Contemporary Natural Philosophy and Philosophies - Part 3

Contemporary Natural Philosophy and Philosophies - Part 3

Guest Editors

Gordana Dodig-Crnkovic
Marcin J. Schroeder

MDPI

Basel • Beijing • Wuhan • Barcelona • Belgrade • Novi Sad • Cluj • Manchester

Guest Editors
Gordana Dodig-Crnkovic
Computer Science and
Engineering
Chalmers University
of Technology
Gothenburg
Sweden

Marcin J. Schroeder
The Faculty of International
Liberal Arts
Akita International University
Akita
Japan

Editorial Office
MDPI AG
Grossspeteranlage 5
4052 Basel, Switzerland

This is a reprint of the Special Issue, published open access by the journal *Philosophies* (ISSN 2409-9287), freely accessible at: www.mdpi.com/journal/philosophies/special_issues/Philosophy_and_Philosophies3.

For citation purposes, cite each article independently as indicated on the article page online and using the guide below:

Lastname, A.A.; Lastname, B.B. Article Title. *Journal Name* **Year**, *Volume Number*, Page Range.

ISBN 978-3-7258-3214-9 (Hbk)
ISBN 978-3-7258-3213-2 (PDF)
https://doi.org/10.3390/books978-3-7258-3213-2

Cover image courtesy of Jonathan McCabe

Contents

About the Editors

Gordana Dodig-Crnkovic

Gordana Dodig-Crnkovic is a Professor of Computer Science at Mälardalen University and a Professor of Interaction Design at Chalmers University of Technology. Holding Ph.D. degrees in Physics and Computer Science, her research explores Morphological Computation, the Study of Information, and Computing Ethics. She has been developing and teaching courses in Formal Languages, Research Methodology, Computing and Philosophy, and Professional Ethics, among others.

Dodig-Crnkovic is a prolific author and editor. She has written and contributed to numerous academic publications, including the books *Investigations into Information Semantics and Ethics of Computing (2006)* and *Information and Computation Nets (2009)*. In addition to these, she has published six volumes edited with Springer and World Scientific, as well as two edited books with MDPI.

A dedicated member of the global academic community, she has been serving on the boards of several research and education networks, including Informatics Europe, IS4SI, and BITrum. She is also a past President of the International Society for the Study of Information, co-editor of the World Scientific Series in Information Studies, and a member of the Springer SAPERE Series Advisory Board. Her editorial contributions extend to several journals, including *Entropy* and *Philosophical Problems in Science.*

Throughout her career, Dodig-Crnkovic has actively contributed to the international research landscape by organizing over 30 conferences and academic events, including ECAP 2005 (MDU, Västerås), IS4SI 2017 (Chalmers, Gothenburg), ECSS 2018 (Informatics Europe, Chalmers), and PT-AI 2021 (Philosophy and Theory of AI, Chalmers). She was a guest editor for fourteen special journal issues. Recognized for her contributions to the study of information and computation, she is a Fellow and Board Member of the International Academy of Information Studies.

Marcin J. Schroeder

Marcin J. Schroeder, Ph.D., has been a member of the academic community for almost 50 years as a faculty member at several universities in Poland, the USA, and Japan, active in several disciplines of study from theoretical physics, mathematics (general algebra), and logic to philosophy, intercultural communication, education, and cultural studies. In recent decades, the focus of his academic activity has been on philosophy and mathematical modeling of information, its dynamics (computation), and its integration (including its role in consciousness) with a special emphasis on the use of the general concept of symmetry. For many years, in addition to teaching in several universities, he held administrative faculty positions as chair of department, dean of academic affairs, director of the Academic Center for Learning Excellence, etc. For his contributions to the establishment and development of the Akita International University in Akita, Japan, he was awarded the honorary title of Professor Emeritus. Since 2016 he has been the founding editor-in-chief of the journal *Philosophies* published by MDPI. He was the president of the International Society for the Study of Information (IS4SI) in the years 2019–2021 and has been a member of the IS4SI Board since 2017, currently as the vice president for research. Since 2022 he has been the academic president of the International Society for Interdisciplinary Studies of Symmetry (SIS). More recently he became, in 2024, a fellow of the International Academy of Information Studies. In addition to presenting and publishing the results of his research, editing multiple research publication initiatives, and serving with his experience on several boards of organizations, he is frequently engaged in the preparation and chairing of academic conferences.

Preface

The evolution of human knowledge has often mirrored the complexity and interconnectedness of the natural world, yet it has also fragmented into isolated silos of thought. In "Contemporary Natural Philosophy and Philosophies—Part 3", we strive to revive the original spirit of natural philosophy: the pursuit of an integrated understanding of nature and humanity's place within it.

This Special Issue presents scholars, scientists, and thinkers across disciplines, fostering dialogues that bridge the divide between philosophy and science, the rational and the intuitive, and the human and the natural world. Inspired by historical traditions and driven by contemporary advancements in fields like artificial intelligence, cognitive science, neuroscience, and complexity science, we synthesize diverse perspectives into a cohesive framework of knowledge.

As a continuation of previous volumes, this collection explores themes such as the informational fabric of reality, the evolving relationship between humans and nature, and the philosophical underpinnings of modern science. It challenges readers to embrace both abstraction and embodiment, coexistence and co-creation, in a holistic understanding of the world we live in.

Through this Special Issue, we hope to contribute to a broader and more interconnected understanding of knowledge—one that resonates deeply in a world facing rapid paradigm shifts and unprecedented challenges. We hope it will inspire further inquiry and collaboration toward a rich contemporary philosophy of nature.

Gordana Dodig-Crnkovic and Marcin J. Schroeder
Guest Editors

philosophies

MDPI

Editorial

Contemporary Natural Philosophy and Philosophies—Part 3

Gordana Dodig-Crnkovic [1,2,*] and Marcin J. Schroeder [3]

1 Department of Computer Science and Engineering, Chalmers University of Technology, 412 96 Gothenburg, Sweden
2 School of Innovation, Design and Engineering, Mälardalen University, 721 23 Västerås, Sweden
3 The Faculty of International Liberal Arts, Akita International University, Akita 010-1292, Japan; mjs@gl.aiu.ac.jp
* Correspondence: dodig@chalmers.se or gordana.dodig-crnkovic@chalmers.se

1. Introduction

In 2018, we initiated a series of three Special Issues dedicated to contemporary natural philosophy in the spirit of the goals of the journal *Philosophies* (See Sections A and B). *Philosophies* journal [1] aims to establish a new unity in diversity within human knowledge, encompassing both "Wissen" (i.e., "Wissenschaft") and "scīre" (i.e., "science"). While "science" exclusively focuses on directly testable explanations and predictions, "Wissenschaft" as the pursuit of knowledge, learning, and scholarship involves all forms of knowledge, including philosophy. Our aim is to promote this broader notion of scholarship that encompasses the understanding and articulation of the learner's role in the knowledge growth process rather than just the final product and its validation. This inclusive approach to knowledge involves both short-term and long-term perspectives and is critical and hypothetical, breaking new ground. It is expected to resonate with basic human value systems, including cultural values.

The contemporary natural philosophy project aims to give importance to humans in the natural world as active subjects and integral parts of nature. It seeks to overcome the compartmentalization of human reality into non-communicating domains by accommodating all forms of knowledge within the network of networks of contemporary natural philosophies. This synthetic network of knowledge promotes coexistence and co-creation between the human and the natural world, where there is room for both rational and intuitive, embodied and abstract, physical and mathematical relations with the world.

As knowledge grows, it tends to spontaneously fragment. We take advantage of existing diversity as both a resource and a starting point for a new synthesis of knowledge. The idea of broad, inclusive knowledge is not new and has been part of natural philosophy from its inception. Scientists such as Newton, Bohr, Einstein, Prigogine, Weizsäcker, and Wheeler were all natural philosophers who embraced a broad understanding of knowledge about nature. However, in modern times, the unifying picture of the natural/physical world is missing. This is because scientific domains have become isolated silos with their own ontologies, methodologies, and epistemologies.

In recent decades, the need for connected and shared knowledge has given rise to new trends toward synthesis. Complexity science, particularly when applied to biology or medicine, helps us understand the importance of connectedness between disparate pieces of knowledge and their frameworks, theories, and approaches. Network science is also emerging as a related field that studies the structures of nodes and edges as connections between actors. These trends toward synthesis and interconnectedness are crucial for advancing our understanding of the world around us.

According to Einstein and Hawkins, problems are not solved within the framework in which they arise but rather in a new framework at the next level of abstraction. This principle guides the approach of this Special Issue, which attempts to construct a new, networked world of knowledge where domain specialists from various disciplines can

Citation: Dodig-Crnkovic, G.; Schroeder, M.J. Contemporary Natural Philosophy and Philosophies—Part 3. *Philosophies* **2024**, *9*, 58. https://doi.org/10.3390/philosophies9030058

Received: 6 April 2023
Accepted: 6 April 2023
Published: 27 April 2024

interact and connect with the wider knowledge-producing and knowledge-consuming communities in an inclusive, extended natural-philosophic manner.

This process of synthesis involves a mutually beneficial relationship between scientific and philosophical investigations. Sciences inform philosophies about the latest knowledge of the world, both natural and human-made, while philosophies scrutinize the ontological, epistemological, and methodological foundations of sciences, providing scientists with questions and conceptual analyses. The goal is to extend and deepen our comprehension of the world, including ourselves as individuals and societies, as well as humankind. Through this inclusive and collaborative approach, we can achieve a more comprehensive and interconnected understanding of the world around us, which is needed in these turbulent times of paradigm shifts caused by the emergence of high-level artificial intelligence.

We would like to give place in this modern natural philosophy to the human in the natural world, both as an active subject and as an integral part of nature, for whom the world comes as an interface (Rössler's "The World as an Interface") [1]. The separation between human and nature, thought and feeling, rational and intuitive, knowledge how and knowledge that, embodiment and abstraction, and physical and mathematical relations to the world have led to the compartmentalization of human reality in various non-communicating domains. All should have a place in this new synthetic network of knowledge of contemporary natural philosophy, in which there is a given place for human in coexistence and co-creation with the natural world.

This Special Issue responds to the call from the journal Philosophies to build a new, networked world of knowledge with domain specialists from different disciplines interacting and connecting with the rest of the knowledge-producing and knowledge-consuming communities in an inclusive, extended natural-philosophic manner. In this process of synthesis, scientific and philosophical investigations enrich each other—with sciences informing philosophies about the best current knowledge of the world, both natural and human-made—while philosophies scrutinize the ontological, epistemological, and methodological foundations of sciences, providing scientists with questions and conceptual analyses. This is all directed at extending and deepening our existing comprehension of the world, including ourselves, both as individuals, societies, and as humankind.

2. Towards a New Synthesis

Historically, scholars have attempted to search for a unity of knowledge originating from a holistic understanding of the world. The examples include Snow's critique of "The Two Cultures" [2] and biologist Wilson's "Consilience: The Unity of Knowledge" [3]. However, the strong development of disciplinary research continued. It was still possible to dig deeper into isolated domains, and the results were still interesting even though a common view was missing. However, new developments in sciences and technology, such as artificial intelligence, neurosciences, cognitive science, and modern medicine called for unified views of emergent levels from microscopic to macroscopic scales. It also connected diverse phenomena of the "body" and mind", the physical and the mental as archetypes of the divide between "two cultures".

The dialogue between sciences and philosophy has become especially interesting regarding the philosophy of science and the question of what constitutes the scientific method, which has become less clear. There are three major methodological challenges: The demise of natural philosophy; "Idol of Numbers" added to Bacon's four Idols of the Mind (Idols of the Tribe, Idols of the Cave, Idols of the Marketplace, and Idols of the Theater) [4]; and isolationism and the self-sufficiency of research disciplines.

3. Connecting the Disparate Knowledge Silos

When modeling a phenomenon, multiple connected theories, seen from a common perspective, contribute to our multifaceted understanding of its structures and temporal behavior.

One very successful approach in this direction was the development of multiscale models for complex physical, chemical, biological, and cognitive systems, including the human brain. Multiscale models [5] combine and connect earlier approaches focused on single scales of time, space, and topology through the integration of data across spatial, temporal, and functional scales.

Another promising path is the reconceptualization (i.e., conceptual engineering) of the basic concepts used to describe different natural and artificial systems—physical, chemical, biological, and cognitive. In this new framework, information is considered the fabric of reality (Deutsch) [6], for an observer, Floridi [7]. The dynamics of information can be modeled as computation, thus forming the basis for the info-computational modeling of a variety of systems, from the physical to the cognitive [8]. According to Kun and Brenner [9], the philosophy of information presents a revolution in philosophy and provides a means of informational metaphilosophy of science, as the philosophy of science. We might also add that information, together with its dynamics (computation), presents a new possibility for the development of the modern philosophy of nature.

4. Topics Covered

For this Special Issue, we particularly encouraged addressing the human aspect of natural philosophy, extended evolutionary syntheses, and the life and capacities of cognition and consciousness from a naturalist point of view—along with the topics already discussed in the previous two issues.

The basic ambition from the beginning was to explore contemporary natural philosophy through the views of researchers investigating broader domains of knowledge based on "the idea of the unity of nature and human as its integral part, from different perspectives of sciences, humanities, and liberal arts from their cultural contexts, including technology".

This resulted in the following list of topics:

What is the current state of the philosophy of nature/natural philosophy?
What might be the role of the philosophy of nature/natural philosophy?
Can the philosophy of nature be based on our best current scientific knowledge? (the thesis of the book "Everything Must Go" [10]);
How can interdisciplinarity/crossdisciplinarity/multidisciplinarity/transdisciplinarity help tie knowledge from different disciplines and interdisciplines at different levels of abstraction in a common intelligible philosophy of the universe with cosmos and chaos, non-living and living parts in it? [11,12]
What would be the new role of research methods in this new high-level take on human knowledge?
Can we imagine any higher authority in matters of truth and existence than the consensus view of our current humanity?
How do the sciences of the artificial [13], AI, relate to the philosophy of nature?
Informational universe—Floridi, Deutsch, Kun—epistemology, and ontology;
"Mechanism" and "materialism" as bases for our understanding of nature;
Nature and mind—the role and character of the mind/cognition/agency in the development of the universe;
Evolving universe—being and becoming in the contemporary philosophy of nature;
Emergent universe;
Connecting a variety of levels of abstraction;
The role of life sciences, with biology and cognitive sciences, in the new natural philosophy;
The role of the observer in the new synthesis;
The role of formal sciences and methods—logics, mathematics, computing, and simulation;
The ecological view of knowledge [14].

5. The Way Ahead

We consider the series of Special Issues only the first step towards a more organized and sustainable collective effort to revive the original fundamental role of natural philosophy, construed as the pursuit of integrated knowledge and understanding of the world.

We plan the continuation of this project in the *Topical Collection on Contemporary Natural Philosophy*. This will allow contributors to submit their work unconstrained by the timelines or deadlines of Special Issues.

Author Contributions: All authors contributed equally to the article. All authors have read and agreed to the published version of the manuscript.

Acknowledgments: The Guest Editors would like to express their gratitude to the authors who contributed to this Special Issue and to numerous anonymous peer reviewers whose work helped in improving the quality of published contributions. We were overwhelmed by the response, both in terms of the number of submissions and their wide range of topics and excellent quality. We hope that this Special Issue will add to the new synthesis in the form of a revived modern natural philosophy.

Conflicts of Interest: The authors declare no conflicts of interest.

Appendix A

The List of Contributions to **Volume 1** of Contemporary Natural Philosophy and Philosophies, available at https://www.mdpi.com/books/book/1331 (accessed on 26 April 2024) [15].

Gordana Dodig-Crnkovic and Marcin J. Schroeder, *Contemporary Natural Philosophy and Philosophies.*
Bruce J. MacLennan, *Philosophia Naturalis Rediviva: Natural Philosophy for the Twenty-First Century.*
Nicholas Maxwell, *We Need to Recreate Natural Philosophy.*
Stanley N. Salthe, *Perspectives on Natural Philosophy.*
Joseph E. Brenner, *The Naturalization of Natural Philosophy.*
Andrée *Ehresmann* and Jean-Paul Vanbremeersch, *MES: A Mathematical Model for the Revival of Natural Philosophy.*
Arran Gare, *Natural Philosophy and the Sciences: Challenging Science's Tunnel Vision.*
Chris Fields, *Sciences of Observation.*
Abir U. Igamberdiev, *Time and Life in the Relational Universe: Prolegomena to an Integral Paradigm of Natural Philosophy.*
Lars-Göran Johansson, Induction and Epistemological Naturalism.
Klaus Mainzer, *The Digital and the Real Universe. Foundations of Natural Philosophy and Computational Physics.*
Gregor Schiemann, *The Coming Emptiness: On the Meaning of the Emptiness of the Universe in Natural Philosophy.*
Koichiro Matsuno, *Temporality Naturalized.*
Robert E. Ulanowicz, *Dimensions Missing from Ecology.*
Matt Visser, *The Utterly Prosaic Connection between Physics and Mathematics.*
Kun Wu and Zhensong Wang, *Natural Philosophy and Natural Logic.*
Lorenzo Magnani, *The Urgent Need of a Naturalized Logic.*
Roberta Lanfredini, *Categories and Dispositions. A New Look at the Distinction between Primary and Secondary Properties.*
Rafal Maciag, *Discursive Space and Its Consequences for Understanding Knowledge and Information.*
Harald Atmanspacher and Wolfgang Fach, *Exceptional Experiences of Stable and Unstable Mental States, Understood from a Dual-Aspect Point of View.*
Włodzisław Duch, *Hylomorphism Extended: Dynamical Forms and Minds.*
Robert Prentner, *The Natural Philosophy of Experiencing.*
Robert K. Logan, *In Praise of and a Critique of Nicholas Maxwell's In Praise of Natural Philosophy: A Revolution for Thought and Life.*

Appendix B

The List of Contributions to **Volume 2** of Contemporary Natural Philosophy and Philosophies, available at https://www.mdpi.com/books/book/3098 (accessed on 26 April 2024) [16].

Gordana Dodig-Crnkovic and Marcin J. Schroeder, *Contemporary Natural Philosophy and Philosophies—Part 2*
Richard de Rozario, *Matching a Trope Ontology to the Basic Formal Ontology*
Ronald B. Brown, *Breakthrough Knowledge Synthesis in the Age of Google*
Andreas Stephens and Cathrine V. Felix, A *Cognitive Perspective on Knowledge How: Why Intellectualism Is Neuro-Psychologically Implausible*
Cathrine V. Felix and Andreas Stephens, *A Naturalistic Perspective on Knowledge How: Grasping Truths in a Practical Way*
Johannes Schmidl, De Libero Arbitrio—*A Thought-Experiment about the Freedom of Human Will*
Cristian S. Calude and Karl Svozil, *Spurious, Emergent Laws in Number Worlds*
Roman Krzanowski, *What Is Physical Information?*
Gordana Dodig-Crnkovic, *Natural Morphological Computation as Foundation of Learning to Learn in Humans, Other Living Organisms, and Intelligent Machines*
Joseph E. Brenner and Abir U. Igamberdiev, *Philosophy in Reality: Scientific Discovery and Logical Recovery*
Marcin J. Schroeder, *Contemporary Natural Philosophy and Contemporary Idola Mentis*

References

1. Rössler, O. *Endophysics: The World as an Interface. Singapore-New Jersey-London-Hong Kong: World Scientific;* World Scientific Publishing Co Pte Ltd.: Singapore, 1998.
2. Snow, C.P. *The Two Cultures;* Cambridge University Press: London, UK, 1959.
3. Wilson, E.O. Consilience: The unity of knowledge. *Issues Sci. Technol.* **1998**, *15*, 90. [CrossRef]
4. Bacon, F. *The New Organon; Book 1. The Works of Francis Bacon;* Spedding, J., Ellis, R.L., Heath, D.D., Eds.; CreateSpace Independent Publishing Platform: Boston, MA, USA, 1863.
5. Weinan, E. *Principles of Multiscale Modeling;* Cambridge University Press: Cambridge, UK, 2011.
6. Deutsch, D. *The Fabric of Reality;* Penguin Books: New York, NY, USA, 1997.
7. Floridi, L. A defense of informational structural realism. *Synthese* **2008**, *161*, 219–253. [CrossRef]
8. Dodig-Crnkovic, G. Nature as a Network of Morphological Infocomputational Processes for Cognitive Agents. *Eur. Phys. J.* **2017**, *226*, 181–195. [CrossRef]
9. Kun, W.; Brenner, J. Philosophy of Information: Revolution in Philosophy. Towards an Informational Metaphilosophy of Science. *Philosophies* **2017**, *2*, 22.
10. Ladyman, J.; Ross, D.; Spurrett, D.; Collier, J. *Everything Must Go: Metaphysics Naturalised;* Clarendon Press: Oxford, UK, 2007; ISBN 9780199276196.
11. Hofkirchner, W. The Quest for a Unified Theory of Information. In *Proceedings of the Second International Conference on the Foundations of Information Science. World Futures General Evolution Studies;* Gordon and Breach: Amsterdam, The Netherlands, 1999.
12. Brier, S. *Cybersemiotics: Why Information Is Not Enough!* University of Toronto Press: Toronto, ON, Canada, 2008; ISBN 9780802092205.
13. Simon, H.A. *The Sciences of the Artificial;* MIT Press: Cambridge, MA, USA, 1996; ISBN 0262691914.
14. Burgin, M.; Zhong, Y. Information Ecology in the Context of General Ecology. *Information* **2018**, *9*, 57. [CrossRef]
15. Dodig-Crnkovic, G.; Schroeder, M.J. *Contemporary Natural Philosophy and Philosophies—Part 1;* MDPI: Basel, Switzerland, 2019.
16. Schroeder, M.J.; Dodig-Crnkovic, G. *Contemporary Natural Philosophy and Philosophies—Part 2;* MDPI: Basel, Switzerland, 2020; ISBN 978-3-03943-536-4.

philosophies

MDPI

Article

Naturalizing Logic: How Knowledge of Mechanisms Enhances Inductive Inference

Paul Thagard

Department of Philosophy, University of Waterloo, Waterloo, ON N2L 3G1, Canada; pthagard@uwaterloo.ca

Abstract: This paper naturalizes inductive inference by showing how scientific knowledge of real mechanisms provides large benefits to it. I show how knowledge about mechanisms contributes to generalization, inference to the best explanation, causal inference, and reasoning with probabilities. Generalization from some A are B to all A are B is more plausible when a mechanism connects A to B. Inference to the best explanation is strengthened when the explanations are mechanistic and when explanatory hypotheses are themselves mechanistically explained. Causal inference in medical explanation, counterfactual reasoning, and analogy also benefit from mechanistic connections. Mechanisms also help with problems concerning the interpretation, availability, and computation of probabilities.

Keywords: induction; inference; logic; mechanisms; naturalism; probability

check for updates

Citation: Thagard, P. Naturalizing Logic: How Knowledge of Mechanisms Enhances Inductive Inference. *Philosophies* **2021**, *6*, 52. https://doi.org/10.3390/philosophies6020052

Academic Editors: Gordana Dodig-Crnkovic and Marcin J. Schroeder

Received: 17 May 2021
Accepted: 17 June 2021
Published: 21 June 2021

1. Introduction

An old philosophy joke (dubiously attributed to Morris Cohen) says that logic texts are divided into two parts: in the first half, on deductive logic, the fallacies are explained; and in the second half, on inductive logic, they are committed. This quip is too hard on inductive inference, which is indispensable in science and everyday life, but it does point to the difference between deduction and induction, which introduces unavoidable uncertainty. This paper argues that an appreciation of mechanisms can help substantially to reduce the problems that attend induction.

Natural philosophy has made substantial progress in integrating epistemology, metaphysics, and ethics with sciences that include physics, psychology, and neuroscience [1]. However, logic might be seen as beyond the reach of naturalism because it provides normative ideals about how people ought to reason, not descriptions of how minds actually work. Nevertheless, advances have been made on the psychology of deduction [2,3], and computational and psychological models of inductive inference have been developed [4,5]. Artificial intelligence has blossomed with applications of deep learning and other kinds of inductive inference [6,7].

This paper explores a different, compatible way of naturalizing inductive inference not just by psychologizing it but also by showing how scientific knowledge of real mechanisms provides large benefits to it. I am not arguing that knowledge of mechanisms is essential to induction, only that some kinds of inductive inference gain substantially from it. Specifically, I show how knowledge about mechanisms contributes to generalization, inference to the best explanation, causal inference, and reasoning with probabilities. Even deduction can be influenced by knowledge about mechanisms in the world.

A Google Scholar search for the term "mechanism" yields more than 7 million responses across fields in the natural and social sciences and more than 200,000 mentions in 2020 alone. Recent philosophy of science has intensely investigated the nature of mechanisms, which can be understood as combinations of connected parts whose interactions produce regular changes [1,8–11]. However, these investigations have neglected the contribution that the understanding of mechanisms makes to the interconnected problems of describing and justifying various kinds of inductive inference. Induction can proceed without mechanisms, but it is more understandable and reliable when inferences are supported by information about mechanisms.

The contribution of mechanisms to good inductive inference is important for practical as well as theoretical reasons. The world is awash in misinformation about issues such as health, climate, and politics. An important way to distinguish information from misinformation is to look at how they differ in their inferential basis. Whereas useful information arises from solid inductive inferences based on mechanisms, misinformation is often based on inferences that either ignore mechanisms or rely on mechanisms that are seriously defective. Hence, an important task for inductive logic is to discriminate strong mechanisms from unreliable ones.

2. Mechanisms

Two of the most pressing problems facing humanity are climate change and viral epidemics. Both problems require inductive inferences to answer important questions, such as whether human industrial activity that produces greenhouse gases is leading to irreversible global warming, and whether new viral diseases such as COVID-19 can be controlled. Fortunately, substantial knowledge has been achieved about the relevant mechanisms as shown in Table 1, which amalgamates the varying terminology used by different philosophers in discussing mechanisms.

Table 1. Mechanisms relevant to global warming and viral epidemics. In the top row, the parentheses show the range of terminology used in philosophical discussions of mechanisms. The other rows describe the operation of climate [12] and viral [13] mechanisms.

	Combination (Whole, System, Structure)	Parts (Entities, Components)	Interactions (Activities, Operations)	Changes	Results (Behaviors, Functions, Phenomena)
Global warming	Solar system including Earth	Sun, solar radiation, Earth's atmosphere, Earth's surface, greenhouse gas molecules	Earth absorbs sunlight. Earth emits energy as infrared light. Greenhouse gases absorb light, retaining energy. Energy heats up the Earth.	Earth warms.	Earth's temperature is permanently increasing. Severe weather and flooding are increasingly common.
Viral epidemic	Human population	Bodies, cells, viruses	Viruses infect cells and reproduce. Viruses spread to other bodies.	Infections spread among bodies	Epidemics and pandemics occur.

Consideration of mechanisms should help with two interconnected problems of induction: description and justification. The description problem is to characterize how people typically make inductive inferences. In logic, this would take the form of a set of rules that are applied to premises to generate conclusions. Cognitive science can take a broader view that goes beyond verbal premises and syntax-driven rules of inference to include a variety of mental representations including pictorial and kinesthetic ones and computational procedures different from logical rules of inference. The justification problem asks whether the inductive procedures so described are legitimized by their production of reliable and useful conclusions. I will argue that including mechanisms in the description of several kinds of inductive inference makes them more useful and justifiable.

3. Inductive Generalization

The simplest and most familiar form of induction is generalization from some to all: for example, from the observation that some aardvarks are burrowing animals to the conclusion that all aardvarks burrow. John Stuart Mill noticed that some such inferences are more plausible than others [14], p. 206:

> *Why is a single instance, in some cases, sufficient for a complete induction, while in others myriads of concurring instances, without a single exception known or presumed, go such a little way towards establishing an universal proposition? Whoever can answer this question knows more of the philosophy of logic than the wisest of the ancients, and has solved the problem of Induction.*

For example, a few examples of burrowing aardvarks might suffice to convince us that all aardvarks burrow, whereas we would need many more cases to be confident that all aardvarks eat peanuts.

By "plausible", I mean that a claim is more coherent with the available evidence than opposing claims, although the degree of coherence may not be enough to establish the claim as definitely accepted [15,16]. Here, coherence can be computationally assessed by maximizing the satisfaction of constraints concerning how hypotheses explain evidence and other hypotheses and concerning competition among incompatible hypotheses.

In the 1980s, research in psychology and philosophy provided an answer to Mill's question based on variability [17,18]. Studies found that people who are told that floridium is a metal are quick to infer from a few cases that a high percentage of floridium burns with a blue flame, whereas people who are told that shreebles are birds are much more reluctant to infer that a high percentage of shreebles are blue. A plausible explanation for this difference is that people are aware that metals have little variability in their combustion properties and that colorful birds such as parrots have a lot more variability in their colors. When inductive generalization is about kinds of things and properties that have little variability, then even a single instance or a few of them can suffice.

This analysis probably does capture psychological differences, but a deeper answer comes from considering mechanisms. In the 1780s, Antoine Lavoisier identified the mechanism of combustion as the combination of materials with oxygen to produce heat and light. Much later, the mechanism was deepened to explain how atoms of elements such as carbon interact with atoms of oxygen to produce heat construed as rapidly moving molecules and with light construed as emission of photons. In the floridium induction, we can presume that interaction of the metal atoms with oxygen produces light with a specific frequency, making it easy to infer that floridium burns with a blue flame.

In contrast, the mechanisms in the shreebles case provide much less assurance about the plausibility of the inductive generalization. The main relevant mechanism for inferring the color of animals is genetics on the assumption that offspring inherit color from their parents. However, often, genes do not produce consistent color in species such as cats, parrots, and humans with several different hair and skin colors. Genes have variants called alleles, and different alleles can produce variations in hair color. For example, in humans, most redheads have a mutation in the gene for the melanocortin 1 receptor that affects hair and skin. Accordingly, in the absence of extensive knowledge about the genetics of shreebles, we should be reluctant to infer from a few cases that all or most shreebles are blue.

Mechanisms also help with paradoxes that afflicted attempts in the 1940s and 1950s to establish accounts of inductive generalization as purely syntactic. Confirmation theory proposed that inductive support for generalizations of the form (x)(Fxșș → Gx) came from instances Fa and Ga. For example, observing a black raven confirms the hypothesis that all ravens are black. However, Carl Hempel noticed that (x)(Fx → Gx) is logically equivalent to (x)(~Gx → ~Fx), so it seems that a black raven also confirms the odd hypothesis that all non-black things are non-ravens [19]. Equally oddly, a white shoe confirms the claim that all ravens are black.

The oddity disappears with recognition that hypotheses that connect a kind with a property, as in all ravens are black, are much more plausible when a mechanism connects the kind to the property. Ravens have genes for color that lack alleles for producing colors other than black, and extremely rare white ravens occur because of albinism resulting from mutations in genes for producing the pigment melanin. Hence, the known genetic mechanisms support the confirmation by a black raven that all ravens are black. In contrast, no mechanisms connect non-black things with non-ravens, so we have no reason to take that hypothesis seriously, despite its syntactic equivalence with all ravens are black. One of the lessons of the failure of logical positivism as a philosophy of science is that scientific reasoning is not just a matter of syntax and should instead consider the physical constitution of the world as understood in terms of mechanisms.

A similar resolution is available for Nelson Goodman's new riddle of induction [20]. Examples of green emeralds seem to support the generalization that all emeralds are green, but they also support the generalization that all emeralds are grue, where things are grue if they are observed before time t and green and blue otherwise. Fortunately, mechanisms provide a valuable contrast between the generalizations that all emeralds are green and that all emeralds are grue. The Gem Encyclopia reports (https://www.gia.edu/seeing-green, accessed on 18 June 2021):

Emeralds are formed when chromium, vanadium, and iron are present in the mineral beryl. The varying presence of these three elements gives emerald its range of color. Chromium and vanadium make an intense green color. Iron gives the stone a bluish tint.

The perceived color of emeralds results first from how their constituent elements (parts) interact with light to reflect light in a specific frequency (around 550 nm), and second from how this frequency of light stimulates receptors in the retina to send signals to the brain that get interpreted as green. Nothing in these two mechanisms points to how time t could be relevant to making emeralds blue. As with Hempel's paradox of the ravens, background knowledge about mechanisms is far more useful to understanding inductive generalization than pure syntax.

Inductive generalization assigns a property to a kind, and many philosophers have recognized that natural kinds support induction better than contrived collections [21]. Natural kinds are sometimes assigned metaphysical essences as being the same in all possible worlds, but that account is useless for science-oriented philosophy. A better account of natural kinds was developed by Richard Boyd, who proposed that biological species are clusters of properties held together by underlying properties that are homeostatic: a stable range of properties is maintained because deviations have a low chance of persisting [22]. Hence, we should think of the induction-promoting value of natural kinds as resulting from their underlying mechanisms.

Inductive generalization does not absolutely require knowledge of mechanisms, as sometimes we can have ample instances of *A* and *B* to support the conclusion that all *A* are *B* even if we lack knowledge of a mechanism connecting *A* to *B*. For example, it was known for centuries that willow bark reduces pain before aspirin was isolated in 1897 and its biochemical mechanism was discovered in 1971. Nevertheless, knowledge of mechanisms is highly useful for grasping the contributions of variability and natural kinds to inductive inference and for understanding the failure of the purely syntactic approach of confirmation theory.

4. Inference to the Best Explanation

A narrow use of the term "induction" covers only generalization from some to all, but the broader use covers any inference that differs from deduction in introducing uncertainty. There are many such kinds of induction ranging from analogy to statistical inference, but one of the most common goes by the name "inference to the best explanation" [23–25]. This name was new in the 1960s, but inference to explanatory hypotheses was recognized by nineteenth-century writers such as William Whewell and Charles Peirce, and precursors can be found as far back as Renaissance astronomers and possibly Aristotle. Peirce introduced the term "abduction" for the generation and acceptance of explanatory hypotheses, and much recent work in philosophy and artificial intelligence analyzes varieties of abductive inference [5,26–28].

The basic form of inference to the best explanation is:

Evidence *E* requires explanation.

Hypothesis *H* provides a better explanation of *E* than alternative available explanations.

Therefore, *H*.

This kind of inference, "IBE" for short, is common in everyday life: for example, when people attribute mental states to other people and when mechanics identify causes of automobile breakdowns. It is also common in the law when jurors conclude that an accused criminal is guilty and in medicine when physicians conclude that a patient has a disease that explains the patient's symptoms.

As with generalization, knowledge of mechanisms is not essential to inference to the best explanation, but mechanisms help reduce the inherent riskiness of IBE. The loosest form of IBE has been dismissed as "modus morons":

If *A* then *B*.

B.

Therefore, *A*.

This form of inference is pathetically weak because there may be many other reasons for *B* besides *A*. One way of tightening it up is to require a causal connection, *A* causes *B*, but there may still be other causes that need to considered. By requiring the best explanation, IBE ensures that some comparative assessment of alternatives has taken place.

Advocates of IBE are usually vague about what constitutes an explanation, which most generally is just fitting something puzzling into a familiar pattern. Useful patterns range from the loose storytelling that is frequent in everyday life to the exact deduction found in mathematical fields such as physics. In biology, medicine, cognitive science, and other fields, explanation is often the description of causal mechanism: for example, when influenza is explained by the infection of cells by viruses.

Mechanistic explanations strengthen IBE in two ways. First, mechanisms provide a much tighter connection between hypotheses and evidence than mere if–then relations or abstract causes. The claim that a patient's symptoms of fever, coughing, and pains are the result of influenza can be fleshed out by many causal details, including that a known virus infected the patient's respiratory system, causing specific bodily reactions. When a mechanism is known, we have good reason for taking a hypothesis as a serious contender for explaining symptoms, in comparison with fanciful mechanism-free hypotheses such as that the patient is possessed by demons. IBE still requires that a proposed hypothesis be evaluated according to whether it explains more than alternative hypotheses, but the use of mechanistic explanations sets a high bar for alternatives through the expectation that they should also be able to provide mechanisms that connect the hypothesis with the evidence.

The second way that mechanisms are important to IBE arises because hypotheses are assessed not only by how much they explain but also by the extent to which they themselves are explained [16,29]. For example, in law, the hypothesis that an accused is guilty of murdering a victim has to explain many aspects of the crime scene such as the accused's fingerprints on the murder weapon. However, the guilt hypothesis also gets support by the provision of a motive for why the murderer killed the victim, for example, out of jealousy. Such legal explanations are based on common-sense knowledge that rely on loose psychological mechanisms based on beliefs and desires, for example, the belief that the victim had seduced the accused's spouse, so the accused desired revenge.

In science and medicine, higher-level explanations that explain hypotheses often point to deeper mechanisms supported by substantial evidence. For example, Darwin's theory of evolution gained support from its ability to explain many observations such as the distributions of species, but it was itself explained by the mechanisms of natural selection and the genetic transmission of inherited traits. Skepticism about the truth of scientific theories is inspired by the observation that many scientific hypotheses have turned out to be false: for example, Aristotle's theory of the aether and the chemical theory of phlogiston [30]. However, this pessimistic induction can be countered by the cautiously optimistic induction that all accepted scientific theories that have been deepened by mechanistic explanations have stood up to scrutiny in the face of additional evidence and the competition from alternative theories [31].

The schema for strong IBE is then:

Evidence E requires explanation.

Hypothesis H provides mechanistic explanations of E that are better than alternative available explanations, including alternative mechanisms.

In turn, the mechanisms underlying H are explained by more fundamental mechanisms.

Therefore, H.

Application of this schema does not completely eliminate the uncertainty of inductive inference but helps to reduce the apparent arbitrariness of IBE. The incorporation of mechanisms helps to overcome the problem identified by Bas van Fraassen that the best explanation might just be the best of a bad lot [32]. If a hypothesis provides a mechanism for the phenomena that constitute the evidence for it, and if this mechanism is itself explained by underlying ones that explain why the parts and their interactions behave as they do, and if both these mechanisms are assessed against alternative explanations, then we have solid grounds for accepting the hypothesis.

The use of mechanisms in support of IBE might seem circular because the existence of mechanisms is itself usually justified by IBE. However, the naturalistic goal is not to provide an a priori justification of inductive inference but rather to identify how induction works when it works well. IBE and inductive inference do not conform to the ideal of deductive inference from indubitable axioms to theorems but require an alternative ideal based on overall coherence among a raft of hypotheses and evidence. Early philosophical ideas about explanatory coherence were vague, but coherence can now be understood mechanistically as a computational process performed by neural networks [33]. The existence of psychologically and neurologically plausible mechanisms for how IBE integrates evidence and hypotheses at multiple levels supports the conclusion that IBE is a good account of much of human induction. Of course, we need to consider alternative hypotheses, and one prominent alternative to IBE is Bayesian inference, as discussed below.

5. Causality and Counterfactuals

One of the most important applications of IBE is to causal claims: for example, the disease COVID-19 is caused by the novel coronavirus; global warming is caused by increasing human emission of greenhouse gases. Such claims go beyond inductive generalizations that all A are B to assert that A causes B. These claims are of practical importance as they suggest that we can deal with undesirable effects such as diseases and global warming by modifying their causes. Mechanisms do not explain causality because they presuppose causal notions lightly disguised by saying that the parts and interactions produce, generate, or are responsible for changes.

Analysis of causal inference depends on what causes are taken to be. Skeptics who claim that causality is a bogus, unscientific idea are freed from having to evaluate causal inferences, but they cannot explain why causal talk is ubiquitous in science, as shown by the millions of Google Scholar citations for "cause". David Hume claimed that causality was just constant conjunction [34], which would reduce causal inference to inductive generalization; however, the distinction between correlation and causation is generally acknowledged. Probabilistic theories of causality that look for causes where $P\ (effect \mid cause) > P\ (effect)$ also try to make causal inference data-driven, but they have trouble with non-observable causes such as subatomic particles. Manipulation theories of causality emphasize how causes can be inferred by identifying interventions that change their effects, but they have trouble with causal relations elsewhere in the galaxy that are beyond human intervention.

I prefer an ecumenical account of causality that avoids definition in favor of identifying standard examples and typical features of causality while noting its explanatory role [1]. Standard examples of cause–effect relations include pushes, pulls, motions, collisions, actions, and diseases. The typical features (looser than necessary and sufficient conditions) of causality are as follows: temporal ordering, with causes before effects; sensory–motor–sensory patterns such as kicking a ball; regularities expressed by general rules; manipulations and interventions; statistical dependencies; and causal networks of influence. Causality explains why events happen and why interventions work.

From this perspective, causality is recognized by inferences to the best explanation that take into account a range of evidence about temporal patterns, correlations, probabilities, and manipulations. Knowledge of mechanisms is not essential to such inferences, but it helps enormously in cases where the interactions of parts connect a putative cause with an effect. For example, the claim that the cause of COVID-19 is infection by the novel coronavirus SARS-CoV-2 is not just correlational, because much is known about how this virus infects cells and disrupts organs such as lungs and blood vessels.

Medical researchers have devoted much attention to analyzing the considerations for inferring the causes of diseases, including the strength of empirical association and background knowledge [35–37]. All of these considerations can be accommodated in computational models based on explanatory coherence [38]. Identifying mechanisms is just one of the considerations that goes into recognizing the coherence of a causal claim, but it provides important backing for claims such as that smoking causes cancer. This hypothesis was accepted in the 1960s before much was known about how cigarette smoke disrupts the normal growth of cells, but it has become all the stronger thanks to understanding of how chemicals are carcinogenic for lung cells. The hypothesis that Zika viruses cause neural defects in infants became more plausible when it was based not just on correlations between Zika infections and birth defects but also on understanding of how the virus infects neurons and produces abnormal growth. In 2021, worries about the occurrence of blood clots in the brains of people who had taken two kinds of vaccines for COVID-19 became more accepted when a mechanism was identified by which the adenovirus-based vaccines cause blood clotting.

Mechanisms are relevant to considering whether a factor *C* is a cause of an event *E* in four situations [39]:

(1) There is a known mechanism by which *C* produces *E*.
(2) There is a plausible mechanism by which *C* produces *E*.
(3) There is no known mechanism by which *C* produces *E*.
(4) There is no plausible mechanism by which *C* produces *E*.

The fourth situation is damning for inductive inference because it suggests that the link between *C* and *E* cannot be given without abandoning well-established science. For example, many paranormal claims such as demonic possession, extrasensory perception, and telekinesis are incompatible with evidence-based physics.

Counterfactuals provide one of the most problematic domains of causal inference. How should we assess such claims as that if the novel coronavirus had not spread to a wet market in Wuhan, then the COVID-19 pandemic would never have happened, or that if the industrial revolution had not occurred, then there would be no global warming? Standard logical treatments of counterfactuals using possible worlds connected by similarity relations are mathematically elegant but scientifically useless.

The AI researcher Judah Pearl developed a much more plausible account of counterfactuals based on causal relations [40]. His account inverts the attempt to analyze causes in terms of counterfactuals: if the cause had not occurred, then the effect would not have happened. Instead, Pearls suggests that counterfactual claims, even though they are not true or false, may yet be plausible or implausible depending on the causal relations in the world. To assess counterfactuals causally, we can work with a causal network and tweak some of the contributory causes to see what happens, either by deleting a cause or by changing the strength of its connection to an effect. Computational methods for such tweaking are available using either Bayesian networks or explanatory coherence networks.

Causal knowledge is not always dependent on mechanisms, but mechanisms enhance causal inference and can also contribute to more plausible counterfactual judgments. Generally, to evaluate a counterfactual claim that if event1 had not happened, then event2 would not have occurred, it helps to ask the following questions. Is there a mechanism connecting event1 to event2? Are there other mechanisms that can produce event2 without event1? For example, consider the counterfactual claim that if Donald Trump had not been infected with the novel coronavirus, then he would not have gotten COVID-19. The mechanisms by which the virus produces the disease are well known, and no other mechanisms produce COVID-19, so the counterfactual about Trump is plausible.

Mechanisms also help with another shaky kind of inductive inference that benefits from causal relations: analogy. At its loosest, analogical inference just notices that two things or events are similar in some respects and infers that they will be similar in another respect. For example, Montreal is similar to Toronto in being a large Canadian city, and Toronto has a subway, so probably Montreal does, too.

Dedre Gentner noticed that analogies are much more useful when they rely on systematic causal relations [41]. If you know the political backgrounds of Canadian cities and how they operate at national, provincial, and municipal levels, then you can construct a causal story about how the decision process that produced a subway in Toronto is likely to have produced a subway in Montreal. Such causal analogies get even stronger from a correspondence between mechanisms operating in the source and target cases. For example, one of the reasons for thinking that the Zika virus causes birth defects is similarity with the mechanism by which measles causes birth defects [38].

For analogical, counterfactual, and causal inference in general, mechanisms are not mandatory. However, they help to reduce uncertainty in inductive inferences that are often error-prone.

6. Probability

Even though probability theory was only invented in the eighteenth century, many philosophers assume that inductive inference should be based on probabilities [42,43]. I find this assumption implausible, because the array of qualitative inferences so far discussed (generalization, IBE, causal, counterfactual, analogical) do not reduce to probabilistic reasoning. Nevertheless, probabilities are indispensable for many kinds of statistical inference: for example, in estimating the effectiveness of vaccines in preventing COVID-19 where data are used to estimate *P (infection | vaccination)*.

At the core of probabilistic inference is Bayes' theorem, which says that the probability of a hypothesis given the evidence depends on the prior probability of the hypothesis times the probability of the evidence given the hypothesis, all divided by the probability of the evidence. In symbols, $P(H \mid E) = P(H) \times P(E \mid H)/P(E)$. As a theorem of the probability calculus, this result is straightforward, but applying it to real cases of inductive inference faces problems concerning the interpretation, availability, and computation of probabilities. Considerations about mechanisms help with all three of these problems.

The syntax of probability theory is uncontroversial thanks to Kolmogorov's axiomatization, but disputes still rage concerning its semantics [44,45]. Should probabilities be construed as frequencies, degrees of belief, logical relations, or propensities? The frequency interpretation seems most consistent with statistical practices, but it has difficulty in establishing what it means for a probability to be a long-run frequency and in applying this notion to the probability of single events. Bayesians assume that probabilities are degrees of belief but face problems about how such subjective beliefs can objectively describe the world and run up against experimental findings that people's thinking often mangles probabilities [46]. Attempts to describe probabilities as logical relations have encountered problems with describing how abstract considerations of logic and evidence can generate probabilities that satisfy the axioms of probability theory while serving as a practical guide to life.

As a result of these problems, I think the most plausible interpretation of probability is the propensity theory, which says that probabilities are tendencies of physical situations to generate long-term relative frequencies. For example, P *(infection | vaccination)* = x is an objective property of the world by which interactions of people, viruses, and vaccines have a disposition to produce over the long run a ratio x of infected people to vaccinated people. However, this interpretation largely ignores the question of the nature of propensities, tendencies, or dispositions.

What does it mean to say that glass has a disposition to break when struck? Fragility is not just a matter of logical relations such as "If the glass is struck, it breaks" or counterfactuals such as "If the glass had been struck, it would have broken." Rather, we can look to the mechanisms by which glass is formed to explain its fragility, including how poorly ordered molecules generate microscopic cracks, scratches, or impurities that become weak points that break when glass is struck or dropped [47]. Similarly, the mechanisms of viral infection, contagion, vaccination, and immunity explain the disposition for people to be protected by vaccines. Mechanisms flesh out the propensity interpretation of probability and point toward a new mechanistic interpretation of probability [1,48].

Karl Popper introduced the interpretation of probabilities as propensities in order to overcome problems faced by the frequency interpretation in applications to single events. He stated that "propensities may be explained as possibilities (or as measures or 'weights' of possibilities) which are endowed with tendencies or dispositions to realize themselves, and which are taken to be responsible for the statistical frequencies with which they will in fact realize themselves in long sequencies of repetitions of an experiment." [49], p. 30.

Similar to forces, propensities point to unobservable dispositional properties of the physical world.

However, Popper did not elucidate the nature of these possibilities, tendencies, or dispositions and did not spell out how they explain frequencies. These gaps are filled by viewing propensities as mechanisms that generate and explain frequencies. Propensities are dispositions to generate frequencies that result from the connections and interactions of the parts in the underlying mechanism. For example, the probability that two dice will roll a total of 12 is 1/36, because the interactions of the dice with their environment and each other will over the long run yield 12 in an approximate proportion of 1/36.

The propensity interpretation, construed mechanistically, works well for statistical probabilities, but it does not apply to the previous kinds of inductive inference considered here. Inductive generalization and inference to the best explanation do not generate conclusions with determinable probabilities because no known propensities generate conclusions such as that all ravens are black and that species evolved by natural selection. Probabilities are just irrelevant to non-statistical judgments [1].

The second problem with Bayesian approaches to inductive inference is that the relevant probabilities are often unavailable, no matter whether they are construed as frequencies, degrees of belief, or propensities. Bayesians usually just present simple examples, but if they worked with examples with tens or hundreds of events or propositions, they would find that Bayesian calculation requires making up vast numbers of conditional probabilities [50]. Paying attention to mechanisms helps to constrain identification of the probabilities that matter in a particular inferential context. For example, understanding the mechanisms for infection, contagion, vaccination, and immunity makes it clear that many extraneous factors can be ignored, such as demonic possession.

Similarly, mechanisms help with the third problem with Bayesian approaches to inductive inference: probabilistic inference has been proven to be computationally intractable in the sense that the amount of computation increases exponentially with the number of variables used [51]. A human brain has thousands or millions of beliefs and large computer data bases can have hundreds or thousands of interrelated variables. Bayesian networks have been developed that prune the potentially explosive networks by introducing a DO operator that makes the networks restricted to causally plausible connections, but the semantics of this operator are ill-specified [52]. Grasping the underlying mechanisms in a situation dramatically prunes the causal relations that provide plausible connections between variables in a Bayesian network, thereby reducing the number of probabilities to be computed.

Thus, knowledge of mechanisms helps with three problems of the Bayesian approach to inductive inference: interpretation, availability, and computation. This assistance is not enough to defend probability theory as a general approach to inductive inference, which would require probabilistic analyses of all the other kinds of induction that I have discussed. However, the legitimate use of probabilities in many important real-life cases of reasoning is enhanced by incorporating knowledge about mechanisms.

7. Evaluating Mechanisms

I have showed the contribution of information about mechanisms to several kinds of inductive inference but have ignored the problem of assessing the quality of mechanisms. To take an extreme example, someone might claim that demonic possession is the mechanism responsible for COVID-19: the parts are demons, organs, and souls, the interactions are that demons invade organs connected to souls, and the results are infected organs and suffering souls. Fortunately, the philosophy of science can assess what makes some mechanisms much more explanatory than others.

Carl Craver and Lindley Darden identify three vices that can occur in representations of mechanisms: superficiality, incompleteness, and incorrectness [9] (ch. 6). Superficial mechanisms merely redescribe the phenomenon to be explained without providing any internal structure, as in Moliere's joke that sedatives put people to sleep because they have dormative virtue. Superficial mechanisms do not seriously compete to be inferred as part of the best explanation of anything. My demon example is not superficial because at least it tries to say something about demons infecting organs and souls to produce symptoms.

Incomplete mechanisms provide only sketches of mechanisms, leaving out crucial parts and interactions. They often have gray or black boxes that need to be filled in. Incompleteness is sometimes unavoidable because of lack of knowledge: for example, when Darwin was unable to explain the inheritance of traits between generations. However, the general aim of science is to fill in the boxes and convert sketches of mechanisms to schemas that provide details about parts, connections, interactions, and causal results. In the evaluation of competing theories, scientists can compare the degree of completeness of the mechanisms they employ in their explanations. My demon example is seriously incomplete because it says nothing about how demons manage to infect bodily organs and how organ changes cause mental suffering.

A mechanism is incorrect if it fails to describe accurately the alleged parts, connections, and interactions, or, equivalently in Craver and Darden's terminology, their entities, organization, and activities. A schema should explain how a mechanism actually works, not just how it might work. At the beginnings of investigation, researchers legitimately can speculate about how a mechanism might possibly work, but evidence should accumulate to suggest that the mechanism is at least plausible in being consistent with background knowledge and ultimately suggest that the mechanism is actually how the world works. We can dismiss my demon mechanism as incorrect because science has found no evidence for the existence of demons and souls, let alone for their activities in causing infections.

In scientific contexts, correctness can be a matter of degree: for example, when good evidence is available for the existence of the parts of the mechanisms but proposed interactions have yet to be empirically established. When alternative mechanisms described by different theories compete, we can compare them with respect to their degree of correctness as well as for their degree of completeness.

An even stronger way in which a mechanism can be incorrect was mentioned above in relation to causal inference. A mechanism that invokes parts and interactions incompatible with legitimate science does not even get to be judged as providing a how-possible explanation. For example, demons have magical capabilities such as taking possession of souls that are incompatible with scientific physics and psychology.

Hence, proposed mechanisms can be evaluated according to their superficiality, completeness, and correctness. We can assess the evidence for the hypothesized parts, connections, and interactions, and for whether the interactions possibly, plausibly, or actually cause the result to be explained.

To sum up the result of this assessment, we can judge a mechanism to be strong, weak, defective, or harmful. A strong mechanism is one with good evidence that its parts, connections, and interactions really do produce the result to be explained. A weak mechanism is one that is not superficial but is missing important details about the proposed parts, connections, interactions, and their effectiveness in producing the result to be explained. Weak mechanisms are not to be dismissed, because they might be the best that can be done currently, as in the early days of investigations of connections between smoking and cancer and between willow bark and pain relief.

More seriously, some mechanisms can be branded as defective because we have good reason to doubt the existence of their parts or interactions, or to doubt the claimed causal production. Doubts about the existence of parts can come from three directions. First, if extensive efforts have failed to find evidence for the parts, then we have reason to believe that they do not exist. The cliché that absence of evidence is not evidence of absence does not apply when extensive attempts to find evidence have failed to provide reason to believe in existence. Extensive attempts to find evidence for demons, unicorns, and gods have provided no evidence for their existence, so belief in non-existence is justified.

Second, hypotheses about proposed parts and interactions can be rejected when the theories that propose them have been superseded by ones that provide better explanations. For example, the phlogiston theory that dominated chemical explanations of combustion for most of the eighteenth century was superseded by Lavoisier's oxygen theory, which proposed different parts and interactions. So, we have good reason to doubt the existence of phlogiston and its interactions with flammable materials. Subsequently, oxygen was isolated from water and other gases, and eventually, oxygen atoms could even be photographed through electron microscopes, so the evidence for the parts and interactions in the oxygen mechanism has become progressively stronger.

Third, the existence of parts and interactions can be doubted because, as already suggested, their operation is inconsistent with established scientific principles. Homeopathic medicine became popular in the early nineteenth century by suggesting that minute quantities of substances that bore some similarity to disease symptoms could be used to cure the disease. This mechanism is defective for explaining diseases because of the general implausibility of causal claims based on minute quantities and similarities.

Some proposed mechanisms are not just defective but actually toxic in that their applications are harmful to human beings by threatening their physical or psychological well-being. The alleged homeopathic mechanism is toxic because people who use ineffective treatments for serious medical problems may fail to get evidence-based treatments that actually work. In the early nineteenth century, homeopathy was probably better than standard treatments based on beliefs about humoral imbalance, which is another defective mechanism lacking in completeness and correctness that was also directly harmful because balance-restoring treatments of bloodletting and purging usually made patients worse.

In summary, mechanisms contribute effectively to inductive inference if they are strong or possibly if they are weak but stronger than available alternatives. Defective and harmful mechanisms block the epistemic and practical effectiveness of inductive inference.

8. Conclusions

The significance of the contributions of mechanisms to inductive inference extends beyond philosophy. Some psychologists have noticed the importance of mechanisms to people's thinking and learning [53,54]. Better understanding of how mechanisms are mentally represented and processed should contribute to further analysis of the advantages of causal mechanisms over relatively superficial knowledge of associations between observable events. Similarly, artificial intelligence has had great successes through the associative inductive method of deep learning, but human-level intelligence will require computers to grasp causality based on mechanisms [6,7]. Both people and computers can learn better if they appreciate the contributions that mechanisms make to inductive inference.

The social significance of the role of mechanisms to inductive inference comes from the need to differentiate misinformation from information. Dealing with climate change and COVID-19 has generated masses of informative evidence, but controversies have also spawned many instances of misinformation such as claims that climate change is a hoax and COVID-19 can be treated by ingesting bleach [55]. Separating information from misinformation requires identifying good patterns of inductive inference that lead to the information and defective patterns that lead to the misinformation. Noting the contribution of mechanisms to justifiable induction is one of the contributions to this separation, as Table 1 summarized for both climate and COVID-19.

It might seem ridiculous that mechanisms could also be relevant to deductive inference, but consider an argument due to Gilbert Harman [29]. Suppose you believe that all aardvarks are gray and that the animal in the zoo is an aardvark. Should you therefore infer deductively that the animal is gray? If you see that the animal is actually brown, you might want to consider instead that it is not an aardvark, or that you were wrong in thinking that all aardvarks are gray. In general, you cannot infer from the premises of a deductive argument that the conclusion is true, because you might need to question some of the premises or even worry about the validity of a particular kind of deductive argument such as disjunctive syllogism. Harman's argument suggests that all deductive inference is actually inductive, so that mechanisms are potentially relevant. You might know, for example, that mutations in color genes are common in aardvark-like animals and thus have further reason to doubt your deductive inference.

I have emphasized that inductive inference does not always depend on mechanisms. Nevertheless, when knowledge of mechanisms is available, it can often be valuable in making inductive inference more reliable. I have shown the relevance of mechanistic information to generalization, inference to the best explanation, causal reasoning, and thinking based on probabilities. Thinking mechanistically makes people smarter and helps to naturalize logic.

Funding: This research received no external funding.

Conflicts of Interest: The author declares no conflict of interest.

References

1. Thagard, P. *Natural Philosophy: From Social Brains to Knowledge, Reality, Morality, and Beauty;* Oxford University Press: New York, NY, USA, 2019.
2. Johnson-Laird, P.N.; Byrne, R.M. *Deduction;* Lawrence Erlbaum Associates: Hillsdale, NJ, USA, 1991.
3. Rips, L.J. *The Psychology of Proof: Deductive Reasoning in Human Thinking;* MIT Press: Cambridge, MA, USA, 1994.
4. Holland, J.H.; Holyoak, K.J.; Nisbett, R.E.; Thagard, P.R. *Induction: Processes of Inference, Learning, and Discovery;* MIT Press: Cambridge, MA, USA, 1986.
5. Thagard, P. *The Cognitive Science of Science: Explanation, Discovery, and Conceptual Change;* MIT Press: Cambridge, MA, USA, 2012.
6. Goodfellow, I.; Bengio, Y.; Courville, A. *Deep Learning;* MIT Press: Cambridge, MA, USA, 2016.
7. Thagard, P. *Bots and Beasts: What Makes Machines, Animals, and People Smart?* MIT Press: Cambridge, MA, USA, 2021.
8. Bechtel, W. *Mental Mechanisms: Philosophical Perspectives on Cognitive Neuroscience;* Routledge: New York, NY, USA, 2008.
9. Craver, C.F.; Darden, L. *In Search of Mechanisms: Discoveries across the Life Sciences;* University of Chicago Press: Chicago, IL, USA, 2013.
10. Craver, C.F.; Tabery, J. Mechanisms in Science. In *Stanford Encyclopedia of Philosophy;* Stanford University: Stanford, CA, USA, 2015.
11. Glennan, S. *The New Mechanical Philosophy;* Oxford University Press: Oxford, UK, 2017.
12. Ranney, M.A.; Clark, D. Climate change conceptual change: Scientific information can transform attitudes. *Top. Cogn. Sci.* **2016**, *8*, 49–75. [CrossRef] [PubMed]
13. Dimitrov, D.S. Virus entry: Molecular mechanisms and biomedical applications. *Nat. Rev. Microbiol.* **2004**, *2*, 109–122. [CrossRef] [PubMed]
14. Mill, J.S. *A System of Logic*, 8th ed.; Longman: London, UK, 1970.

15. Connell, L.; Keane, M.T. A model of plausibility. *Cogn. Sci.* **2006**, *30*, 95–120. [CrossRef] [PubMed]
16. Thagard, P. Explanatory coherence. *Behav. Brain Sci.* **1989**, *12*, 435–467. [CrossRef]
17. Nisbett, R.E.; Krantz, D.; Jepson, C.; Kunda, Z. The use of statistical heuristics in everyday inductive reasoning. *Psychol. Rev.* **1983**, *90*, 339–363. [CrossRef]
18. Thagard, P.; Nisbett, R.E. Variability and confirmation. *Philos. Studies* **1982**, *42*, 379–394. [CrossRef]
19. Hempel, C.G. *Aspects of Scientific Explanation*; The Free Press: New York, NY, USA, 1965.
20. Goodman, N. *Fact, Fiction and Forecast*, 2nd ed.; Bobbs-Merrill: Indianapolis, IN, USA, 1965.
21. Bird, A.; Tobin, E. Natural Kinds. In *Stanford Encyclopedia of Philosophy*; Stanford University: Stanford, CA, USA, 2017.
22. Boyd, R. Realism, anti-foundationalism and the enthusiasm for natural kinds. *Philos. Studies Int. J. Philos. Anal. Trad.* **1991**, *61*, 127–148. [CrossRef]
23. Harman, G. *Thought*; Princeton University Press: Princeton, NJ, USA, 1973.
24. Lipton, P. *Inference to the Best Explanation*, 2nd ed.; Routledge: London, UK, 2004.
25. Thagard, P. The best explanation: Criteria for theory choice. *J. Philos.* **1978**, *75*, 76–92. [CrossRef]
26. Josephson, J.R.; Josephson, S.G. (Eds.) *Abductive Inference: Computation, Philosophy, Technology*; Cambridge University Press: Cambridge, UK, 1994.
27. Magnani, L. *Abductive Cognition: The Epistemological and Eco-Cognitive Dimensions of Hypothetical Reasoning*; Springer: Berlin, Germany, 2009.
28. Peirce, C.S. *Collected Papers*; Hartshorne, W.P., Burks, A., Eds.; Harvard University Press: Cambridge, MA, USA, 1958.
29. Harman, G. *Change in View: Principles of Reasoning*; MIT Press/Bradford Books: Cambridge, MA, USA, 1986.
30. Laudan, L. A confutation of convergent realism. *Philos. Sci.* **1981**, *48*, 19–49. [CrossRef]
31. Thagard, P. Coherence, truth, and the development of scientific knowledge. *Philos. Sci.* **2007**, *74*, 28–47. [CrossRef]
32. Van Fraassen, B. *The Scientific Image*; Clarendon Press: Oxford, UK, 1980.
33. Thagard, P. *Coherence in Thought and Action*; MIT Press: Cambridge, MA, USA, 2000.
34. Hume, D. *A Treatise of Human Nature*; Selby-Bigge, L.A., Ed.; Clarendon Press: Oxford, UK, 1888.
35. Hill, A.B. The environment and disease: Association or causation? *Proc. R. Soc. Med.* **1965**, *58*, 295–300. [CrossRef] [PubMed]
36. Hennekens, C.H.; Buring, J.E. *Epidemiology in Medicine*; Little, Brown and Co.: Boston, MA, USA, 1987.
37. Dammann, O. *Etiological Explanations*; CRC Press: Boca Raton, FL, USA, 2020.
38. Dammann, O.; Poston, T.; Thagard, P. How do Medical Researchers Make Causal Inferences? In *What Is Scientific Knowledge? An Introduction to Contemporary Epistemology of Science*; McCain, K., Kampourakis, K., Eds.; Routledge: New York, NY, USA, 2019; pp. 33–51.
39. Thagard, P. *How Scientists Explain Disease*; Princeton University Press: Princeton, NJ, USA, 1999.
40. Pearl, J. The algorithmization of counterfactuals. *Ann. Math. Artif. Intell.* **2011**, *61*, 29–39. [CrossRef]
41. Gentner, D. Structure-mapping: A theoretical framework for analogy. *Cogn. Sci.* **1983**, *7*, 155–170. [CrossRef]
42. Howson, C.; Urbach, P. *Scientific Reasoning: The Bayesian Tradition*; Open Court: Lasalle, IL, USA, 1989.
43. Olsson, E. *Against Coherence: Truth, Probability, and Justification*; Oxford University Press: Oxford, UK, 2005.
44. Hájek, A. Interpretations of Probability. In *Stanford Encyclopedia of Philosophy*; Stanford University: Stanford, CA, USA, 2019.
45. Hájek, A.; Hitchcock, C. *The Oxford Handbook of Probability and Philosophy*; Oxford University Press: Oxford, UK, 2016.
46. Kahneman, D.; Tversky, A. (Eds.) *Choices, Values, and Frames*; Cambridge University Press: Cambridge, UK, 2000.
47. Kuhn, J. Part 2: Why does glass break? *Corning Museum of Glass*. 2015. Available online: https://blog.cmog.org/2015/06/03/part-2-why-does-glass-break/ (accessed on 18 June 2021).
48. Abrams, M. Mechanistic probability. *Synthese* **2012**, *187*, 343–375. [CrossRef]
49. Popper, K.R. The propensity interpretation of probability. *Br. J. Philos. Sci.* **1959**, *10*, 25–42. [CrossRef]
50. Thagard, P. Causal inference in legal decision making: Explanatory coherence vs. Bayesian networks. *Appl. Artif. Intell.* **2004**, *18*, 231–249. [CrossRef]
51. Kwisthout, J.; Wareham, T.; van Rooij, I. Bayesian intractability is not an ailment that approximation can cure. *Cogn. Sci.* **2011**, *35*, 779–784. [CrossRef] [PubMed]
52. Pearl, J.; Mackenzie, D. *The Book of Why: The New Science of Cause and Effect*; Basic Books: New York, NY, USA, 2018.
53. Johnson, G.S.; Ahn, W.K. Causal networks or causal Islands? The representation of mechanisms and the transitivity of causal judgment. *Cogn. Sci.* **2015**, *39*, 1468–1503. [CrossRef] [PubMed]
54. Keil, C.F.; Lockhart, K.L. Beyond cause: The development of clockwork cognition. *Curr. Dir. Psychol. Sci.* **2021**, *30*, 167–173. [CrossRef]
55. Thagard, P. Mechanisms of misinformation: Getting COVID-19 wrong and right. Forthcoming.

philosophies

Article

The Philosophy of Nature of the Natural Realism. The Operator Algebra from Physics to Logic

Gianfranco Basti

Faculty of Philosophy, Pontifical Lateran University – Vatican City 00120; basti@pul.va

Abstract: This contribution is an essay of formal philosophy—and more specifically of formal ontology and formal epistemology—applied, respectively, to the philosophy of nature and to the philosophy of sciences, interpreted the former as the ontology and the latter as the epistemology of the modern mathematical, natural, and artificial sciences, the theoretical computer science included. I present the formal philosophy in the framework of the category theory (CT) as an axiomatic metalanguage—in many senses "wider" than set theory (ST)—of mathematics and logic, both of the "extensional" logics of the pure and applied mathematical sciences (= mathematical logic), and the "intensional" modal logics of the philosophical disciplines (= philosophical logic). It is particularly significant in this categorical framework the possibility of extending the operator algebra formalism from (quantum and classical) physics to logic, via the so-called "Boolean algebras with operators" (BAOs), with this extension being the core of our formal ontology. In this context, I discuss the relevance of the algebraic Hopf coproduct and colimit operations, and then of the category of coalgebras in the computations over lattices of quantum numbers in the quantum field theory (QFT), interpreted as the fundamental physics. This coalgebraic formalism is particularly relevant for modeling the notion of the "quantum vacuum foliation" in QFT of dissipative systems, as a foundation of the notion of "complexity" in physics, and "memory" in biological and neural systems, using the powerful "colimit" operators. Finally, I suggest that in the CT logic, the relational semantics of BAOs, applied to the modal coalgebraic relational logic of the "possible worlds" in Kripke's model theory, is the proper logic of the formal ontology and epistemology of the natural realism, as a formalized philosophy of nature and sciences.

Keywords: quantum field theory; Kripke model theory; physical causality principle

Citation: Basti, G. The Philosophy of Nature of the Natural Realism. The Operator Algebra from Physics to Logic. *Philosophies* **2022**, *7*, 121. https://doi.org/10.3390/philosophies7060121

Academic Editors: Gordana Dodig-Crnkovic and Marcin J. Schroeder

Received: 15 May 2022
Accepted: 30 September 2022
Published: 26 October 2022

1. Introduction: From Logic to Physics and Vice Versa

1.1. A Methodological Premise: Mathematical Logic and Philosophical Logic

The final aim of this contribution is to develop the formal ontology and epistemology of the *natural realism* (NR), as a formalized philosophy of nature and a formalized philosophy of science. They are interpreted, respectively, with the former as the ontology, and the latter as the epistemology of the modern natural and artificial science, the theoretical computer science (TCS) included, using the *category theory* (CT) as metalanguage of logic and mathematics. In this framework, the NR formal ontology is a categorical interpretation of the so-called *ontic structural realism* (OSR) approach to the philosophy of quantum physics (see [1,2]), or more generally of the *ontic interpretation* of the ψ-wave function (see [3] for an updated discussion).

Now, the proper *modal relational semantics* of the NR-formal ontology is that in it (the complex Boolean structures of), the propositional formulas of a descriptive ontology of the physical systems/processes can be validated "by homomorphisms up to isomorphisms" directly onto (the complex algebraic structures of) the mathematical models of the physical systems to which the ontological formulas descriptively refer. This depends, ultimately on the extension of the *operator algebra formalism* from physics to logic (see Sections 3.2 and 5.1), and then on the algebraic relational interpretation of the meaning function $[\cdot]$ in CT logic, for which the extension of a complex formula φ of the propositional calculus making it true, i.e. $[\varphi]$, is not defined by operations onto set-subset partial orderings such as in the set-theory (ST) logic, but primarily by operations onto a complex algebra (algebra-subalgebras) structure, in the common framework of the *operator algebra* formalism, extended from the mathematical physics to the Boolean logic, i.e., the so-called *Boolean algebra with operators* (BAO) (see [4,5] and below Section 5.1.4). All this is synthesized into the motto that "meaning is homomorphism" because meaning is based on a structure preserving mapping or homomorphism from the algebraic complex structure of a physical object in its mathematical model, onto the algebraic complex structure of the logic of a predicative sentence, in the descriptive language of ontology, which just because of this homomorphism is "referring to" or "signifying" this physical object [6].

Effectively, in this way, I want to emphasize the relevance of R. Goldblatt's suggestion synthesizing the main difference between the CT and the ST metalanguage in the slogan "arrows instead of epsilon" (see [7], pp. 37–74). Specifically, in ST, we suppose Russell's *set-elementhood* principle expressed in the *Principia* [1] for avoiding in axiomatic ST Frege's and Cantor's antinomies, and then we are supposing the predicate logic making of the *set-membership* relation $\in$ is a primitive in ST. On the contrary, in CT, we can formalize Peirce's pioneering intuition of a triadic algebraic construction of the predicate domains, making *morphisms* (arrows) the primitive of CT, with the consequent categorical notion of the set as *hom-set* (see Section 3).

The CT metalanguage is particularly suitable, therefore, for formalizing the constructive power of nature in constituting *dynamically* new domains of predication as it is required by an *evolutionary* approach not only in biology but also and primarily in cosmology. This is based on the universal mechanism of the (infinitely many) *spontaneous symmetry breakings* (SSBs) of the quantum fields at their ground state (i.e., the so-called *quantum vacuum* (QV) condition) in the *quantum field theory* (QFT), conceived as fundamental physics. This holds, both at the *microscopic* level of relativistic quantum physics of the *standard model* (SM) of elementary particles (see Section 3), and at the *macroscopic* level of the *condensed matter* physics of the chemical and biological systems (see [8–10] for a synthesis).

In this CT framework, the subcategory in the category **Set** of the *non-well-founded* (NWF) sets, violating the "set-elementhood" principle (see Note 1) because it satisfies P. Aczel's *anti-foundation axiom* [11] by which set *self-membership* is allowed, is particularly suitable for our aims. Specifically, for modeling in CT logic and mathematics the notion of *emergence* of new physical systems as a result of as many SSBs of the QV, i.e., as many *phase coherence domains* of the quantum fields at their ground state, which can be modeled in NWF-set theory as new "self-containing wholes", irreducible to the simple "combinatorics" of elements according to the famous expression "more is different" that was coined by the Nobel Prize Ph. Anderson precisely for characterizing any *phase transition* in fundamental physics [8].

Finally, both in ST and CT logics, the distinction holds between the *mathematical* and the *philosophical* logics that in its modern form is due to the American logician Ch. I. Lewis in his criticism of the application of the *extensional, truth-functional* mathematical logic of the *Principia* to the analysis of the philosophical, especially metaphysical, theories [12], thus criticizing *ante litteram* the core of Wittengstein's *Tractatus.* The philosophical logic is, indeed, the *modal logic* (ML), the logic of necessity and possibility, of "must be", and "may be" of which Lewis first proposed an axiomatic version by adding new modal *symbols* (essentially, the *necessity* $\Box$ and the *possibility* $\Diamond$ operators) and *axioms,* respectively, to the alphabet and to the axioms of the standard propositional calculus of mathematical logic to define for the first time in the history of logic a formal *modal calculus* (MC) [13]. Therefore, by combining in a proper way the modal axioms, we can obtain as many *modal systems* as the proper syntax of different philosophical theories (see [12–14], for a complete presentation of the axiomatic approach to the MC, and Section 5.2.3 below for a partial exemplification). In this way, the distinction, and at the same time the relationship between *mathematical* and *philosophical* logics, started to take its actual form using the rigor of the axiomatic method.

Indeed, saying that ML is the logic of necessity and possibility—a distinction per se that is meaningless in mathematical logic—means, using S. Kripke's many-worlds *modal relational semantics* [15,16], that in the *modal model theory*, we are dealing with truth or falsity of propositions not concerning only one state-of-affairs, or "actual world", as in the standard Tarskian model theory in mathematical logic [17], but also with truth or falsity in other *possible states-of-affairs* or "possible worlds" that possess *some relation* with the actual one. An approach that, also intuitively, is compliant with an evolutionary cosmology, based on the physical *causality principle* of the special relativity (SR) "light-cone" that holds both in general relativity (GR), and QFT (see below Section 1.2), and where, therefore, "cosmogony is the legislator of physics", according to J. A. Wheeler's intriguing statement about quantum gravity in cosmology [18]. Consequently, in ML, a proposition will be *necessary* in a world, if it is true in *all possible worlds related* to that world, and *possible*, if it is true *at least in another world*, relatively to the former one. This implies, of course, that in ML, the logical connectives (propositional predicates) are not *truth-functional*, at least in Frege's sense related to the usage of the *truth-tables* for the propositional connectives/predicates ("not", "and", "or", "if... then", ...) [2].

To sum up, the *different meanings* of the modal operators correspond to as many *different semantics* and then to as many *truth criteria*, ruled by suitable axioms, for the interpretations of the MC, by which formalizing in a proper way, and then comparing, different philosophical theories, their consistency, and their effectiveness in solving the problems for which they were developed and defended by the respective supporters. Now, the main semantics of the MC generally admitted in ML are the following:

1. The *alethic logics*, where the meaning of the modal operators is *possibly/necessarily true* in descriptive theories of the world states, in the different senses of the *logical*, and the *ontological* (physical and metaphysical) truth. Specifically, without confusing the *logical* (linguistic, abstract) and the *ontic* (causal, real) possibility/necessity, and their relationships. Historically, this distinction is the core of the classical Aristotelian philosophy and it was reintroduced in the contemporary analytic philosophy debate by S. Kripke at the end of the XX cent (see [19] and Section 6.2). Of course, the *onto*-logical alethic interpretation of MC is the proper logic of the *formal ontology*.
2. The *epistemic logics*, where the meaning of the modal operators possible/necessary is related to different levels of knowledge *certainty*, and then to the distinction between *opinion/science* (*dóxa/epistéme*, in the Platonic language of the classic philosophy) [20–22]. Therefore, the necessity operator is interpreted in epistemic contexts as the "knowledge operator" **K**, and the possibility operator is here interpreted as the "belief operator" **B**. The possible worlds concerned here are the *believed representations* of the world relatively to a knowing (conscious) singular/collective *communication agent*, *x*. Additionally, the passage from "believing for *x* that *p*", **B**(*x*,*p*), to "knowing for *x* that *p*", **K**(*x*, *p*), depends on the satisfaction of a *foundation clause* **F**, i.e., $\mathbf{K}(x,p) \Leftrightarrow \mathbf{B}(x,p) \wedge \mathbf{F}p$, in the sense that the *sound* (true) beliefs or scientific knowledges are those founded in the real world. Of course, the clauses **F** will be different for different epistemologies, and for different underlying ontologies, which in this way can be rigorously compared and discussed (see [20] and Section 1.4).
3. The *deontic logics*, where the meaning of the modal operators possible/necessary is related to different levels of ethical/legal *obligation*, and then the necessity/possibility operators of MC must be interpreted as the deontic operators of *obligation* **O**, and *permission* **P** [20,23,24]. The possible worlds concerned here, namely, the "ideal worlds" of the *ought to be*, as distinct from the "real world" of the *to be*, are those related to the ethical values or "goals" to be pursued. Or, more precisely, they are related with the axiological *optimality/maximality* criteria of "goodness" for actions to be satisfied according to the different ethical/legal systems. This means imposing ethical/legal constraints or "obligations" for the *effective pursuing* of the goals in the "real world"

by the human agents in terms of ethical optimality/maximality goodness constraints being satisfied [3]. Where, of course, the distinction between *moral* and *legal* obligations, and then between the *individual* and the *common* good(s) is fundamental [20]. From the standpoint of the history of philosophy, the distinction between the "alethic" and the "deontic" semantics of ML gives a formal foundation to the so-called "Hume problem" of the distinction between the "world of facts" ("to be": alethic logic) and the "world of values" ("ought to be": deontic logic), well known to the Middle Age logic but lost during the Renaissance and recovered by Hume. Moreover, in the case of the deontic obligatoriness being distinct from the logical necessity, the "possible worlds" x concerned are the *optimal states s of the world* (so introducing the "optimality operator" **Op** of the axiological logic (the "logic of values")), for a *given (individual, collective) subject x*, i.e., **Op** (x, s) [4]. Therefore, the ethical obligatoriness expressed by the moral/legal norm p, i.e., **Ob** p, ruling the behavior for pursuing effectively in the real world a given optimal state s by x, i.e., **Ob** $p(x,s)$, satisfies the following axiomatic scheme: $\mathbf{Ob}p := (\mathbf{Op}\,(x,s) \wedge c_a \wedge c_{ni}) \leftrightarrow \mathbf{Ob}\,p(x,s)$, where the two clauses c_a and c_{ni} express, respectively, the "condition of acceptance" by the individual/collective subject x of the optimal ordering **Op,** and the "condition of non-impediment" for x of effectively pursuing s in the real world [20].
4. Finally, in the MC semantics, it is possible to also formalize *intensional objects* and *predicates*, and not only intensional interpretations of modal operators, as we did till now, sometimes denoted as *individual concepts* ([14], p. 332). Generally, indeed, the "possible worlds" are modeled as classes of objects satisfying given modal rules. For this reason, MC is normally formalized in ST using **NBG** as its metalanguage but with the remembered restrictions and distinctions characterizing the different modal object domains [25]. However, it is also possible to model possible worlds by considering, for defining the truth evaluation functions of the modal semantics, the *individuals* within *a partition* of possible worlds of the universe (i.e., of the set of all possible worlds) considered. In this way, in the validation procedure, the *contingent identity* can also be considered, that is, *the identity of individuals* satisfying different predicates in different possible world partitions. In this sense, the ML semantics, because of its high flexibility, appears to be able to formalize the *intensional* (with "s") *logics* also in the sense of the *subject–object intentional* (with "t") *relationship* of the phenomenological inquiry [26]. Specifically, it expresses the *singular/plural first-person ("I"/"we") language* of individual/collective *intentional agents*, i.e., the "belief systems" of the different individuals and cultural groups in a society. This means that—against the dominating "relativism"—using the intensional logic formalization, it becomes possible to compare different visions of the world, in ontology, ethics, epistemology…, as far as each group, each "we", makes the effort of formalizing what they "intend" with their respective doctrines, i.e., in their "intensional logics". Then, according to the synthetic but effective account of John Searle, we can summarize by saying that the *intensional* (with "s") logic is also the proper logic of the cognitive, subjective *intentionality* (with "t") [27].

We can conclude, therefore, that the main distinction between *philosophical logic* and *mathematical logic* reduces to that between *modal* (*intensional)* and *extensional* logics, respectively [20,21], against the reductionist program of the early Neo-Positivistic approach to the philosophical analysis. Moreover, we must recall that the "philosophical logic" is not the same as the *philosophy of logic,* that is, the philosophical enquiry about the foundations of the formal (mathematical and philosophical) logic.

Finally—and this brings us back to the formal core of this paper—in addition to the early Lewis' *axiomatic* approach, and Kripke's *relational* approach to MC and ML, both based on the ST metalanguage, today, the more fruitful approach to MC and ML is the *algebraic* approach to Kripke's modal relational semantics that applies both to mathematical and philosophical languages in the framework of CT metalanguage. The algebraic approach is, indeed, based on a categorical modal interpretation of BAOs. For this taxonomy of the different ML approaches, see [28] and Section 5 of this paper.

1.2. The Logical Issue of Whichever Formal Ontology and Epistemology of Natural Sciences

For our aims, the relevance of a categorical formalization of ML emerges clearly when we reflect on the main issue of whichever formal ontology and epistemology of the natural sciences. For this, we can refer to the teaching of W. V. O. Quine, and more specifically to his criticism of the axiomatic approach to ML developed by Ch. I. Lewis in its pretension of being the proper logic of ontology and metaphysics:

> What the resulting Lewis' systems describe are actually modes of *statement composition*—revised conditionals of a non-truth-functional sort—rather than implication relations between statements. If we were willing to reconstrue statements as names of some sort of entities, *we might take (metaphysical) implication as relation between those entities rather than between the statements themselves*; and correspondingly for equivalence, compatibility, etc. ([29] p, 32. Italics are mine).

In a word, what Quine is rightly vindicating as a proper foundation of the modal logic of the metaphysical *implication* (premise-conclusion) in a *formal* ontology and/or metaphysics is the necessity that the modalities of the *logical* relations between statements be able *to denote* ("to name") in some proper way the modalities of the *real* (causal) relations among the extra-linguistic entities, to which an ontological/metaphysical statement pretends to refer. However, Lewis' modal logic system is not able, in principle, to satisfy this requirement!

As I synthesized elsewhere [30–32] and we discuss at length in this contribution, the more direct and elegant way to satisfy Quine's deep requirement is to justify in a naturalistic ontology the *functorial dual equivalence* $\overset{\rightleftarrows}{\underset{\Omega^*/\Omega}{\longleftarrow}}$ in a categorical setting, between the *logical entailment* for which "it is impossible that the premise is true, and the consequence is false" $(\neg\diamond(\alpha^*\wedge\neg\beta^*))$ on the *logical* side of the descriptive statements of the ontological language, and the dual *causal modal entailment* "it is impossible the effect without its cause" $(\neg\diamond(\beta\wedge\neg\alpha))$ on the *ontic* side of the physical objects to which the descriptive statements refer. Here, the latter must be considered in some proper way as the semantic extension on which it validates *dually* the propositional formulas of the former.

As we see, this *dual relationship* between the *logical* and the *causal* entailments is the core of the Aristotelian theory of the *demonstrative syllogism* (premise $\rightarrow$ conclusion), where the soundness of its premise is founded dually by homomorphism on the conclusion of the *causal syllogism* (cause $\leftarrow$ effect). This is a theory that can only be justified in the categorical framework of the theory of the *functorial bounded morphism* between Kripke models, respectively, on the ontic and on the logical sides of the NR formal ontology, as discussed in Section 5 (see Sections 5.1.3 and 5.2.3). We return to this specific point in the conclusive Section 6 of this work when we examine our theoretical proposal developed in this paper in a historical perspective. To conclude, it is worth emphasizing that this distinction between the causal and the logical necessitations (entailments) was reproposed in recent times by Kripke in his seminal *Naming and Necessity* book [19], even though it received its proper formal justification only in the CT framework of a coalgebraic semantics of Kripke model theory (see Section 5.2.3).

What indeed immediately excited the interest of scholars in Kripke's proposal is the evidence that the causal entailment, in Kripke's many worlds *relational semantics*, appears to be the ML version of the *causality principle* in fundamental physics based on the so-called "light-cone" of special relativity (SR). In fact, and it is important to recall this in our context, this causality principle holds, both for the three quantum interaction force fields of the relativistic QFT, and for the gravitational force field of general relativity (GR), as one of the most distinguished theoretical physicists of our time, the 1979 Nobel Laureate in Physics Steven Weinberg (1933–2021), also recently pointed out. He, indeed, in his last published book dedicated to the *Foundations of Modern Physics*, in the paragraph about the *causality* in fundamental physics, stated (see Figure 1):

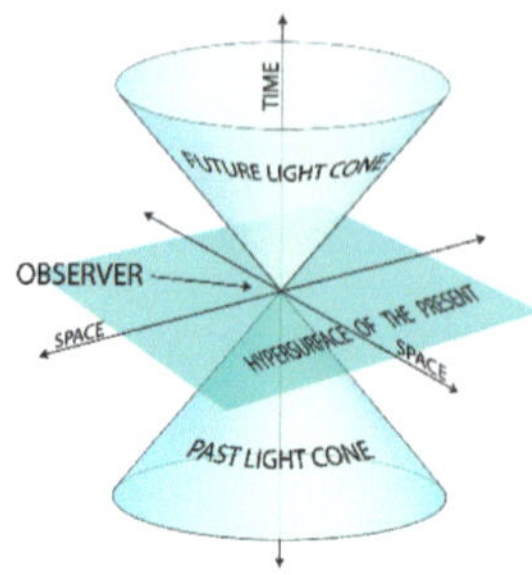

Figure 1. Intuitive representation of Minkowski's light-cone of special relativity, which is defined on complex numbers, where the future-directed states are defined on real numbers and the past-directed states are defined on imaginary numbers.

> We saw (. . .) that no Lorentz transformation acting on a body at rest could give it a speed greater than *c*, the speed of light. We can derive a stronger result, that no influence whatever can travel faster than light. This is not just a confession of technological inadequacy, but a consequence of an assumption of causality, that *effects always come after causes*" ([33], pp. 121–122 (italics mine)).

Intuitively—but overall formally (see Section 5.2.3)—it is evident that Kripke's many-world relational semantics is the proper model theory of the causal light-cone granting *a dynamic partition criterion* among the possible world-states in terms of their "causal accessibility" from/to a given past/future physical event. Moreover, this evidence acquires a precise formal justification in the categorical formalization of the relativistic quantum physics (QFT) when we reflect upon the evidence that the "causal relations" from/to past/future events satisfy the dual definition of *morphisms* (arrows) from/to an *initial/terminal object*, characterizing, respectively, the categories of *algebras* and *coalgebras* in CT logic and mathematics (see Section 2.4 and especially Definition 7. and Note 12).

Particularly, it is worth emphasizing from the ontological standpoint that, when we consider the "future light-cone" on the cosmological scale of the *universe evolution* by which it populates progressively itself of ever more complex objects and structures in the "hot big-bang hypothesis", because of the strongly non-linear character of the causal processes involved (symmetry breakings) [10], the *logical/mathematical unpredictability* of the "effects" (future events) as to their common "cause" implies that the only morphisms that logically make sense are those from the effects as to their common cause. It, therefore, categorically plays the role of the common *terminal object*, to which all the "arrows" relating the effects to their cause are directed. This notation emphasizes the coalgebraic nature of the "future" light-cone, and then the "coinductive" nature of the "causal entailment" (see Section 5.2).

Only from this simple reflection does the *mathematical* and *logical* relevance of the *category of coalgebras (coproducts)* functorially represented appear, which we discuss in Section 2.4, but also their *physical* relevance, which we discuss in Section 4.5. Indeed, the (Hopf) coproducts and coalgebras play a fundamental role in the QM and QFT calculations over lattices of quantum numbers. In fact, in this case the coproducts—in terms of summations for calculating the total energy of a superposition of particles (fields) in a quantum state (phase)—are the fundamental way for knowing how many and which type of particles are superposed in a quantum state. Effectively, this is how many and which type of matter fields (of which the "elementary particles" or "fermions" are their quanta) stay in a coherent phase. In this case, for coming back to the causal light-cone, its "local accessibility relations" among physical states (phases) in the physical space-time become as many *phase transitions allowed* among quantum fields in QFT. We present in Sections 3 and 4 a sketch of the categorical formalization of the representation theory of these phase transitions (or unitarily inequivalent representations of the quantum fields dynamics) in QFT, according to its different interpretations and models.

Therefore, coming back to our philosophical discussion, a possible significant solution of Quine's conundrum about the same possibility of a formal ontology on a naturalistic basis (effectively, a formal ontology and epistemology of QFT as fundamental physics) is given in the framework of the CT logic. Namely, according to a categorical *(co-)algebraic relational semantics* of the *meaning function* $[\cdot]$ mapping a formula $[\varphi]$ of the Boolean propositional calculus into its coalgebraic *extension* φ, *validating* "making φ true". This relational semantics was inaugurated by Jónsson's and Tarski's application to Boolean logic of the *operator algebra* formalism, already extensively applied in quantum and classical physics, i.e., the so-called *Boolean algebra with operators* (BAO), in the framework of the celebrated Stone's *representation theorem for Boolean algebras* (RTBA) (see Section 5.1 and Appendix B). Indeed, the *topologies* of RTBA in logic and quantum physics *are ultimately the same,* so that RTBA is the theoretical foundation of any possible bridging between physics and logic, and then of any possible formal ontology of quantum physics.

Without anticipating here all the passages of the argumentation given in Sections 4 and 5 of this paper, we can emphasize two essential points. Before all, it is possible to demonstrate in CT logic the completeness of Kripke's relational semantics using a coalgebraic interpretation over trees on NWF-sets defined in the Stone spaces, in the framework of the *functorial dual equivalence* between the category of Stone coalgebras and the category of the modal Boolean algebras with operators (MBAO), $\mathbf{SCoalg(\Omega)} \simeq \mathbf{MBAO(\Omega^*)}$ [6]. Secondly, it is possible to extend this categorical dual equivalence to the category of the Hopf coalgebras in physics for the Bogoliubov functor $\mathcal{B}$ $\mathbf{qHCoalg(\mathcal{B})}$ [34].

Effectively, this relational semantics is only a significant application of the more general principle characterizing the CT logic, according to which "a statement α is *true* in/about a category C if and only if its dual α^{op} (i.e., obtained from α by reversing all the arrows and their compositions) is *true* in/about the opposed category C^{op}" (see Section 2.3).

On the other hand, all this offers a categorical solution not only for Quine's *ontological* conundrum but simultaneously for the other modern conundrum of the justification of the *soundness* of the premises (sufficient conditions) in the hypothetical reasoning, afflicting the epistemology of modern sciences from Galilei to Popper (see Section 6.2). Indeed, Karl R. Popper (1902–1994) so synthesized the problem in his masterpiece *The logic of scientific discovery* (1935):

> If we distinguish, with Reichenbach, between a 'procedure of finding' and a 'procedure of justifying' a hypothesis, then we have to say that the former—the procedure of finding a hypothesis—cannot be rationally reconstructed. Yet the analysis of the procedure of justifying hypotheses does not, in my opinion, lead us to anything which may be said to belong to an inductive logic. For a theory of induction is superfluous. It has no function in a logic of science ([35], p.307).

Evidently, in the light of what we just said, we try to demonstrate in this contribution that there exists in CT logic a rational procedure for justifying the *soundness* of the hypothesis in the hypothetical deductive method of modern sciences.

1.3. A Scheme of this Paper

In the Section 2, I summarize some elements of CT as a formal metalanguage of the mathematical and logical theories in a systematic comparison with set theory (ST), with which philosophers (and physicists) are generally more acquainted than with CT. Particularly, I emphasize that CT is particularly suitable as a formal metalanguage of the *operator algebra* and then of the *topological* approach in mathematics and physics, and logic and computer science because of the development of the so-called BAO by Jónsson and Tarski [4,5] in the framework of the celebrated Stone's *representation theorem for Boolean algebras* (RTBA) [36] (see below Section 5.1 and Appendix B).

In Section 3, I summarize some elements of the QM formalism, particularly the completion of the original Von Neumann formalization of QM, via the so-called *Gelfand–Naimark–Segal (GNS) construction* that inaugurated the operator algebra approach to QM. Afterward, I present some basic notions of the QFT formalism in the framework of special relativity theory (SR), according to the original Dirac's interpretation of QFT as a *second quantization* (SQ) with respect to QM.

In Section 4, I summarize the core of the extension of the QFT system representation theory to the modeling of quantum dissipative systems (or dissipative QFT) persistently in far-from-equilibrium conditions because of passing through different phases. This is based on the *Bogoliubov transform* mapping between different phases of fermionic and/or bosonic quantum fields, both in the relativistic QFT (at the physical *microscopic* level) and QFT of the condensed matter physics (at the *macroscopic* level of the chemical and biological systems). Because of the necessary *non-commutative* character of the Hopf algebra *coproducts* in calculations over lattices of quantum numbers in the case of open systems [5], the mathematical formalism of dissipative QFT implies the necessity of the *algebra doubling*, and then of the doubling of the *state (phase) spaces*, and finally of the same *Hilbert spaces* for recovering the canonical (closed) *Hamiltonian representation* of the total system.

This is obtained by also inserting in the Hamiltonian—through the method of the algebra doubling—the thermal bath degrees of freedom, with which any quantum dissipative system is necessarily *entangled*, so to grant a far from equilibrium *energy balance*, and the "closed" character of the resulting system as required by the Hamiltonian "canonical" representation.

The main mathematical result of this approach is then the so-called principle of the *doubling of the degrees of freedom* (DDF), by which it is possible for the system itself "to decide dynamically", which is the proper finite number of the degrees of freedom of the statistical expectations in the Hamiltonian for a faithful representation of the system dynamics.

This means that the ground state of the quantum fields in the dissipative QFT, i.e., their *QV condition*, because it is necessarily at a temperature $T > 0$, allows *different phase-coherence domains*, non-interfering with each other, to coexist in the same balanced (0-summation free energy) ground state of the quantum fields. Each of these phase coherences of the quantum fields at their ground state—according to the fundamental *Goldstone theorem*—corresponds to a *spontaneous symmetry breaking* (SSB) of QV, from which new properties in physical systems emerge. Each SSB corresponds indeed to the spontaneous instauration of *long-range correlations* among quantum fields at their ground state (QV), and it is therefore univocally indexed by the *unique value* $\mathcal{N}$ of the condensate of the so-called *Nambu–Goldstone* (NG) bosons, i.e., the quanta of the long-range correlations among the quantum fields.

Because of the stability of these collective modes of quantum fields that do not require any further energy contribution since all coexist at the same balanced ground state (0-sum energy) of a dissipative system, it is possible to justify a *dynamic partial ordering* of them. All this is the core of the *QV-foliation* principle that can be formalized in CT using the *colimit* operation (see Appendix A), which therefore appears to be the fundamental tool used by nature for generating *complex systems* and for justifying at its fundamental physical level the notion of *memory* in biological and neural systems [34,37,38].

Effectively, in biology and neurosciences, the QV-foliation allows the proposal of an original solution of the debated issue of the *long-term memories* in mammals' brains, modeled as *dissipative brains* entangled (balanced) with their environment (i.e., with the rest of the body, and through it, with the outer environment), in the framework of the *intentional* interpretation of cognitive tasks [39,40]. Intentionality has its biological foundation, therefore, in the *homeostasis* characterizing all living systems as dissipative systems, according to A. Damasio's original proposal [41]. In this way, this neurophysiological application becomes one of the main empirical supports of QFT as the fundamental physics of biological systems.

In Section 5 of this paper, therefore, to arrive at the presentation of the coalgebraic foundation of the Kripke modal relational semantics as the proper logic of the NR formal ontology and epistemology, we start from a synthetic illustration of the momentous Stone's *representation theorem for Boolean algebras* (RTBA: see also Appendix B). From this the consequent development of BAOs derives, and then a *relational semantics* based on the algebraic interpretation of the meaning function in CT logic. In the CT approach to ML, this is based on the *dual equivalence* between the category of the coalgebras of NWF-sets defined in Stone spaces, **SCoalg**, and the category of the modal Boolean algebras with operators **MBAO**, for the contravariant application of the Vietoris functor $\mathcal{V}$. This constitutes the core of the coalgebraic justification of Kripke's modal relational semantics in CT logic [6,42,43]. Now, starting from the evidence that the Stone spaces in logic are the same topological spaces of the C*-algebras of Hilbert spaces in physics (see [44,45], and Appendix B), it is possible to define the Kripke relational semantics of NR-formal ontology directly in the category of physical coproducts for the contravariant application of the Bogoliubov functor $\mathcal{B}$ [34]. The Vietoris coalgebraic construction, indeed, grants—as the Bogoliubov functor $\mathcal{B}$ does *dynamically* via the DDF construction in the category of coalgebras for QFT systems—a *selection criterion of admissible sets* on which the semantics of the Boolean modal algebras are defined, analogously to the ultrafilters of Stone's RTBA. In both cases, indeed (the *physical* one (Bogoliubov) and the *logical* one (Vietoris)), the set indexing is performed by the *colimit operation* over categories of coproducts on NWF-sets. For this reason, we can write the categorical dual equivalence that is the core of the modal logic of the NR-formal ontology: $\mathbf{SCoalg}(\mathcal{B}) \simeq \mathbf{MBAO}(\mathcal{B})^*$, just as in logic, we write $\mathbf{SCoalg}(\mathcal{V}) \simeq \mathbf{MBAO}(\mathcal{V})^*$ [6].

Particularly, it is worth emphasizing that in the case of Kripke relational structures defined on NWF-sets, only *local modal truths* are allowed by the powerful notion of *functorial bounded morphism* between Kripke models, respectively, on the logic ($\mathfrak{M}$) and on the physical ($\mathfrak{M}'$) side, i.e., $\mathfrak{M} \overset{\leftrightarrows}{\leftarrow} \mathfrak{M}'$, as we illustrate in Sections 5.1.3 and 5.2.3. This semantics is indeed exactly what we need for formalizing a descriptive ontology of an evolutionary cosmology where "cosmogony is the legislator of nature".

Finally, the concluding Section 6 is dedicated to two fundamental metaphysical and epistemological issues to which the NR formal ontology could suggest a solution. At the beginning of the Modern Age, Immanuel Kant in his famous booklet *Prolegomena to any Future Metaphysics that will be able to come forward as a Science* [46] published in 1783, even though it was originally conceived as an Introduction to Kant's masterpiece *Critique of the Pure Reason*, stated that the future of a naturalistic metaphysics as science will pass necessarily through a new foundation of the *causality principle* in physics and metaphysics. Indeed, the modern Galilean and overall Newtonian physics, of which Kant's *Critique* wanted to constitute the epistemology, confuted the Aristotelian and Scholastic causal view of nature, and the causal justification of its laws. At the same time, it confuted the core of the Aristotelian epistemological realism, for which the *logical relations* among *objects* in reasoning (i.e., in the language of mind), *depend on*, and then *refer to* because *abstracted from* the *causal (real) relations* among *things* in nature. We summarize in which sense the categorical duality between the causal and logical modal entailments presented since the beginning of this *Introduction* (see Section 1.1), and formally justified in the rest of this paper, is in continuity with the *relational natural realism* of the NR formal ontology (see Section 1.4) in the framework of the CT logic presented in this contribution.

1.4. A Taxonomy of the Different Formal Ontologies in Western Thought

As a conclusion of this *Introduction* devoted to illustrating the theoretical and historical background of my proposal of a renewed philosophy of nature as a formal ontology of the natural sciences, let us sketch briefly which are *the main formal ontologies* in the history of Western thought to immediately locate my proposal in this schematic survey.

Today, the term "formal ontology" is widely used in the computer science environment, particularly in the so-called knowledge engineering realm for the development of *semantic databases*. In this sense, ontologies refer to the fundamental conceptual categories by which different linguistic groups organize their knowledges about the objects of their specific environments, that is, their representations of reality. It is often forgotten, however, that this usage of the term "formal ontology" in computer science refers implicitly to the origins of this term in the phenomenological philosophy [47,48].

Historically, indeed, Edmund Husserl introduced the terms "formal ontology" in the contemporary philosophical jargon for signifying the *ante-predicative foundation* of predicates in formal logic that he developed in his *transcendental logic,* based on the notion of the *intentional transcendental subject* [26] [7]. Specifically, against the *formalism* of René Descartes' *cogito,* and Immanuel Kant's *Ich denke überhapupt* ("I think in general"), which made the *self-conscious evidence* of the pure thinking the *conceptualist* foundation of the logical truth. In his criticism of the epistemic formalism of Descartes and Kant, Husserl, following his teacher Franz Brentano [49], vindicated that any psychical act as such (believing, thinking, willing, sensing. . .) is *evidently* characterized by an intrinsic *aboutness* or "reference to an object". In this sense, the pure *cogito* cannot exist or the Kantian "I think in general" since "I/we think (believe, will, sense. . .) always *something*", i.e., a given object. Conversely, no object can exist in logic or mathematics, according to Husserl, without supposing an implicit reference to a knowing (individual/collective) *subject.* To sum up, the modern principle of evidence has an intrinsic *intentional constitution,* based on the *transcendental relationship subject–object.*

Effectively, in the *Third Logical Investigation,* Husserl defends this *ontological* foundation of the *logical* truths because knowledge can access *real* beings/things only as objects-for-a-subject". Particularly, in the "Introduction" to this *investigation,* Husserl refers to the notion of *formal ontology* as the *"pure (a priori) theory of objects as such"* (see [50], p.3).

Indeed, this reference to the ontology, because of his criticism of the *formalism* typical of the modern "reshaping" of mathematics by the *axiomatic method* ([51], pp. 21–23), constitutes the main motivation of Husserl's phenomenological method since the very beginning of his career. Namely, since his PhD work (1891) in mathematics concerning the "calculus of variations", Husserl introduced the notion of *Inhaltlogik.* This is the "logic of contents", or "intensional logic", as he denoted it, for correcting the formalistic, purely syntactic nature of the calculus in modern extensional logic and mathematics [52], according to Frege's *Begriffsschrift* [53].

Now, in this light, it is important to compare Husserl's and Peirce's criticisms they independently made about Ernst Schröder's first volume of his treatise on the *Algebra of Logic* [54], published in 1890, which was the first historical proposal of a mathematical logic, before Frege's logistic or predicative one, based on his logic of classes [55]. A comparison between Husserl and Peirce is relevant for us because both agree independently about the insufficiency of Schröder's *dyadic* algebra of logic for justifying a satisfactory theory of *signifying* in logic. However, while Husserl vindicated the necessity of the reference to *an intentional subject* for giving the algebraic formulas of Schröder's calculus the capacity of signifying something [56], Peirce, in his famous review paper on Schröder's book, *The Logic of Relatives* [57], introduced the necessity of an irreducible *algebraic triadic relation* for "signifying" the dyadic relation between subject–predicate in the linguistic tokens. In other papers, Peirce denoted this third term of the "semiotic" (signifying) relation as an *interpretant,* with a neologism invented for excluding—against any conceptualist view on the foundations of logic—any necessary reference to a knowing subject, or *interpreter,* for justifying the capacity of signifying a predicative formula in logic. In Peirce's words:

> This definition [of semiotic relation] no more involves any reference to human thought than does the definition of a line as the place within which a particle lies during a lapse of time ([58], p. 52).

On this *triadic algebra of relations* is based, therefore, Peirce's semiotic notion of *sign* as a being-for (*esse per,* in Scholastic Latin) a third term, by which the dyadic relation of being-to (*esse ad*) between the two terms subject–predicate of a predicative relation acquires the capacity of *signifying.* On this theory, Peirce's famous theory of the three *semiotic categories* of "firstness", "secondness", and "thirdness" [59] is also based. These are *ante-predicative algebraic categories,* in the sense that any classical predicative theory of *logic* categories (i.e., intended as the most general and then irreducible predicates of a given language) supposes these three semiotic *algebraic* categories.

We see in Section 5.1 how, through the axiomatization of Peirce's naïve algebra of relations into an *axiomatic calculus of relations,* by A. Tarski (see below Note 11) and then the development by the same Tarski of a BAO with its algebraic relational semantics (see

Section 5.1 and Appendix B), Peirce's pioneering work is in the background of whichever ontology of the *relational natural realism*, my NR-formal ontology included.

Given this necessary historical background, let us now illustrate shortly a taxonomy of the *main formal ontologies* proposed in the history of Western thought. This synthesis is inspired by a similar one developed by my colleague and friend Nino B. Cocchiarella [60,61], a logician and philosopher of logic, now Emeritus at the Philosophy Dept. of the Indiana University at Bloomington (USA). What I share with him—apart from some significant differences—is the general idea that the main ontologies of whichever philosophy and culture can be interpreted, in formal philosophy, like many *theories of predication*, as far as predication is not reducible to the only class/set membership relation $\in$. The main theories of predication are, indeed, in the history of logic, the *nominalism*, the *conceptualism*, and the *realism*, which historically can be viewed like many *theories of universals*. By "universal" we intend, again with Cocchiarella, "what can be predicated of a name", according to Aristotle's classical definition (*De Interpretatione*, 17a39).

To sum up [60–62], we can *synchronically* distinguish along the centuries of the (Western) history of thought (generally distinct into Ancient, Middle, and Modern Ages) at least *three types of ontologies*, with the last one subdivided into two others (see Figure 2). For each of these subdivisions, I quote indicatively in parenthesis some authors, who belong indifferently to one of the three main ages of the Western tradition [8].

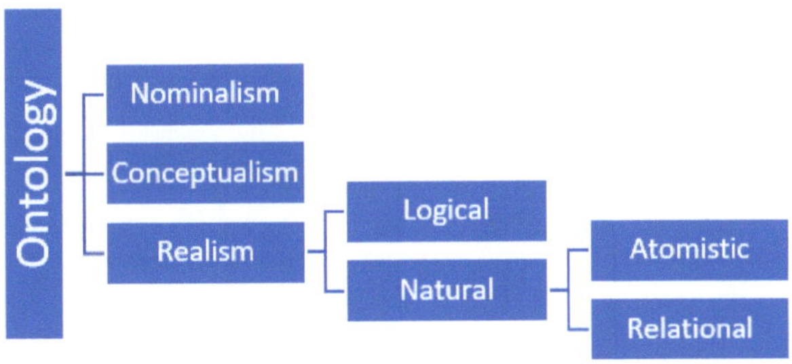

Figure 2. Scheme of the main ontologies in the Western tradition.

1. *Nominalism*: the predicable universals are reduced to the predicative expressions of a given language that, *by its conventional rules*, in the referential usages of predicative sentences, determines the truth conditions of the ontological propositions (Sophists, Roscellinus, Ockham, Hobbes, Quine, etc.).
2. *Conceptualism*: the predicable universals are expressions of *mental concepts*, so that the *laws of thought*, in the referential usages of predicative sentences, determine the truth conditions of the ontological propositions (Descartes, Leibniz, Kant, Husserl, Stein, etc.).
3. *Realism*: the predicable universals are expressions of *properties and relations* existing independently of the linguistic and/or mental capacities in:
    a. *The logical realm*: we have, therefore, the ontologies of the so-called *logical realism*, where the *logical relations*, in the referential usage of predicative sentences, determine the truth conditions of the ontological propositions, independently of human linguistic and mental capacities (Plato, Guillaume de Champeaux, Frege, Russell, Fraenkel, Gödel, etc.).
    b. *The physical realm*: we have then the ontologies of the so-called *natural realism*, or "naturalism". In turn, naturalism can be of two types:
        - *Atomistic*: without natural kinds, where the *logical-mathematical laws* with their *empirical fulfilment* by measurements on physical events, in the referential usages of predicative sentences, determine the truth conditions of the ontological propositions (Democritus, Newton, Laplace, Wittengstein's *Tractatus*, Carnap, etc.).

- *Relational*: with "natural kinds", where the *real relations* (causes) among "things" in nature determine the *logical relations* among "objects", in the referential usages of predicative sentences in language, and then determine the truth conditions of the ontological propositions (Aristotle, Aquinas, Poinsot, Peirce, Kripke, NR, etc.).

2. Some Elements of the Category Theory and Its Relational Semantic in Logic

As we anticipated, this section is devoted to acquainting philosophers with the basic notions of CT, discussed in their relationships with the correspondent notions in ST. The strong interdisciplinary character of formal philosophy is even more evident when we consider the actual *algebraic formalization* of ML in the context of CT logic and mathematics [28], by which the very same algebraic relational structures appear to be at the *common roots* of the mathematical and the philosophical logics. For my synthetic exposition, I refer essentially to [63], which is addressed explicitly to introduce physicists and philosophers into CT, while I refer to [64] and [65] as two CT textbooks addressed mainly to professional mathematicians and computer scientists.

2.1. The Ante-Predicative Definition of Category in Category Theory

As we recalled since the beginning of this contribution, the proper formal character of the CT metalanguage as to the (standard) ST metalanguage consists in not taking $\in$ of the set-membership as a primitive, so to limit the *constructive* approach in logic and mathematics to the *inductive* one, extended to infinite sets (transfinite induction), based on Von Neumann's "cumulative hierarchy of *ranks* of ordinals" and then on Zermelo's "well-ordering theorem" because of the "foundation axiom" in **ZF(C)** set theory [66]. In this sense, given the strict dependence of ST on the predicate logic, which is the deep reason underlying the fact of taking $\in$ as a primitive, CT can be defined as an algebraic ante-predicative theory on the foundations of logic and mathematics. Therefore, in the following exposition, I compare systematically some basic CT notions with the corresponding set-theoretic ones, with which we are more acquainted, to emphasize the differences and contact points. Of course, this is without supporting any non-sensical opposition between ST and CT in the foundations of logic and mathematics.

Indeed, it is well-known that it is possible to interpret CT at the foundational level within **NBG** set theory, even though not within **ZF** because of the presence of "large" categories requiring "classes" with a cardinality greater than V ("large cardinals") and then supposing Gödel's "generalized continuum hypothesis". Nevertheless, what is evident is that CT, initially meant to organize certain fields of mathematics in a systematic way (such as algebraic topology and homological algebra), categories soon became objects of study in their own right ([45], p. 805).

What I want to emphasize in this work is that the CT metalanguage allows not only the working mathematician, as S. Mac Lane suggested [64], but also the working philosopher, as S. Abramsky first suggested [67,68], to discover and formalize axiomatically *structural* similarities between theories; in our ontological case, between logical and physical theories, in which an exclusive "predicative" interpretation of the category notion that takes the $\in$ of the membership relation as a primitive would be forbidden as an inconsistent "category jump".

Indeed, in CT, the *primitives* are:

1. *Morphisms* or *arrows*, *f*, *g*,—intended as a (purely relational) generalization of notions such as "function", "operator", "map", etc.
2. The *identity arrow*, such that, for any object *A*, there is an identity arrow or reflexive morphism $\mathrm{Id}_A = \mathbf{1}_A\colon \mathrm{A} \to \mathrm{A}$ [9].
3. Two *maps* or *operations* from arrows to objects, dom(·), codom(·), assigning a domain or *source* and a codomain or *target* to each arrow.
4. The *compositions of arrows*, written as *g*, *f*, or *f* o *g*, in which the codomain of *g* is the domain of *f*, that is, for any three objects *A*, *B*, *C* in the theory, there exists a morphism

composition $f \circ g$, that is, $A \xrightarrow{g} B \xrightarrow{f} C = A \to C$, satisfying a *transitive* property among arrows.

These primitives must satisfy two axioms regulating compositions and identities among morphisms by which domains and codomains match appropriately:

Axiom 1. *(Associativity Law):* $h \circ (g \circ f) = (h \circ g) \circ f$.

Axiom 1. *(Identity or Unity Law):* $f \circ \mathrm{Id}_A = f = \mathrm{Id}_B \circ f$.

Therefore:

Definition 1. *(Category, C): Any structure-preserving collection of «arrows» (or «morphism»), «objects», «compositions», and the two «mappings» dom(f), cod(f), assigning to each morphism f its domain-codomain of objects, and satisfying associativity and identity, constitutes a category C in CT.*

In this way, it becomes possible to locate the algebraic, ante-predicative notion of category among the other algebraic structures more used by the working mathematicians with their defining axioms, according to the following Table 1.

Table 1. Main algebraic structures with respect to their defining axioms.

	Closure	Associativity	Identity	Invertibility	Commutativity
▪ Semigroupoid	*Unneeded*	Needed	*Unneeded*	*Unneeded*	*Unneeded*
▪ Category	*Unneeded*	Needed	Needed	*Unneeded*	*Unneeded*
▪ Groupoid	*Unneeded*	Needed	Needed	Needed	*Unneeded*
▪ Magma	Needed	*Unneeded*	*Unneeded*	*Unneeded*	*Unneeded*
▪ Quasigroup	Needed	*Unneeded*	*Unneeded*	Needed	*Unneeded*
▪ Loop	Needed	*Unneeded*	Needed	Needed	*Unneeded*
▪ Semigroup	Needed	Needed	*Unneeded*	*Unneeded*	*Unneeded*
▪ Inverse Semigroup	Needed	Needed	*Unneeded*	Needed	*Unneeded*
▪ Monoid	Needed	Needed	Needed	*Unneeded*	*Unneeded*
▪ Group	Needed	Needed	Needed	Needed	*Unneeded*
▪ Abelian Group	Needed	Needed	Needed	Needed	Needed

Furthermore, if we add the algebraic notion of *homomorphism* as a *structure-preserving mapping* between algebraic structures—not to be confused with the notion of *homeomorphism* denoting an *isomorphism* (i.e., an invertible homomorphism) between topological spaces—

we can give the following examples of typical categories in mathematics useful for our aims, each characterized by specific objects and specific arrows:

- **Set** (sets and functions);
- **Grp** (groups and homomorphisms);
- **Mon** (monoids and epimorphisms), where "monoids" are "one-object categories" and "epimorphisms" are the categorical counterpart of "surjective functions" in ST;
- **Top** (topological spaces and continuous functions/paths);
- **Vect**$_k$ (vector spaces defined on a numerical field k and linear functions).

Moreover, in CT, the formal tool for *calculating* and *demonstrating* and then to grant *universality* and *truthfulness* to CT constructions are the *commutative diagrams* of the algebraic calculus of relations. In this way, to continue with Abramsky, the "arrow-theoretic" way of reasoning consists essentially in a *diagrammatic way of reasoning* [63] (p. 10) [10]. Following step by step his useful exemplification, it is asserted that the equations $g \circ f = h$ and $g \circ f = k \circ h$ correspond to the *commuting triangle* and *commuting square diagrams*, respectively, which are the basic commutative diagrams in CT, i.e.,

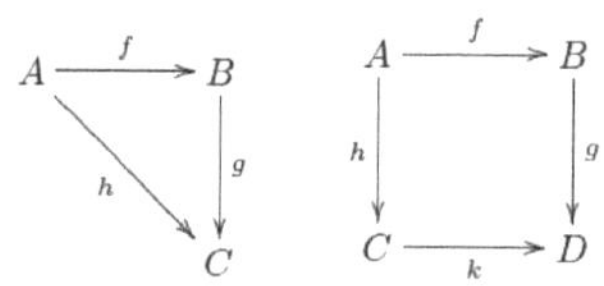

Similarly, the two equations asserting the "associative" and the "identity" laws above, i.e.,

$$h \circ (g \circ f) = (h \circ g) \circ f;\ f \circ \mathrm{Id}A = f = \mathrm{Id}B \circ f$$

characterizing a *category* C in universal algebra (e.g., *groups* also satisfy "closure" and "invertibility" axioms) can be expressed by the two diagrams below, respectively:

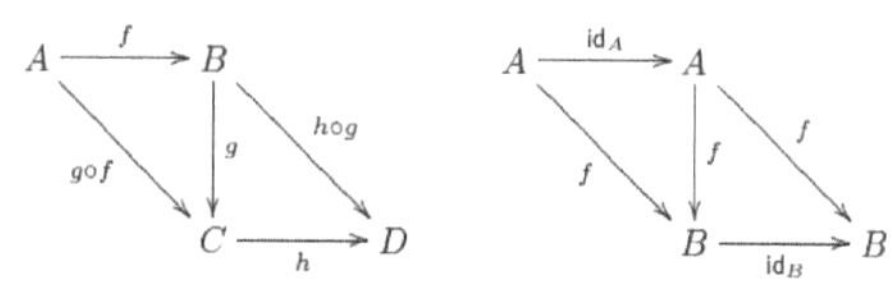

The definition of the commutative diagram can be the following one, which is a simplified version of the rather cumbersome one given by Abramsky and Tzevelekos ([63], p. 11):

Definition 2. *(commutative diagram): A commutative diagram in a category C is a directed graph, whose nodes are objects in C, and edges are morphisms in C. This diagram commutes, if any, two paths with a common source and target that are equal, where at least one of them has a length greater than 1. Specifically, given paths:*

$$A \xrightarrow{f_1} C_1 \xrightarrow{f_2} \ldots C_{n-1} \xrightarrow{f_n} B \text{ and } A \xrightarrow{g_1} D_1 \xrightarrow{g_2} \ldots D_{m-1} \xrightarrow{g_m} B$$

if max(numb) > 1, then $f_n \circ \ldots \circ f_1 = g_m \circ \ldots \circ g_1$. *This commutativity property immediately grants the uniqueness of the diagram concerned, and then the universality of a diagrammatic demonstration.*

Remark 1. *From the historical standpoint, the commutative diagrams in CT are the formalization in the framework of Tarski's axiomatic algebraic calculus of relations* [69] *of the naïve intuition of Peirce, who first introduced diagrams as a calculation tool in the earliest stage of the algebra of relations he inaugurated. Moreover, the evidence that the commuting triangle satisfying the equation* $g \circ f = h$ *is the more fundamental diagram confirms Peirce's intuition that the triadic relations, and not the dyadic ones, are the irreducible relations in algebra. Specifically, they are the basic structure of "semiotics", i.e., of any "signifying" structure in logic and mathematics* [11].

2.2. The Categorical Definition of Sets as Hom-Sets

If all this justifies Abramsky's intriguing statement that "we will refer to any concept which can be defined purely in terms of compositions and identities as *arrow-theoretic*" ([67], p. 3), this perspective change is made explicit when we consider the categorical notion of set as *hom-set*. Indeed:

> Category theory can be seen as a "generalized theory of functions", where the focus is shifted from the pointwise, set-theoretic view of functions to an abstract view of functions as *arrows* ([67], p. 8).

In fact, given in a category $\mathcal{C}$ the two collections of arrows (or morphisms), Ar($\mathcal{C}$), and objects, Ob($\mathcal{C}$), characterizing the category definition, we can also define the arrow-theoretic notion of *hom-set* for a category C, where the prefix *hom-* stays for *homomorphism*, i.e., a structure-preserving mapping between pairs of objects, as the arrow-theoretic interpretation of a "function". Specifically, for each pair of objects A, $B \in$ Ob ($\mathcal{C}$), we define the set:

$$\mathcal{C}(A,B) := \{f \in Ar(\mathcal{C}) | f : A \to B\}.$$

We therefore refer to $\mathcal{C}(A,B)$ as a "hom-set", where distinct hom-sets are *disjoints* ([63], p. 9).

2.3. The Notions of Functors as Morphisms between Categories and Natural Transformations as Morphisms between Functors

Another fundamental notion of CT we have already used is the notion of *functor F*, that is, a «morphism between categories» ([63], p. 28):

Definition 1. *(Functors). A functor F: C → D is given by:*

- *An object-map assigning an object A of D to every object A of C.*
- *An arrow-map assigning an arrow Ff: FA → FB of D to every arrow f: A → B of C, in such a way that compositions and identities are preserved:* $F(g \circ f) = Fg \circ Ff$; $Fid_A = id_{FA}$.

In this way, a functor justifies a *homomorphism* or a «structure-preserving mapping» between the categories $\mathcal{C}$ and $\mathcal{D}$. Of course, for each category $\mathcal{C}$, there exists an *endofunctor* mapping a category onto itself: $\mathcal{C} \to \mathcal{C}$ and an *identity functor* Id_C by which all identities (objects) in C are given, i.e., ([63], p. 31):

$$\mathrm{Id}_{\mathcal{C}} : \mathcal{C} \to \mathcal{C} := A \mapsto A, f \mapsto f$$

For our aims, other significant types of functors are:

- The *inclusion functor I:* $\mathcal{C} \hookrightarrow \mathcal{D}$ between a category $\mathcal{C}$ and its sub-category $\mathcal{D}$. Of course, this is achieved by taking the identity map both for object-maps and arrow-maps.
- The *forgetful functor U:* **Mon** → **Set,** which sends monoids to their set of elements, "forgetting" the algebraic structure, and sends a homomorphism to the corresponding function between sets.

Moreover, the application of each functor can be covariant if it also preserves between the two categories, in addition to all the objects, the directions of morphisms and the orders of compositions. On the contrary, the application of a functor G is contravariant if it preserves all the objects but reversing all the directions of the morphisms (i.e., from $A \to B$, to $GA \leftarrow GB$), and the orders of their compositions (i.e., from *f* o *g* to *Gg* o *Gf*). In this case, the target category of the functor is the opposite of the source category. In a word:

Definition 1. *(Contravariance). Let C, D be two categories. A contravariant functor G from C to D is a functor G:* $C^{op} \to D$ *(or equivalently* $C \to D^{op}$*).*

Finally, another fundamental notion of CT is the notion of *natural transformation*, i.e., of morphisms between functors that are fundamental for a categorical *representation theory*.

Definition 1. *(Natural transformations). Let $F, G : \mathcal{C} \to \mathcal{D}$ be functors, either both covariant or both contravariant. A natural transformation $t : F \to G$ is a family of morphisms in $\mathcal{D}$ indexed by objects A of $\mathcal{C}$*

$$\{t_A : FA \to GA\}_{A \in Ob(\mathcal{C})}$$

such that for all f: $A \to B$, the following diagram commutes:

$$\begin{array}{ccc} FA & \xrightarrow{Ff} & FB \\ \Big\downarrow{\scriptstyle t_A} & & \Big\downarrow{\scriptstyle t_B} \\ GA & \xrightarrow[Gf]{} & GB \end{array}$$

This condition is known as naturality. If each t_A is invertible and then it is an isomorphism, t is a natural isomorphism:

$$F \overset{\cong}{\to} G$$

i.e., F and G are naturally isomorphic written $F \cong G$ ([63], p. 36).

Of course, as far as a natural isomorphism between functors F and G is given, the isomorphism between the relative categories, is given too, both in the covariant $\mathcal{C} \cong \mathcal{D}$ and the contravariant case $\mathcal{C} \cong \mathcal{D}^{\mathrm{op}}$. From this, the definitions of equivalence and dual equivalence, respectively, are derived between categories ([63], p. 40):

Definition 1. *(Equivalence between categories). Two categories $\mathcal{C}$ and $\mathcal{D}$ are equivalent, $\mathcal{C} \simeq \mathcal{D}$* if there are functors F: $\mathcal{C} \to \mathcal{D}$ and G: $\mathcal{D} \to \mathcal{C}$, *and natural isomorphisms with the identity functors of the two categories:*

$$\mathrm{G} \circ \mathrm{F} \cong \mathrm{Id}_{\mathcal{C}}, \quad \mathrm{F} \circ \mathrm{G} \cong \mathrm{Id}_{\mathcal{D}}$$

If the two functors F, G are contravariant, we have the dual equivalence between the relative categories, i.e., $\mathcal{C} \simeq \mathcal{D}^{op}$.

The notion of opposite categories being functorially defined leads us to the categorical interpretation of the *principle of duality* that has a secular tradition in the history of logic (think only of the duality between $\wedge$ and $\vee$ in the De Morgan laws and then in a Boolean lattice), mathematics (think only of the duality between a function $f(x)$ and its inverse $f^{-1}(x)$), and physics (think only of the duality between a function f and its Fourier transform $\hat{f}$. See also [70] for a survey about the notion of duality in mathematics and physics). Now, one of the more significant applications in quantum physics of a natural transformation between contravariant functors—emphasizing the radiographic power of CT in mathematics—concerns the categorical interpretation of the *GNS-construction* for a family of Hilbert spaces—effectively, for a sub-category of the category **Hilb** of the Hilbert spaces, those satisfying the Stone–Von Neumann theorem of the "finitely many unitarily equivalent representations of a quantum system" in QM—based on the *double contravariant application of the Gelfand functor* ([45], p. 807), as we discuss in Section 3.2.

Finally, this hint to the GNS-construction in the mathematical formalism of QM, which historically inaugurated the operator algebra approach in quantum physics, introduces us to the application of the *functorial dual equivalence* between opposed categories to CT logic, as far as we were made to extend the operator algebra formalism from physics to logic because of the fundamental Stone's RTBA (see Section 5.1.2), which allowed Jónsson and Tarski to define the powerful construction of BAOs [4,5] (see Section 5.1.3). In fact, the dual equivalence between statements means that in CT logic, a statement α is *true* in/about a category $\mathcal{C}$ *if* its dual α^{op} (i.e., obtained from α by reversing all the arrows and their compositions) is *true* in/about the opposed category $\mathcal{C}^{\mathrm{op}}$. This means that in CT logic, truth is *invariant* for the reversal of arrows and of the arrow composition orders ([63], p. 40).

2.4. The Dual Equivalence between the Categories of Algebras and Coalgebras

In this subsection, we briefly introduce the fundamental categorical duality between algebras and coalgebras because coalgebraic structures are becoming ever more significant in several fields of modern sciences, from mathematics—think only of the notion of *colimit* as a categorical counterpart of the notion of *direct limit* in mathematical analysis (see Appendix A, and especially [71,72] discussed in it)—to physics, logic, and computer science [6,73,74]. Indeed, in standard ST, the set-membership primitive is strictly related to (Cartesian) *products* among sets and then with algebraic structures $A \times A \rightarrow A$. The coalgebraic structures $A \rightarrow A \times A$ are, on the contrary, characterized by the dual operation of *coproducts*, effectively disjoint sums or set disjoint unions.

Indeed, to limit ourselves to the more interesting cases for us, direct products, categorically defined, correspond ([63] p. 21):

- In **Set**, to Cartesian products.
- In **Pos,** to Cartesian products with a pointwise order.
- In **Top,** to Cartesian products with a topological order.
- In $\mathbf{Vect}_k$, products are direct sums.
- In a poset, seen as a category, products correspond to the *greatest lower bounds.*

Now, coproducts are the dual notion as the products in the sense that, formally, "coproducts in C are just products in C^{op}, interpreted back in C" ([63] p. 23). In fact, coproducts, categorically defined, correspond to direct sums, that is ([63] p. 24):

- In **Set,** to disjoint unions.
- In **Top,** to topological disjoint unions.
- In $\mathbf{Vect}_k$, direct sums are coproducts
- In a poset, seen as a category, coproducts correspond to the *least upper bounds.*

The other starting point for illustrating the categorical duality algebras-coalgebras is the duality, with which we already met in illustrating the two past/future light-cones of the causality principle in fundamental physics, between *final* objects that are *initial* and *terminal* objects, respectively [12]:

Definition 1. *(Initial and terminal objects). An object I in a category C is "initial" if for every object A in C, there exists a unique arrow from I to A, which we write as $\iota_A : I \rightarrow A$. An object T in a category C is "terminal" if for every object A in C, there exists a unique arrow from A to T, which we write as $\tau_A : A \rightarrow T$* ([63], pp. 17–18).

Of course, initial and terminal objects are *dual* in the sense that if A is initial in C, e.g., in the category of algebras **Alg,** it is terminal in C^{op}, e.g., in the category of coalgebras **Coalg,** and vice versa. Indeed, algebras are characterized by products and initial objects, and coalgebras by coproducts and terminal objects.

Let us now illustrate arrow-theoretically the dual categorical characterization of algebras $A \times A \rightarrow A$ and coalgebras $A \rightarrow A \times A$ for the contravariant application of the same functor Ω. Following Y. Venema ([6], pp. 394–395), we recall that an *algebra* $\mathbb{A}$: $A \times A \rightarrow A$ over a signature Ω is a set A with an Ω-indexed collection $\left\{ f^{\mathbb{A}} \middle| A^{ar(f)} \rightarrow A \right\}$ of operations, i.e., polynomial functions indexed by their ariety *ar*, that is, by the number of their arguments. These operations may be combined into a single map constituting the signature of a given algebra $\mathbb{A}$ i.e., $\alpha : \sum_{f \in \Omega} A^{ar(f)} \rightarrow A$, where $\sum_{f \in \Omega} A^{ar(f)}$ denotes the *coproduct* (or *sum* or *disjoint union*) of the sets $\left\{ A^{ar(f)} \middle| f \in \Omega \right\}$. It is easy to verify that a map f: $A \rightarrow A'$ is a homomorphism between the algebras $\mathbb{A} = \langle A, \alpha \rangle$ and $\mathbb{A}' = \langle A', \alpha' \rangle$ if the following diagram commutes:

$$\begin{array}{ccc} A & \xrightarrow{f} & A' \\ {\scriptstyle\alpha}\uparrow & & \uparrow{\scriptstyle\alpha'} \\ \Omega A & \xrightarrow{\Omega f} & \Omega A' \end{array}$$

where it is obvious the signature Ω is interpreted as the *polynomial* set functor $\sum_{f\in\,\Omega}\mathcal{I}^{ar(f)}$, emphasizing that Ω operates on functions between sets. From this, it is possible to generalize to the notion of the algebra category for a given endofunctor Ω, i.e., **Alg(Ω).**

Definition 1. *(Category of algebras for an endofunctor Ω). Given an endofunctor Ω on a base-category C, an Ω-algebra is a pair* $\mathbb{A} = \langle A, \alpha \rangle$ *where* $\alpha : \Omega A \to A$ *is an arrow in C. A homomorphism from the Ω-algebra* $\mathbb{A}$ *to the Ω-algebra* $\mathbb{A}'$ *is an arrow f:* $A \to A'$*, such that* $f \circ \alpha = \alpha' \circ (\Omega f)$*. We denote the induced category as* **Alg(Ω)**.

Similarly, but dually for coalgebras, $A \to A \times A$.

Definition 1. *(Category of coalgebras for an endofunctor Ω). Given an endofunctor Ω on a category C, an Ω-coalgebra is a pair* $\mathbb{A} = \langle A, \alpha \rangle$ *where* $\alpha : A \to \Omega A$ is an arrow in C. A homomorphism from the Ω-coalgebra A to the Ω-coalgebra A' is an arrow f: A → A', such that $f \circ \alpha = \alpha' \circ (\Omega f)$, and for which the following diagram commutes:

$$\begin{array}{ccc} A & \xrightarrow{f} & A' \\ {\scriptstyle\alpha}\downarrow & & \downarrow{\scriptstyle\alpha'} \\ \Omega A & \xrightarrow{\Omega f} & \Omega A' \end{array}$$

It is evident that "the collection of coalgebra homomorphisms contains all identity arrows and it is closed under the arrow composition. Hence, the Ω-coalgebras with their homomorphisms form a category: **Coalg(Ω)**" ([6], p. 394), where the category C is called the base category of **Coalg(Ω)**.

In this way, the clear similarities between the structures of the algebra and the coalgebra categories can be made formally precise in CT. In fact, from observing the two diagrams above, the basic idea, which also explains the name "coalgebra", is that a coalgebra $\mathbb{C} = \langle C, \gamma : C \to \Omega C \rangle$ over a base category C might also be seen as an algebra in the opposite base category C^{op}, i.e., [6], p. 417:

$$\mathbf{Coalg}(\Omega) = (\mathbf{Alg}(\Omega^{op})^{op}$$

Specifically, the category of Ω-coalgebras is *dually isomorphic* to the category of algebras over the functor Ω^{op} (i.e., acting like Ω but on the opposite category C^{op}). From this, the *dual equivalence* between the categories **Coalg(Ω)** and **Alg(Ω^op)**, for the contravariant application of the same functor Ω, immediately derives, i.e.,

$$\mathbf{Coalg}(\Omega) \simeq \mathbf{Alg}(\Omega^{op})$$

To conclude, it is fundamental to recall with Venema himself ([6], p. 395) a *fundamental difference* between the categories of algebras and coalgebras functorially defined. Indeed, while in the category of coalgebras we are dealing with *arbitrary* set functors that can be whichever type of *homomorphic mapping*, in the category of algebras, we are constrained to dealing with functors that are *polynomials*. This difference is made clear when we reflect on the fundamental functorial coalgebraic construction of the *colimit operation* (see Appendix A), which plays a fundamental role in the categorical formalization of the notion of the "QV-foliation" in QFT and then for the formalization of the notions of *emergence* in the NR-formal ontology and the philosophy of nature, particularly in its application to a topology of NWF-sets (see Section 4.6 and [38]).

2.5. Non-Wellfounded Sets in the Category of Coalgebras as Causal Sets

All the *standard* axiomatic set theories, **ZF**, as far as sharing the *membership relation* $\in$ taken as a *primitive*, also share the *axiom of extensionality* and the *well-founded* character of set membership, granted by the *axiom of foundation* in its different versions. For instance, Zermelo's *axiom of regularity* grants the well-founded character of sets by not allowing a set to contain itself, so forbidding infinite chains of set inclusions. The axiom of regularity states that every non-empty set A contains an element that is disjoint from A. In its FOL formulation, it reads: $\forall x(x \neq \varnothing \rightarrow \exists y(y \in x \wedge y \cap x = \varnothing))$. In this way, by the prohibition of unbounded chains of set inclusions, set *total ordering*, and finally Zermelo's *well-ordering theorem* are granted too—even though definitively by adding the *axiom of choice* AC in **ZFC** (see [66], pp. 320–321 and pp. 360–372 for further explanations).

As Adam Rieger recalls in his monograph about NWF-set theories ([75], pp. 181–182), this means that, given well-ordering, the inductive constructive mechanism of new sets in all well-founded set theories is in terms of the construction of sets that at each stage S are formed as a collection consisting of sets formed at stages before S (see Figure 3 left). This is an inductive procedure that in **ZF** is extended to infinite sets by including in it Von Neumann's construction of the *cumulative hierarchy* of ordinal numbers as *ranks* of *well-founded* sets, i.e., as ranks or *stages* of a hierarchy of sets having a *minimal* element, to make axiomatically consistent Cantor's transfinite induction [76].

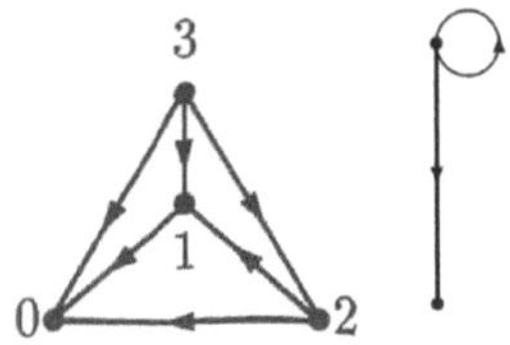

Figure 3. (Left) Set-tree representation of the number "3" as the root of a well-founded oriented graph of subsets satisfying von Neumann's number construction in ZF: $0 = \{\} = \varnothing$; $1 = \{\varnothing\}$; $2 = \{\varnothing, \{\varnothing\}\}$; $3 = \{\varnothing, \{\varnothing\}, \{\varnothing, \{\varnothing\}\}\}$. Note that in such a representation, the empty set is the node to which all the morphisms are pointing while the unary set is always the central node (from [11], p. 3). **(Right)** Representation of a non-well-founded oriented graph where the self-containing set $\{\{\cdot\}\}$ (reflexive graph) is allowed by the anti-foundation axiom, and where the symbol $(\cdot)$ emphasizes that the construction holds for sets of whichever cardinality (from [75], p. 9).

All this implies that in well-founded set theories, "when we are forming a set z by choosing its members, we do not yet have the object z, and hence we cannot use it as a member of z". Or, more synthetically, a set is a collection of previously given objects. In this sense, as Rieger says, referring this time to G. Boolos [77], it is evident that a set must *include* itself as a subset, like the same symbol of set inclusion $\subseteq$ signifies, but this is not the same as saying that a set *contains* itself as a *member*. In well-founded set theories, to satisfy the set-elementhood principle (see Note 1), each set can be a member/element only of another higher rank set. In a word, *self-inclusion is not self-membership*! In other terms, it is perfectly consistent in set theory writing: $\exists x(Sx \,\&\, x \in x)$, where Sx stays for "x is a set", but if we take $\in$ as meaning strictly "'it is a member/element of' it is very, very peculiar to suppose it true". This peculiarity is precisely what characterizes Peter Aczel's NWF-set theory with its *anti-foundation axiom* [11]. In it, set self-membership, in the sense of a *self-containing set* $\{\{\cdot\}\}$ (see Figure 3, right), is allowed and then infinite chains of set inclusions are allowed too, so that no set total ordering but only set *partial orderings* are allowed in NWF set theory [11].

On the other hand, as again Rieger ([75], p. 178) but also as Aczel himself [11] recall, the Russian mathematician Dmitry Mirimanoff was the first in 1917 [78,79] to introduce the distinction between *ordinary* sets not admitting infinite descending membership chains (and then satisfying Zermelo's foundation axiom) and *extraordinary* sets admitting such infinite chains (not satisfying the foundation axiom) without *per se* being antinomic [13]. However, in the light of the well-founded set theory and the role played in it by the transfinite induction in the construction of Von Neumann's *cumulative hierarchy* applied in **ZF** to the universe V (*proper class*) of sets—and in **NBG** also to V extensions with a cardinality higher than V because of Gödel "generalized CH"—it is hard not to agree with Von Neumann's statement of the "superfluous" character of non-well-founded sets [80].

Rieder, in his survey, emphasizes the recent revival of interest in NWF-set theories ignited by Peter Aczel's NWF-set theory based on the strong *anti-foundation axiom* [11], for its wider applications in TCS. It models, indeed, *parallel and concurrent computations* and *data streaming*, and it is applied to the categorical formalization of Kripke's *model theory of modal logic* (see [75], pp. 184–185, and overall [81] for a synthesis).

However, what completely escapes Rieger's (and Von Neumann's) treatment of NWF-set theories is that the proper formalization of Aczel's NWF-set theory requires the formal apparatus of the CT metalanguage to be fully expressed and justified. This dependence of NWF-set theory on CT formalization with the notion of set as hom-set (see Section 2.2) is, on the contrary, the starting point from which Aczel moves (see [11], 71–102). Before all, for justification of the powerful *final coalgebra theorem* for NWF-sets (see [11], 81–90 and [82]), this demonstrates that all the trees of NWF-sets share the same root as a common terminal object in the category of coalgebras (see Definition 7.).

This theorem, indeed, *mutatis mutandis*—where the main difference is that no *set total ordering* is here admitted but only an *infinite arbitrary branching* of trees of posets—plays the same role in the NWF-set theory based on the anti-foundation axiom that Zermelo's well-ordering theorem plays in well-founded set theories (see on this regard [83] and Section 5).

On this regard, the core difference in a categorical setting between (1) well-founded sets, admitting set total and well-ordering, and (2) NWF sets, where only set partial orderings are admitted, is synthesized in a very effective way by Aczel himself in the following way (see [11], Chapter 6, especially p. 77):

1. In the recursive induction (the transfinite induction included) of well-founded set theories the *continuous set operators*, i.e., satisfying the CT primitive of the morphism composition, have only one *least* and one *greatest* fixed points, i.e., the *empty set* $\varnothing$ and the *universal collection* V (Von Neumann's *proper class*), respectively.
2. In the recursive constructions of NWF-sets, where several arbitrary partial orderings are admitted, *the continuous set operators have many fixed points*—effectively, many possible lower and upper bounds of different recursive algebraic-inductive *upward directed* $\{\uparrow\}$, and coalgebraic-coinductive *downward directed* $\{\downarrow\}$ poset construction procedures (see below the application to *concurrent computations* in TCS for modal BAOs in Section 5.2.2 and Appendix A for applications to mathematical analysis in the CT framework).

More intuitively from a logical and epistemological standpoint, what is typical of standard ST based on set total ordering and well-ordering is the formalization of an inductive procedure, and then a *generalization procedure*. When we generalize, indeed, the recursive construction of ever more inclusive collections makes sense, i.e., sets of higher cardinalities that are typical of Boolean algebras, which have the property of recursively constructing the numerical sets on which their operations are defined.

On the contrary, the dual coalgebraic construction of *coinduction* (see [72] for a coalgebraic interpretation of the mathematical continuum and Section 5.2.2), based on *set-trees "unfolding"* from a common root according to *reciprocally irreducible unfolding paths* of the different posets, is aimed at epistemologically formalizing a *specification procedure*. In epistemology, this is typical of the logical/ontological theory of the *natural kinds* (genus/species) on a causal basis, as Kripke first emphasized [19].

Using a biological intuitive example of the natural kind logic, a "genus" (e.g., "the mammals") does not "include" $\subseteq$, in a proper set-theoretical sense, its different species (e.g., "elephants", "dolphins", "squirrels", "humans", …) like a set its subsets but simply "admits" $\ni$ them. Indeed, the different species evolved as different, reciprocally irreducible branches of the *ascendant-descendants' evolutionary trees* from their common "mammalian-root" (that is, from some (hypothetical?) common ancestor of all mammals). Effectively, $\ni$ is significantly also the symbol of the coalgebraic "co-membership relation" that is *dual* to the "membership relation" $\in$ in the CT relational semantics of the *modal* BAOs (see Sections 5.1 and 5.2). Indeed, as it is trivially evident from this biological example, no well-ordering relationship, and much less *no common metrics* justifying a common ordering relation $\leq$, is shared by the different species of mammals.

Now, as we see immediately, in QFT, this distinction genus-species also applies to physical objects such as the three different "generations" (not "sets"!) of fermions and gauge-bosons of the SM hat have in SSBs of the quantum fields at their ground state (="quantum vacuum condition") their common "branching mechanism" in an evolutionary cosmology (see Section 4). In other terms, this logic and mathematics is compliant with the evolutionary quantum-relativistic cosmology, based on the universal mechanism of the *symmetry breaking*, by which our universe progressively "populated itself" of ever more complex systems and structures (see [10] and Section 4).

Not casually, to formalize set-theoretically these strongly non-linear processes related to the causal light-cone in the universe evolution, some authors, e.g., R. D. Sorkin, proposed the so-called *causal set theory* as the proper set theory of quantum cosmology, with quantum gravitation included [84]. What characterizes Sorkin's trees of causal sets is indeed that they admit only *partial order relations* (i.e., reflexive, transitive, anti-symmetric, and locally finite order relations $\leq$) among sets, where the order relations are interpreted like the many *causal relations* in the Lorentzian manifold of the causal light-cone. The non-acceptable price to be paid for justifying the causal set theory in Sorkin's version is the supposition of the *discrete character* of the space-time manifold of the relativistic universe, which would mean renouncing the formal apparatus of GR in cosmology, and, finally, the same topological approach to the theoretical quantum and relativistic physics, "string theory" included (see [85] for a synthesis).

On the other hand—and this is the deep reason of Sorkin's theory, it would be non-sensical, if not contradictory at all, to suppose the set total or well-ordering in a causal set theory used for modeling the strongly non-linear, unpredictable character of the universe evolutionary branching processes, with each based on the universal mechanism of the *symmetry breaking* (phase transitions) with respect to the preceding universe states of the universe dynamics.

It is evident, however, that the same result of the limitation to posets when we speak of causal sets in physics can be obtained by modeling them in the framework of the NWF-set theory in a *categorical coalgebraic setting*, in a way that is perfectly compliant with the topological formalism of operator algebra and the string theory (continuous set operators) in relativistic and quantum physics and cosmology. Indeed, while in Sorkin's causal set theory for limiting the constructions to posets it is necessary to suppose the discrete character of the spatial-temporal manifold on which they are defined, in the NWF-set interpretation of causal set trees, this is neither necessary nor allowed. In it, indeed, we might limit the construction to causal posets simply because the NWF-sets naturally satisfy reflexivity and not totality in their ordering relations. Finally, their categorical coalgebraic setting is perfectly compliant with the interpretation of the causal event as *a terminal object* in the category of coalgebras for all the other events (effects) referring to it as to their common cause, and then belonging to the (subcategory of the) future-oriented light-cone with respect to their cause (see Section 2.4, Appendix A and overall [72]).

In fact, one of the more significant results of our categorical approach to QFT is the demonstration that the coalgebraic sub-category of the *doubled Hilbert spaces* **DHilb**—differently from the category of Hilbert spaces **Hilb** to which it belongs—satisfies the powerful *cocompleteness theorem* and then a *compactness condition* of the underlying topological space (see Section 4.8). In this modeling, indeed, each pair of doubled Hilbert spaces corresponds to a different dissipative quantum system, modeled as the result of SSB of QV and then a phase transition among quantum fields at their ground state, with a necessary change in metrics that each phase transition implies in physics.

Historically, indeed, in the Aristotelian ontology of the natural kinds that applies to all physical entities and not only to the biological ones, each species (individual) with respect to the common genus (species) to which it belongs in the proper sense of "from which it *causally* derives" or "from which it is *generated*" adds a "specific *difference*". Thus, one of the greater and more influential Aristotelian and logically skilled philosophers and theologians of the Middle Age, Thomas Aquinas (1274–1323), in his *Commentary* to Aristotle's *Metaphysics* book, stated on this regard that:

> the predicate 'there exists' is said 'as many times as these differences are' (see [86], *Sententia libri Metaphysicae*, VIII, ii, 1694) [14].

In modern terms, this means that in the logic and ontology of the natural kinds, the existence of an object—either an individual, or a collection of individuals—is not related to the "set-elementhood principle" for which each object x for existing must be a member (element) of another set (class) of a higher ordinal rank (or "higher type": see Note 1). In the limit, it must be an element of the universal class V, as the consistent usage of the "existential quantifier" exemplifies in ST predicate logic [15]. On the contrary, the modal existence predicate (not quantifier!) $E(x)$ in natural kind logic [60] is justified wherever a new identity relation Id_x and then a new unitary relation 1_x "*causally* emerges as a self-containing new whole" from within a collection of previously given objects. This logic and ontology, however – and in this I completely disagree with Cocchiarella –, can be formalized only in the CT metalanguage, where the assignment of a domain/codomain of objects to each morphism (predicate) depends on the primitive of the $\mathrm{dom}(\cdot)/\mathrm{cod}(\cdot)$ maps, and not on the membership relation taken as a primitive (Section 2.1). For this reason, "existence" can be a predicate $E(x)$ and not a simple quantifier $(\exists x)$ like in ST logic.

In the natural kind logic based on coalgebras of NWF sets, in other terms, it is like what happens for the "non-normal classes" of the Russell paradox. Specifically, the *non-normal classes*, according to Russell's definition, "contain themselves as members", as it is exemplified in the well-known linguistic "Richard paradox". The class of all "polysyllables", indeed, *contains* the predicate "being polysyllable", i.e., the denotation of the identity shared by all the elements of the class, because "polysyllable *is polysyllable*", differently from the class of the monosyllables, given that "monosyllable *is not monosyllable*". The logical relevance of NWF-sets is that by the self-containing sets $\{\{\cdot\}\}$ of whichever cardinality, we can justify this notion of a "self-containing collection of elements" in a consistent axiomatic set-theory within CT based on Aczel's "anti-foundation axiom". On the other hand, it is trivial but perhaps significant to recall that NWF-sets are not classes but proper sets because if they do not satisfy set total and well-ordering, they satisfy *partial ordering* relations.

This means that NWF-sets make sense wherever we must model in a suitable ST the "emergence" of a *new property* Id_x shared by all the elements x of a self-containing set and/or a *new object* 1_x, which are irreducible to the simple combinatorics ("summation") of previously given elements because it is a new emerging self-containing "whole", as happens wherever we must deal in physics and more specifically in QFT with *phase transitions*: "more is different"! [8]. Think, for instance, of the well-known phase transition between the non-ferromagnetic and ferromagnetic phases of a metal (see Figure 4). In the QFT interpreted as the fundamental physics of the condensed matter systems, this process corresponds to the *symmetry breaking* of the quantum fields of the atoms of the metal in its non-ferromagnetic phase, characterized by the fact that the atom "magnetization vectors" (the momenta of the magnetization dipoles) are pointing in whichever direction (randomly aligned, so satisfying a spherical symmetry).

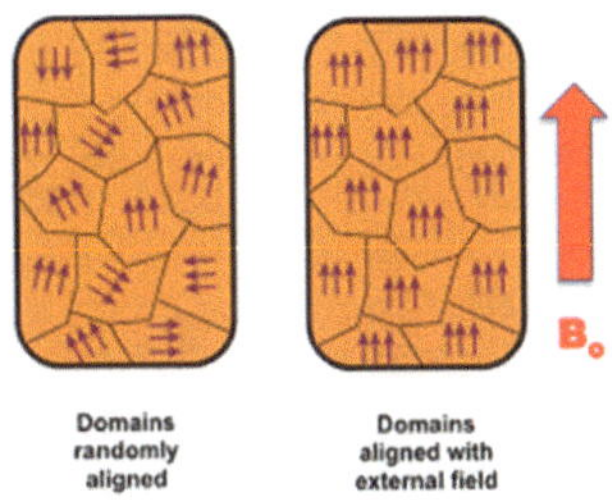

Figure 4. Intuitive representation of the phase transition between the non-ferromagnetic (**left**) and ferromagnetic phases (**right**) of a metal.

By the action of an external field, the phase transition occurs, by which the magnetization vectors are suddenly aligned along only one direction (symmetry breaking), from which the new collective property/predicate Id_x emerges dynamically (causally) of "being magnet", effectively as a new *ordered phase-coherence domain* of the quantum fields. A collective property of the metal that can be lost dynamically by a phase decoherence of the quantum fields because of a change in the boundary conditions, that is, by raising the temperature beyond a given threshold.

Because of the universal character of the symmetry breaking mechanism in quantum physics and cosmology, we dedicate the next Sections 3 and 4 of this contribution to illustrate this on the physical-mathematical side of QFT. Then, we dedicate Section 5 to a discussion of the correspondent logic and ontology of this physics, the NR-formal ontology, all modeled in the unifying formal metalanguage of CT.

3. Some Elements of the Quantum Mechanics Formalism in a Categorical Setting

As a premise to this Section, a synthetic view of the so-called *standard model* (SM) of the quantum elementary particles (extended also to the *gravitons* of quantum gravity, now only hypothetical) is shown in Figure 5. SM (without gravitons) is actually one of the two sources, together with the *cosmological standard model* (CSM) of the general relativity theory (GR), of the evolutionary quantum-relativistic cosmology (see [10], for a synthetic overview). Effectively, SM is one of the more meaningful results achieved by QM and QFT during the XX cent., of which formalism I in this section provide some elements of a historical reconstruction to help the philosopher's understanding.

3.1. The Stone–Von Neumann Theorem in the Quantum Mechanics Formalism

As everybody knows, the main difference between the classical (Newtonian) mechanics and the quantum mechanics (QM) is that, because of the *uncertainty principle,* we cannot have "deterministic" but *irreducibly* only "statistical" representations (measurements) of the two *canonical variables*—position x and momentum p—identifying univocally the *physical state* of a particle in the *state space* of a mechanical system. In other terms, in QM, the two canonical variables are no longer representable as independent of each other such as in classical mechanics. This means that in QM, when we represent the time evolution of a particle dynamics in its *state space,* we cannot have the two canonical variables as its orthogonal (independent) dimensions *commuting* with each other [16]. Indeed, in QM, because of the uncertainty principle:

$$\Delta p \Delta x \geq \hbar/2 \tag{1}$$

where $\hbar$ is the Planck constant, they are *conjugate variables,* i.e., their measurements depend on each other, so that they *do not commute* such as in the classical mechanics. In fact, the uncertainty principle means that a higher precision in determining the position x means *necessarily* a lower precision in determining the momentum p, and vice versa.

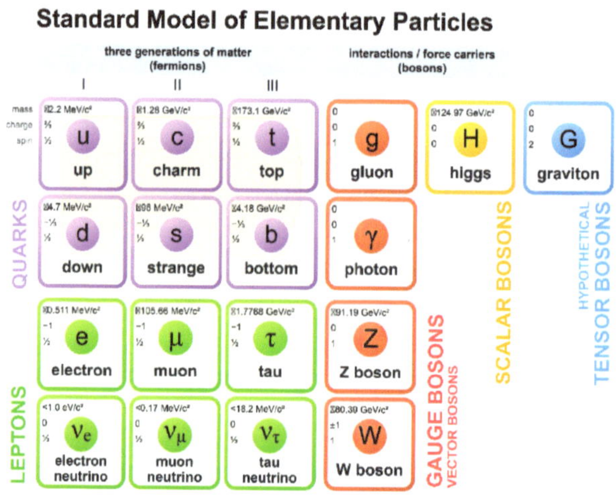

Figure 5. An updated scheme of SM extended to the hypothetical *gravitons*. The particles, interpreted in QFT as quanta of the relative (material and interaction) fields (see text), are subdivided into: (1) *Three generations (families) of "fermions"*, i.e., the quanta of the respective "matter fields", obeying the *Fermi-Dirac statistics*, i.e., because all have a fractional "spin", they occupy distinct levels of energy at their ground (minimum energy) state, such as electrons in the atom. Each generation is constituted by two massive "quarks" and two massive but lighter "leptons". The first generation of fermions, manifesting themselves at the universe lower energies (temperatures), are particles constituting the atoms of our ordinary experience (quarks "up" and "down" constituting protons and neutrons ("baryons") in the atom nucleus, plus the "electrons" and the "electronic neutrinos" of atoms). (2) *Three types of massive and massless "bosons"*, i.e., all obeying the *Bose–Einstein statistics*. Specifically, because all having an integer spin, at the ground state, they all occupy the same energy lower level. They are (a) the "gauge (carrier) bosons" of the three quantum force fields: "strong" (massless *gluons*), "electromagnetic" (massless *photons*), "weak" (massive *Z, W bosons*); (b) the "scalar bosons" of the *Higgs fields* (massive *Higgs bosons*); and (c) the "tensor bosons" of the *gravitation field* (massless *gravitons*) of the quantum gravity theory, now only hypothetical.

The *non-commutative* character of the quantum canonical variable measurements seemed, therefore, at the beginning of XX sec., to preclude the usage of a geometric representation of the dynamics of a quantum system. Effectively, in his famous work on the foundations of geometry *Grundlagen der Geometrie* (1899) [87], David Hilbert, for the first time in the history of thought, gave a complete axiomatic version of the Euclidean geometry, demonstrating that geometrical spaces are *commutative varieties*. However, Hilbert himself suggested the solution of the conundrum for the nascent quantum physics formalism by his following works on the *functional analysis*. In it he introduced the notion of the *infinite dimensional* Euclidean spaces endowed with an *inner structure* ("inner product": see Note 19), afterward developed and denoted by J. Von Neumann as *Hilbert spaces*. Indeed, the spatial-temporal position of particles in QM is no-longer "pointwise", but it corresponds to a "field", i.e., to a "volume" of the possible spatial position distribution for each "instant" of time because of the uncertainty principle. This is a notion that Hilbert applied to his further discovery of the *spectral theory* in the algebra of matrices [88].

Because Hilbert's abstract spectra also apply unexpectedly well to the study of the observed atomic spectra of the hydrogen atom electromagnetic emissions, and then to Werner Heisenberg's *matrix mechanic* formulation of QM, all this led J. Von Neumann—at the time, the assistant of Hilbert at Göttingen—to write in 1932 his famous book on the foundations of the mathematical formalism of QM [89]. Indeed, at the beginning of the 1930s, as any historical reconstructions of the birth of QM recall, there existed two different theoretical interpretations (models) of the nascent QM, both having at that time its essential observational basis in the spectra of the light emission of the "excited" hydrogen atom. The two models recalled are: (1) *Werner Heisenberg's* matrix *mechanics* formulation of QM [90], completed in 1925 with the contributions of Max Born and Pascual Jordan [91,92]; and (2) *Erwin Schrödinger's statistical wave mechanics* formulation of QM, published in its complete form in 1926 [93]. Now, in his momentous book, Von Neumann demonstrated the *algebraic isomorphism* and then the *mathematical equivalence* of these two early formulations of QM. To achieve this result, the essential step was the demonstration that it is sufficient to interpret quantum states as *vectors of a Hilbert space* $\mathcal{H}$, and the *quantum observables* x (position) and p (momentum) as *operators* acting on this vector space and constituting in turn another Hilbert space $\mathcal{H}^*$, *isomorphic and dual* as to $\mathcal{H}$.

Effectively, the possibility of using the Hilbert vector spaces satisfying the commutativity condition in QM depends on the fact that the statistical wave function f of each canonical variable, x or p, in QM *commute* with the *Fourier transform* $\hat{f}$ of the statistical wave function of the other canonical variable, i.e., $\hat{p}_x$ or $\hat{x}_p$, respectively. From this, the fundamental notion in quantum physics of the *canonical commutation relation*, CCR, derives, each identifying univocally a quantum state in QM, i.e.,

$$[\hat{x}, \hat{p}_x] = i\hbar \tag{2}$$

where i is the imaginary unit, given that the Fourier transform and then the Hilbert vector spaces are defined on the complex numerical field $\mathbb{C}$. In this way, it is easier to understand, at least intuitively, why a Hilbert space is connoted as a *functional space*. In it, indeed, the orthogonal dimensions of the space, on which the "coordinates" of the quantum states are defined, are the outputs of a particular class of statistical distribution functions, those obeying the CCR principle. What is fundamental to recall for our theoretical aims is that the core of Von Neumann mathematical formalism is related to the momentous theorem, formulated simultaneously by Von Neumann himself and by the American mathematician Marshall Stone between 1930 and 1932 [94–97]. Specifically, the so-called *Stone–Von Neumann theorem*. Stone is the same Marshall Stone of the RTBA in *logic*, published in 1936, which we already recalled in this contribution, and we discuss in Section 5.1.2. The Stone–Von Neumann theorem indeed demonstrated the *uniqueness* of a *finite* number of *unitarily equivalent CCRs* and then of a family of Hilbert spaces to faithfully represent a quantum system in QM.

Finally, because of the *superposition principle* in QM, while a *pure state* regarding the statistical wave function of a single particle corresponds to a *ray* in the Hilbert space, a *vector* in the Hilbert space corresponds to the *quantum state* as superposition of several wave functions [17], i.e., it corresponds to a *mixed state*, relative to many particles "occupying" the same quantum state that are anyway indistinguishable because of the uncertainty principle to constitute an irreducible "whole".

3.2. The Completion of Von Neumann's Formalism by the Gerlfand–Neimark–Segal Construction and Its Categorical Interpretation

As Klaas Landsman emphasizes very well in his recent textbook [45] on classical and quantum mechanics re-interpreted within the unified formalism of the operator algebra, Von Neumann's early formalization of QM did not satisfy the ideal of *absoluteness* pursued by Von Neumann himself in his construction. Effectively, indeed, his fundamental demonstration of the *reversed isomorphism* $V \to V^* \to V^*$ between a finite Hilbert (vector) space $\mathcal{H}$ and the dual Hilbert space $\mathcal{H}^*$ of its observables (operators) [18] holds only "locally" for a given Hilbert space, and not "generally" for a *whole family* of Hilbert spaces, so to make Von Neumann's early formalism not per se applicable to the "relativistic QM", or QFT, that is, a quantum theory, in which the special relativity principles hold.

Therefore, to complete our survey, it is necessary to recall the momentous *Gelfand–Naimark–Segal (GNS) construction* that systematically extended the Fourier *duality* between a function f and its Fourier transform $\hat{f}$, from the infinite *function spaces* of functional analysis to the infinite *vector spaces* such as per se the Hilbert spaces. Effectively, indeed, the relevance of the GNS-construction for the completion of the QM formalism is strictly related to the attempt started by Von Neumann himself to model the algebras of the physical observables, i.e., the operators over Hilbert spaces. Historically, this is related to the abstract characterization given in 1943 by Israel Gelfand and Mark Naimark of the so-called *C*-algebras* [19] in a momentous paper [98], in which they refer to Hilbert spaces but without any reference to *operators* on a Hilbert space. This extension, in addition to the term and the notion of "C*-algebra", was introduced in 1947 by Erwin Segal [99] to signify the norm-closed subalgebras of the Banach algebra of a Hilbert space $B(H)$, namely the space of bounded operators on some Hilbert space $\mathcal{H}$, where C of C* stays for "closed". Therefore, Segal defined a *commutative* C*-algebra of a Hilbert space as a "uniformly closed, *self-adjoint* algebra (the symbol *, in C*) of bounded operators on a Hilbert space", where self-adjoint elements are those satisfying the condition $x = x^*$ (see Note 19).

Theoretically, this had a fundamental double consequence on the QM formalism. On the one hand, it grants that an *algebra of observables*, i.e., a finite number of operators on the Hilbert space, as far as *irreducible*, is sufficient for faithfully representing a quantum system. On the other hand, it mathematically grants the existence of *a finite orthonormal basis* of the Hilbert space for each quantum system. Thus, it generalizes the Von Neumann demonstration of the "reversed" *isomorphism* (= dual equivalence) between a Hilbert space $\mathcal{H}$ and its dual operator space $\mathcal{H}^*$ to a whole *family* of Hilbert spaces. Following Landsman's reconstruction in the framework of CT formalism (see [45], pp. 807–808), in the case of vector spaces such as the Hilbert spaces, the GNS-construction generalizes the "local" or, using the CT jargon, the "unnatural" isomorphism $V \to V^* \to V$ discovered by Von Neumann, to a whole subcategory of the category of Hilbert spaces **Hilb.** Namely, these satisfy the Stone–Von Neumann theorem in the representation theory of a quantum system in QM.

Indeed, by the *double contravariant application* of the *Gelfand transform*, so constituting the *functor* of this subcategory of Hilbert spaces, we obtain, by the consequent *natural transformation* (see Definition 5.), the *natural* isomorphism: $V \to V^* \to (V^*)^* \to V$. Therefore, the algebraic GNS-construction determined, from the second half of the last century on, the development of the *operator algebra* formalism as the proper formalism of quantum physics in the framework of Hilbert space modeling, which we can formalize in CT in a particularly effective way.

3.3. *The Interpretation of QFT as "Second Quantization" with Respect to QM*

As we know, *quantum superposition* is a fundamental principle of QM that has no correspondence in classical mechanics. It states that any two or more quantum states, as represented by statistical wave functions ψ, can be "added together" or *superposed* to obtain another valid quantum state. The physical analogue of this property is the "constructive" or "destructive" interferences among two or more wave forms, which in the classical "double-slit experiment" of QM reveals the *wave-like* behavior of all quantum objects. Mathematically, the superposition principle is a consequence of the *linearity* of the Schrödinger equation Ψ, for which any linear combination of solutions of the system will also be a solution.

The superposition principle was originally suggested by Paul Dirac to extend the QM formalism to model the quantum behavior not only of single particles but also of many particles "occupying" the same quantum state to determine a sort of "ontological shift" in quantum physics. Namely, this is from considering particles as the fundamental objects to considering *oscillating fields* $f(x_N, t)$, that is, different spatial distributions varying in time of N particles, as the fundamental object of inquiry in QFT.

In this framework, we must consider "particles" as *quanta* of the respective oscillating *material* fields, just like photons are quanta of the *interaction* electromagnetic field. The same Dirac suggested the name of "Second Quantization" (SQ) to this approach to QFT [100] because, in the case of non-relativistic systems where the number of particles is *fixed* and *finite* but too large to use the Schrödinger wave function, it gives a formal alternative for the computations in QM. Effectively, SQ taken in this sense is an "algorithm", as C. M. Becchi states in his useful survey about the SQ formalism, to which we mainly refer in this subsection [101]. In the QM case, indeed, the approach is immediately compatible with the finite number of the degrees of freedom depending on the Stone–Von Neumann theorem just recalled.

However, when we pass into the *relativistic* realm of the quantum field theory (QFT), there exists the problem that the *number of the degrees of freedom* of the system cannot be considered finite any longer but *in(de)finite*. Effectively, this is the core of the famous *Haag's theorem* in QFT [102], which admits different solutions for this problem. We briefly illustrate two of them. Namely, (1) the SQ solution for QFT systems staying mainly *in only one phase* ("closed" QFT systems), which is the object of this subsection, and (2) the solution for QFT systems *passing through different phases* ("open" or "dissipative" QFT systems) in Section 4.

For our aims, following the reconstruction of the SQ method applied both to QM and QFT given in [101], it is important to consider *two new concepts* consequently introduced by Dirac in the general formalism of QM and QFT [100] and that, as such, have a general significance in quantum physics. Indeed, Dirac developed the construction of SQ originally for the treatment of the superposition among finitely many identical *bosons* since his early work in 1927. It concerned the emission and absorption of the electromagnetic radiation, and it was afterward extended to *fermions* by a successive work of Pascual Jordan and Eugene Wigner in 1928 [103] (see Figure 5).

These two notions, using the further contribution of the Russian mathematician Vladimir Fock [104], are: (1) the *Fock space*, as a particular type of Hilbert space generated by the *tensor product* of several one-particle Hilbert spaces to satisfy the superposition principle; and (2) the dual *creation-annihilation operators* acting on the Fock space. Because these operators are "Hermitian conjugates" to each other, they are denoted as $A, A^{\dagger}$, respectively. Effectively, these operators do not act directly on a state of a N-bosons (or fermions) Hilbert space, but to use them, we need to extend the (Hilbert) state space $\mathcal{H}_S^{(N)}$—where the subscript S stays for the type of symmetry (bosonic or fermionic) the algebra satisfies (see below)—to the *direct sum* of N state-spaces. Each of them, in turn, is the *tensor product* of N single-particle Hilbert spaces to constitute a "Fock space" $\mathcal{F}(\mathcal{H})$. Intuitively, we can see at the "Fock space", like a "book (summation of pages)", whose "pages" are "Fock states", namely, Hilbert spaces, each of which is a tensor product of N one-particle Hilbert spaces [20].

Effectively, the original intuition behind the Dirac treatment of the SQ algorithm in QM [100] is the strict relationship existing between a composite system and the *tensor product* of the related Hilbert spaces. In other words, "the state space of an assembly of systems is identified with the tensor product of the state spaces of each system" [101].

At this point, always following [101], we can informally define the "Fock space" notion as the "sum" of a set of Hilbert spaces representing zero-particle states, one-particle states, two-particle states, and so on. Therefore, using Dirac *matrix bra-ket symbolism* to denote the wave-function of a quantum state $|\psi\rangle$, the SQ algorithm for N-identical superposed particles works in the following way. We start from the 1-particle state $|1\rangle$. We apply the "annihilation operator" and obtain the 0-particle state or the *quantum vacuum* (QV) state $|0\rangle$. Then, we apply one-time, two-times, three times ... N-times the "creation operator", and we obtain the 1-particle state again $|1\rangle$, and then the 2-particles $|2\rangle$, 3-particles $|3\rangle$,... N-particles $|N\rangle$ quantum states, where N therefore denotes the particle *occupation number* of each state of the resulting Fock space. On this basis, given the definitions of the "tensor product" $\otimes$ and "direct sum" $\oplus$ between vector spaces (see Note 20), we can give a more formal characterization of the Fock space $\mathcal{F}(\mathcal{H})$ as the (Hilbert) direct sum of tensor products of N copies of single-particle Hilbert spaces $\mathcal{H}$, i.e.,

$$\mathcal{F}_\nu(\mathcal{H}) = \mathcal{H}^{(0)} \oplus \mathcal{H}^{(1)} \bigoplus_{N=2}^{\infty} \mathcal{H}_{S_\nu}^{(N)}, \text{ where } \bigoplus_{N=2}^{\infty} \mathcal{H}_{S_\nu}^{(N)} = (S_\nu(\mathcal{H} \otimes \mathcal{H})) \oplus (S_\nu(\mathcal{H} \otimes \mathcal{H} \otimes \mathcal{H})) \oplus \ldots \quad (3)$$

As we see in the formula, the number of the composing Hilbert spaces—and then of the system *degrees of freedom*—can go, in principle, to infinity, so subtracting any algorithmic (i.e., finitary) value to Dirac's SQ construction. Moreover, in Equation (3) defining the notion of Fock space, the further operator S_ν appears. It makes the tensor *symmetric* or *anti-symmetric*, depending on whether the Hilbert space representing N particles is obeying *bosonic* ($\nu = +$) or *fermionic* ($\nu = -$) statistics. Indeed, the *Pauli exclusion principle* in QM states that two or more *identical fermions* cannot occupy the same quantum state *simultaneously*. For instance, in the case of the electrons in atoms, for which Wolfgang Pauli originally formulated his principle, this means it is impossible that in an atom with many electrons, they have the same values for all *four quantum numbers* identifying their quantum states.

Specifically: (1) n (roughly, the quantized electron energy); (2) the "azimuthal quantum number"$\updownarrow$; (3) the "magnetic quantum number" $m_\updownarrow$; and (4) the "spin quantum number" m_s. Therefore, if two electrons stay in the same "orbital" or "energy level" of an atom at its ground state, the first three quantum numbers $n, \updownarrow, m_\updownarrow$ must be the same, and then their m_s must be different to satisfy Pauli's *exclusion principle*. Specifically, the electrons must have opposite half-integer spin projections of $\frac{1}{2}$ and $-\frac{1}{2}$, respectively. On the contrary, particles with an integer spin, that is, *bosons*, are not subject to the Pauli exclusion principle. In this way, any number of bosons can simultaneously occupy the same quantum state.

A more rigorous statement of the same principle that applies directly to the explanation of the S_ν symbol in Equation (3) concerns the permutations among identical particles, such as those characterizing the tensor product of a finite number of single-particle Hilbert space defining a Fock state in the Fock space (see Equation (3) and Note 20). The total wave function is *anti-symmetric* (i.e., it changes its sign) if the particles are fermions while the sign remains unchanged and then the total wave function is *symmetric* if they are bosons. For this reason, we say that fermions satisfy sets of *canonical anti-commutation relations* (CARs) while bosons satisfy sets of *canonical commutation relations* (CCRs) (see Section 3.1).

When we pass to the relativistic QFT, the situation changes. On the one hand, the Fock space with its decomposition (a sort of factorization, indeed) of the composite Hilbert space into a set of component Hilbert spaces—and particularly, the 0-occupation state of the QV-state $|0\rangle$—acquire a precise physical meaning, losing their purely algorithmic flavor. Indeed, to use Becchi's words in his review of the SQ approach [101], in the relativistic QFT, "particles are produced and absorbed", i.e., continuously "created" and "annihilated", from/to the QV.

On the other hand, all this means that we cannot suppose that the Stone–Von Neumann theorem of QM holds *generally* also in QFT, as the Haag theorem demonstrated [102]. Indeed, per se, in QFT, we are faced with an indefinite number of degrees of freedom—and/or with *an indefinite number of unitary inequivalent representations of the CCRs (CARs)* —and not with *a finite number of unitary equivalent representations of CCRs (CARs)*, as the Stone–Von Neumann theorem states for a single quantum system in QM (see Section 3.1).

The solution proposed firstly by Dirac himself [102], developed systematically during the further 50 years into an SQ approach to QFT [105], and discussed at large in [101] is to model QFT fields as *freely oscillating in the vacuum* to model them as *non-interacting* systems (i.e., energetically closed systems). This makes it possible to also model QFT systems in the standard framework of the *statistical mechanics* of systems that are at *equilibrium* in the *asymptotic (infinite limit) condition*. This means, overall, the possibility of also applying the *perturbative methods* in QFT (like in QM) to model field interactions, having "cut away" all the undesired interactions, so that we can always use the canonical *Hamiltonian representation*, which holds only for *closed* dynamic systems, also for QFT system *dynamics* to define its *total energy*.

The two steps followed by Dirac and leading him to the construction of the *quantum electro-dynamics* (QED)—that is, the quantum theory of the electromagnetic interactions–consists in firstly modeling bosonic fields (photon fields in Dirac QED), and secondly fermionic (electron) fields as *freely oscillating in the vacuum*.

Starting from the *bosonic fields*, both problems emerging in such a QFT modeling (the so-called "zero-energy point" problem and the necessity of using an indefinite number of quantum harmonic oscillators in QFT because of the Haag theorem (see Section 4.1)) can be systematically solved by supposing the reference to (the knowledge of) the only Hamiltonian of the system [101]. Indeed, the supposition of quantum fields freely oscillating in the vacuum means that we are modeling a QFT system in a canonical Hamiltonian way, that is, as a "closed" system.

Similar considerations can be used to define the Fock space for *fermionic fields*, so that "in the general situation we have a mixed bosonic and fermionic Fock space with a unique vacuum state" [101]. Moreover, in the case of QFT for fermionic fields, we have the further advantage, as Dirac first noticed for the electron theory [100], that their normal modes are those whose energy is not bounded from below. Indeed, in the ground state of such models, to satisfy Pauli's exclusion principle, their normal modes "under a certain level (the zero-energy Fermi level) are occupied, while those above this level are empty. This corresponds to the picture of the *Fermi Sea*" [101].

Surely, the main successes of QFT based on the SQ approach and finally on the systematic usage of the perturbative methods of statistical mechanics in fundamental physics are innumerable. The same SM in its actual form is surely the most significant of them. On the other hand, the successes of SM are strictly related to the powerful calculus tool of the so-called *Feynman diagrams*.

The intuitive mechanistic model in the background of the ordinary divulgation of the Feynman diagrams is the following. The particles (fermions) interact by exchanging reciprocally force quanta (bosons, e.g., photons) isolated in the vacuum, such as when two ice-skaters move up/away from each other, by exchanging a basketball, being completely isolated from, i.e., having no interaction with, the crowd of the other skaters populating the icy lake. This is an ideal but *nonrealistic situation* indeed. In fact, Feynman diagrams are rightly considered as the more effective application of the perturbative methods of statistical mechanics to QFT because in these diagrams, all the undesired interactions (undesired interaction branches) are systematically cut away. So, *only* the significant interactions (those satisfying the Hamiltonian in the framework of Feynman's "path integral" formalism) are considered (see [106] for a popular exposition of the theory by Feynman himself).

4. The Extension of QFT to Modeling Dissipative Systems in a Categorical Setting

4.1. The Theoretical Problem at Issue with the Haag Theorem

At the end of his survey on the SQ formalism in QFT to which we referred to in the previous section [101], Becchi quotes the "alternative statistics" related to the *Bogoliubov transform* that also applies to "open" quantum systems and that we illustrate systematically in this section using the CT formalization.

As we know, the fundamental component of Von Neumann's standard formalism of QM [89] is the so-called "Stone–Von Neumann theorem" recalled in Section 3.1, for which a finite number of *unitarily equivalent CCRs (CARs) and then Hilbert spaces* is sufficient for representing a system in QM. In the SQ approach to QFT, this holds as far as we are representing one or more superposed quantum particle fields but staying in *only one phase*, i.e., as freely oscillating in the vacuum, and then satisfying the Hamiltonian of energy for closed systems.

Now, the Haag theorem demonstrated that this representation—the so-called "Dirac's picture"—does not hold *generally* in QFT because, as far as we interpret a QFT system as an *interacting (not closed) "many-body system"*, i.e., we consider the crowd of the other skaters on the icy lake surface following the just quoted metaphor, it can pass through *different phases*. Indeed, the Haag theorem demonstrated formally that in QFT—in the infinite volume limit of the functional analysis—there exists an infinite number of *unitarily inequivalent representations of the commutation (bosons) and anti-commutation (fermions) relations*, all compatible with the QV ground state $|0\rangle$. More precisely, in QFT, there exists a mismatch between the *field dynamics* (the Heisenberg matrix equations) and its representation (the Hilbert space of physical states).

In fact:

> the same dynamics (i.e., the same set of Heisenberg equations) may lead to different solutions when unitarily *inequivalent* representations (Hilbert spaces of physical states) are used in computing the matrix elements. *The choice of the representation to describe our system is thus of crucial importance in solving the dynamics:* the same dynamics may be realized in different ways (i.e., in different unitarily inequivalent representations). The choice of the representation may be considered as *a boundary condition* under which the Heisenberg equations have to be solved (see [9], p. 55).

This "boundary condition" could *epistemologically* be the supposition that we know the Hamiltonian of the system, as Becchi taught us in his reconstruction of the SQ approach in QFT (see Section 3.3), but it could also *physically* be the "thermal bath" with which a quantum system is continuously exchanging energy in a *balanced* way, in a dissipative interpretation of QFT as "many-body physics". In such a case, the Hamiltonian is not supposed but *dynamically generated* by the system itself!

4.2. The Thermal Interpretation of QV

Generally, in QM, we could imagine the QV ground state as the physical state in which all the quantum fields at their ground state "are zero". Effectively, following, for instance, J. Maldacena's divulgation, the fields take random values around zero because of the uncertainty principle. However, these field fluctuations "happen at short distances" while "in the vacuum, at long distance they average out to zero, so that we recover the classical result where the fields are all zero" [107]. In other terms, by such a mathematical construction, there would be no ultimate distinction between QV and the *mechanical vacuum* of the classical and statistical mechanics. In it, no energy contribution is from the vacuum and then the system can be considered as isolated, i.e., the quantum fields are freely oscillating in the vacuum.

To oversimplify, in this approach, we are considering the unavoidable QV fluctuations in the light of the overall averages of Boltzmann's "molecular chaos" of statistical mechanics, and then in the light of the *second principle of thermodynamics* for "closed" or "isolated" systems. However, it is possible to move a step forward, and consider, more realistically, the unavoidable fluctuations of the QV at the ground state in the light of the *third principle of thermodynamics.* This step forward was made by the Japanese physicist Hiroomi Umezawa during the 1990s of the XX cent., leading him to interpret QFT as a "thermo-field dynamics" (TFD) [108,109]. This interpretation was originally developed by Umezawa for modeling condensed matter physics systems, and specifically the neurodynamics of the brain interpreted as a "dissipative system" or "open" system.

Indeed, the *third principle of thermodynamics* states: "The entropy of a system approaches a constant value as the temperature approaches the absolute 0 °K (−273 °C)". It was the Nobel Laureate (1921) Walter Nernst who first discovered that for a given mole of matter (namely, an ensemble of an Avogadro number of atoms or molecules), for temperatures close 0 °K, T_0, the variation of entropy ΔS would become infinite (by dividing by 0). Nernst therefore demonstrated that to avoid this catastrophe, we must suppose that the molar heat capacity C is *not constant* at all but vanishes in the limit $T \to 0$ to make ΔS finite, as it must be. This means, however, that near the absolute 0 °K, there is a mismatch between the variation of the system inner content of energy and the supply of energy from the outside. We can avoid such a paradox, which would violate the *first principle* of the energy balance in any physical system, by supposing that this mysterious *inner supplier of energy is the vacuum* for whichever physical system. Conversely, this implies that the absolute 0°K is unreachable for whichever physical system. In other terms, *there exists an unavoidable fluctuation of the elementary constituents of matter,* at any level of matter organization.

Therefore, an immediate consequence of the *third principle of thermodynamics* is the association of whichever mole of matter with an oscillating *matter field*, and then in QFT, the consequence of a *thermal interpretation* of the QV as *the universal energy reservoir*, with a *temperature* $T > 0$ °K. This means that no physical—classical or quantum—system is conceivable as "isolated" or "energetically closed", since it is necessarily "open" to the unavoidable vacuum fluctuations in its background. At the same time, this means that, in QFT interpreted as the fundamental physics, at any level of matter field organization (see the notion of "QV foliation"), the QV at its ground state $|0\rangle$ corresponds to the fluctuating ground state, with temperature >0 °K characterizing a given "heap" of matter fields. The ontological conclusion for the fundamental physics is that we cannot any longer conceive the physical systems, either at the microscopic, mesoscopic, or macroscopic levels, as isolated in the mechanical vacuum, such as in the classical representation:

> The vacuum becomes a bridge that connects all objects among them. No isolated body can exist, and the fundamental physical actor is no longer the atom, but the field, namely the atom space distributions variable with time. Atoms become the "quanta" of this matter field, in the same way as the photons are the quanta of the electromagnetic field ([110], p. 1876).

For this discovery, eliminating the notion of the "inert isolated bodies" in the mechanical vacuum of the Newtonian mechanics at the fundamental level, Walter Nernst is a chemist who is one of the founders of the modern quantum physics.

To complete this historical reconstruction, we must recall that Umezawa's "thermal" interpretation of QFT as fundamental physics of biological and neural systems cannot give a suitable memory mechanism to brains as dissipative systems because in this approach, each new memory "script" overwrites the precedent one. This limit is overcome by Giuseppe Vitiello's further formalization of the principle of *QV-foliation* based on the theoretical—and experimental, e.g., in neuroscience—evidence that *different, stable phase coherences of the quantum fields* at the same "balanced" (0-energy summation) ground state with $T > 0$ *can coexist* in dissipative quantum systems, without interfering with each other.

Moreover, they can be *locally* ordered over each other, being univocally *addressable* ("labeled") because they are univocally and "dynamically" *indexed* to become an effective tool used by nature for "constructing" *complex systems* and effective *memory sub-systems* in biological and neural dissipative systems (see Sections 4.6 and 4.7). The illustration of Vitiello's approach and his group to QFT is covered in the rest of this section (see the more comprehensive synthesis of this approach in [9]). Afterward, we offer a categorical formalization of the QV-foliation using the powerful colimit operators in Section 4.8.

4.3. The Bogoliubov Transform

The Haag theorem showed us that QV at its *ground state* is compatible with infinitely many unitarily inequivalent representations of the canonical commutation relations (CCRs), for bosons, and the canonical anti-commutation relations (CARs) for fermions and then of the relative Hilbert spaces.

The fundamental mathematical modeling in QFT of the process of creation/annihilation from QV of bosons and fermions as quanta of their relative fields was offered by the Russian physicist and mathematician Nikolay Bogoliubov in 1958 [111]. He demonstrated that, given a pair of CCRs (mathematically defined on a hyperbolic function basis) for a pair of creation/annihilation operators for a boson on the Hilbert space, i.e., $\left[\hat{a},\hat{a}^{\dagger}\right] = 1$, and another pair of operators for another boson, i.e., $\left[\hat{b},\hat{b}^{\dagger}\right]$, there exists a transformation (the Bogoliubov transformation) mapping the former pair into the latter.

The same holds for the CARs (i.e., defined on a circular function basis) between pairs of fermionic creation/annihilation operators, i.e., $\{\hat{a},\hat{a}^{\dagger}\} = 0$, $\{\hat{a},\hat{a}^{\dagger}\} = 1$. In other terms, Bogoliubov demonstrated that there exists an *isomorphism*, either of the *CCR algebras* for bosons, or of the *CAR algebras* for fermions, by which different "degenerate states" of the QV at the ground state $|0\rangle$ are modeled, as previewed by the Haag theorem and corresponding to a process of creation/annihilation of bosons or fermions, respectively.

What is fundamental for our aims is that the application of the Bogoliubov transform implies a *phase shift* of the respective fields. In other terms, differently from QM (and QFT in its SQ formalization), the Bogoliubov transform allows QFT to calculate over *phase transitions* that per se imply he consideration of QFT systems as *open systems* because of the passing through of different phases. In this sense, Becchi spoke about "an alternative statistics" as to the SQ one that, as we know, can model per se QFT systems in only one phase.

4.4. The Goldstone Theorem

The "dynamic richness" of QV and its "degenerate states" is the result, in fundamental physics, of another momentous theorem that historically played a fundamental role in the construction of SM: *the Goldstone theorem*. Indeed, this theorem, together with the original idea of P. Higgs about the existence of the so-called *Higgs field* and its quantum, the *Higgs' boson*, led to the explanation of the *electro-weak symmetry breaking* that is a fundamental ingredient of SM [112–114]. Effectively, the Goldstone theorem demonstrates the existence of infinitely many *spontaneous symmetry breakings* (SSBs) of the QV ground state $|0\rangle$, corresponding to the spontaneous instaurations (i.e., without any energy waste) of *long-range correlations* among quantum fields at their ground state, and then corresponding to as many *phase coherence domains* among them (see Figure 6).

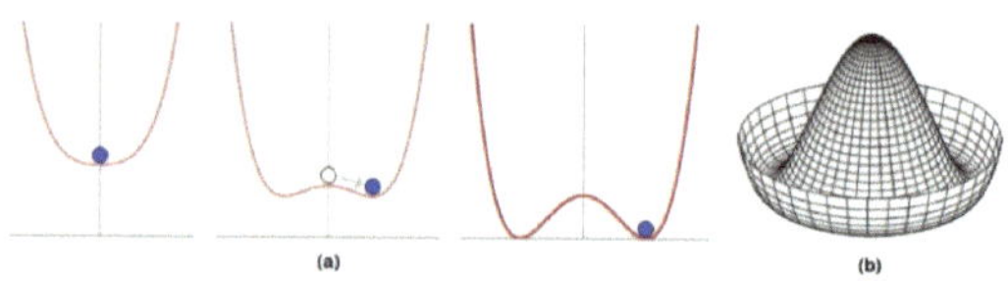

Figure 6. Intuitive representation of the Goldstone theorem. (**a**) The principle of SSB. (**a**) Left: the ground state of the system (minimum of the energy potential function) is not at 0-energy equilibrium. (**a**) Center: This induces the possibility that, despite the overall symmetry being conserved, the system (blue ball) can "choose" between different (two, in the example) states, each "locally" breaking the global symmetry. (**a**) Right: Indeed, the system stays in one of the two available ones at the same ground state to determine an SSB. (**b**) Effectively, at the QV ground state, there are infinitely many possible SSBs to determine the classical "sombrero hat" energy potential diagram of the Goldstone theorem.

Therefore, the infinitely (not denumerable) many QV conditions (or "QV degenerate states") compatible with its ground state ultimately exist because of the instauration of these *long-range correlations* among quantum fields at their ground state. Moreover, the SSB principle holds not only in the relativistic (microscopic) domain of QFT but also applies to non-relativistic many-body systems in condensed matter physics of chemical and biological systems (neural systems included), and even at the cosmological scale.

In other words, this discovery of the dynamically generated *long-range correlations* among quantum fields deeply changed the fundamental physics (QFT) of the elementary particles (at the microscopic level) and the condensed matter physics (at the macroscopic level). Indeed, these correlations are mediated by the *Nambu–Goldstone bosons* (NG-bosons) as quanta not of the energy exchanges (interactions) such as the gauge-bosons of the SM (see Figure 5) but as quanta of the *long-range correlation waves* among quantum fields propagating in the vacuum to be associated to the different *coherent modes of oscillating in phase* of the quantum fields [114,115], i.e., the phenomenon of the so-called *quantum entanglement* [21].

Effectively, the short note in which Higgs synthetically expressed his revolutionary idea [113] was published in 1964, two years after the usage made of it by Salam, Goldstone, and Weinberg to solve the electro-weak SSB in [114]. Indeed, Higgs in his note refers to this paper. However, Higgs illustrated the core of his idea in a lecture given at Princeton in 1961, at which both Weinberg and Goldstone were present. The "Higgs field" hypothesis, indeed, solved the fundamental conundrum of the electro-weak unification. Specifically, it solved the problem of how the *Z-W bosons,* the gauge-bosons of the weak force, can acquire their tremendous mass (of the order of the mass of an iron atom!), so violating the Goldstone theorem for which the gauge-bosons must be massless, just like the *photons* and the *gluons*—the gauge-bosons of the electromagnetic and the strong forces—are (see Figure 5).

Indeed, the Goldstone theorem is also strictly related to Yoichiro Nambu's application of the SSB principle in 1960–61 to model the *chiral symmetry breaking* in *quantum chromodynamics* [115–117]. Specifically, in the QFT of the "strong force" with its three different charges ("colors"), through which the quarks interact. What is amazing for non-physicists such as us is that the largest part (more than 99%) of the mass of protons, and then of atoms, and finally of our bodies depends on the chiral symmetry breaking by which protons (and neutrons) can acquire the largest part of their masses by the "strong interactions" of the quarks constituting them. This is despite "gluons"—that is, the gauge-bosons of the strong force field, see Figure 5—being per se massless such as photons. This sends us back again to the role of the "Higgs field" [113], applied this time not to the electro-weak interaction symmetry breaking but to the electro-strong interaction "chiral" symmetry breaking (see [107] for its intuitive illustration). For this fundamental work, Nambu was awarded the Nobel Prize in Physics in 2008. In this way, SSBs of the Goldstone theorem and the Higgs field play an essential role in the construction of the "local gauge theories" of the three fundamental quantum (strong, electromagnetic, weak) forces of SM.

The quanta of the long-range correlation waves are named *Nambu–Goldstone (NG) bosons.* Now, the NG-bosons are with mass even though always very small (if the symmetry is not perfect in finite spaces), or *without mass at all* (if the symmetry is perfect, in the abstract infinite volume of functional analysis). The lesser the inertia (mass) of the correlation quantum, the greater the distance on which it can propagate, and hence the distance on which the correlation (and the ordering relation they determine) constitutes itself (see Note 21).

To sum up, the main novelty introduced by the Goldstone theorem in this QFT picture is that each of the QV degenerate states constitutes an SSB of QV at its ground state. Each SSB, in turn, corresponds to the "spontaneous" instauration of *long-range correlations* among force fields in QV. Therefore, they can display *collective behaviors* that make their treatment in terms of *individuals* meaningless. This implies a deep *ontological paradigm shift* in modern fundamental physics.

This shift can be summarized as follows:

1. Firstly, all this means that each massive or non-massive "elementary particle" of the SM, both fermions (quarks, neutrinos, and electrons) and gauge-bosons (gluons, photons, *Z-W* bosons), and the Higgs-boson, are considered in QFT as *quanta* of their respective fields (see Figure 5). This ontological stance is consistent with the passage to the mathematical formalism of the so-called *string theory,* where a particle is not represented by a "point" and its motion as an "unidimensional trajectory" in the state space, but it is represented as a *vibrating string* and its motion as a *bidimensional brane,* where the intensity of the string vibration is proportional to the energy of the associated field.
2. Secondly, in QFT, an uncertainty relation holds and then a *particle-wave duality,* similar to Heisenberg's one of QM relating the *statistical* uncertainty between the momentum and position of particles (see Equation (1), above). Effectively, in QFT, in the light of the Goldstone theorem, the uncertainty and then the particle-wave duality concerns the number of the field quanta n, with respect to the field phase ϕ, namely:

$$\Delta n \Delta \varphi \geq K \tag{4}$$

where K is a quantization constant related to the type of long-range correlation involved. If ($\Delta n = 0$), ϕ is undefined, so that it makes sense to neglect the waveform aspect in favor of the individual, particle-like behavior. On the contrary, if ($\Delta\phi = 0$), n is undefined because an extremely high number of quanta are oscillating together according to a well-defined phase, i.e., within a given phase coherence domain. In this case, it would be nonsensical to describe the phenomenon in terms of individual particle behavior since the *collective modes* of the field prevail. We already presented a condensed matter physics example of this phenomenon, that is, the ferromagnetic phase (see Figure 4).

3. Thirdly, another fundamental unifying notion, not only with respect to quantum physics but also with respect to quantum biology and quantum computing as far as both are based on QFT, is the notion of the *NG-bosons*. Of course, they are not "gauge bosons", quanta of energy, such as the bosons of the four fundamental forces of the SM (see Figure 5), since they are quanta of the coherent modes of being in phase of the quantum fields. Therefore, they appear in all the equations of QFT related to the instauration of long-range correlations among quantum fields. In this way, it is a further consequence of the Goldstone theorem that any long-range phase coherence among quantum fields related to SSB of QV at its ground state has its "fingerprint" in the unique *countable* value $\mathcal{N}$ of a *given condensate of NG-bosons*. Now, despite "these correlation quanta" being real particles, observable with the same techniques (diffusion, scattering, etc.) of the other particles, nevertheless, because their mass is in any case negligible (or even null), *their condensation does not imply a change in the energy state of the system*. This means that, if the symmetric state is a possible ground state (a minimum energy state or a degenerate "vacuum" of a QFT system), the coherent state, after the symmetry breakdown, also remains in *a state of minimal energy* to be *stable* in time. In the macroscopic terms of classical kinematics, it is representable as a *stable (chaotic) attractor* of the overall dynamics [118]. Or, more properly, in the formalism of QFT, this phenomenon is the core of the principle of *foliation of QV* at its ground state for different values $\mathcal{N}$ of a given condensate of NG-bosons, as a "robust principle of 'construction' and 'memory' used by nature to generate ever more complex systems [9,34], as we already recalled and will explain formally using the powerful colimit operation of the CT mathematics (see Section 4.8).
4. Fourthly, the Goldstone theorem and the SSB principle related to the instauration of long-range correlation applies both to the relativistic QFT of atomic and subatomic physics at the *microscopic level*, and it applies to many-body physics of condensed states of matter at the *macroscopic level*. This constitutes a fundamental *analogy*—to use Nambu's words [116,117]—between these two levels of matter organization. Indeed, the long-range correlations, related to the instauration of phase coherence domains among the involved matter fields and their quanta, all imply a *dynamic re-definition of the metrics* characterizing the system dynamics and its properties. In this sense, the macroscopic phenomena of condensed matter physics related to system phase transitions have their own proper explanation at the *microscopic* quantum level [9].

To sum up, if any phase transition in physics is characterized by an *order parameter* that is the physical magnitude—generally, a statistical density distribution, whose sudden change at the phase boundary characterizes the phase transition [22]—in dissipative QFT, a given condensate of NG-bosons plays the role of a *dynamic control parameter*. Changing its numerical value $\mathcal{N}$ means that the quantum fields can be subject to different dynamic regimes, with different collective properties, and hence with different collective behaviors and functions, *at the same ground state* of the quantum fields.

From the standpoint of theoretical physics, the demonstration of what we intend by saying that condensates of NG-bosons act as a "control parameter" of the phase transition among different phase coherence domains in the same QV-ground state at a temperature $T > 0$ can be found in ([9], pp. 166–169). In the equations explained in these pages, indeed, it is demonstrated that the simple emission-absorption of "a few of" (a finite number of) NG-bosons can also induce a phase transition, *without any energy waste*! Ontologically, they are "quanta of *form*" (ordering relations) and not "quanta of *matter*" (mass-energy) such as fermions and gauge-bosons.

In this way, NG-bosons in condensed matter physics acquire different names, according to the different topological phases of matter they control.

For instance, in solid state physics (mechanics), the NG-bosons are named *phonons*, as far as they are quanta of the collective coherent modes of mechanical (elastic) oscillations (vibrations) of the molecules. In this case, indeed, the symmetry breaking concerns the *Galilean spherical symmetry* in the propagation of the mechanical vibrational motion of molecules, according to which these vibrations propagate *casually* (i.e., satisfying a spherical symmetry) in whichever direction of the 3D-space. The breaking of such a symmetry determines either their *longitudinal* coherent propagation, corresponding macroscopically to the phase transition to the "liquid state" of the collective behavior of the moles of some material (e.g., the liquid state of a water flow), or the *longitudinal and transverse* coherent propagation, corresponding to the "solid state". In this latter situation, in the case of a rigid crystalline lattice of oscillating atoms/molecules, their coherent oscillation modes determine the regular distribution according to a periodic law of the particles in the lattice to determine dynamically the regular geometric structure of a crystal (e.g., the beautiful geometries of the icy state of water in a snowflake).

Another example is the phase transition to the *magnetic phase* of some metals, which we already presented intuitively in Figure 4, where NG-bosons are named *magnons*. In this case, indeed, the broken symmetry is the rotational symmetry of the *magnetic* dipole of the electrons, and the macroscopic phenomenon of "magnetization" consists of the correlation among all (most) electron spins, so that they "choose", among all the directions, the one correct for the magnetization vector.

Finally, in biological matter and water, in which only the biological molecules are active (this is the deep reason why more than 80% of our bodies is made of water, and more than 90% of our molecules are water), the NG-bosons are named *polarons*. Indeed, the broken symmetry in this case is the rotational symmetry of the *electrical* dipoles characterizing water and organic molecules. The coherent modes of propagation of dipole currents are indeed the dynamic "secret" by which distant bio-molecules "can feel each other" in the cell inner-outer "watery" environment to constitute the physical basis of the ordered collective modes, which ultimately consist of both the complex structures of biological matter and the ordered sequences of chemical reactions of a biological function [23].

To conclude, please note again this ontological consequence for fundamental physics, which we already introduced descriptively at the end of Section 2.5. It is similar to if the matter dynamics by long-range correlations and the relative *phase coherences of the quantum fields* corresponding to as many SSBs of QV defined a *new property and/or function and/or predicate of matter*: "being liquid", "being solid", "being magnet", "being organic", etc. In a proper sense, which will emerge clearly when we discuss more deeply the common *coalgebraic semantics over NWF-sets*, both of QFT in physics and f Kripke's coalgebraic modal semantics in logic, it is similar to if a physical phase coherence constitutes the *logical/mathematical domain* of a given function/predicate.

This physical process can only be properly formalized in the (ante-predicative) CT metalanguage. In it, the constitution of the domain/codomain of objects for a given morphism (function/predicate) depends on the primitive of the $\mathrm{dom}(\cdot)/\mathrm{cod}(\cdot)$ mappings and not on the set-elementhood principle of the (predicative) ST metalanguage, taking the $\in$-membership predicate as a primitive (see Note 1, and Section 2.1).

4.5. The Non-Commutative Coalgebraic Modeling of QFT Dissipative Systems

Having given a descriptive survey of some main notions of QFT, let us deepen the understanding of some aspects of the mathematical formalism involved in this subsection. In the quantum physics mathematical formalism, the *Hopf algebras* play an essential role in performing mathematical computations on a lattice of quantum numbers $\mathbf{k}$ associated with given quantum variables (e.g., the energy $\mathbf{E}$ or the momentum J of particles).

Mathematically, a Hopf algebra H is a "bi-algebra" because it is characterized by two types of operations, i.e., *coproducts* of the coalgebra: $H \to H \times H$, and *products* of the algebra: $H \times H \to H$. Both products and coproducts *are commuting*, and defined over a *field* K with a K-linear *map* S: $H \to H$, or *antipode*, sending dually commuting *coproducts* over commuting *products*, and *counits* ε over *units* η, and vice versa, so that the diagram of Figure 7 commutes.

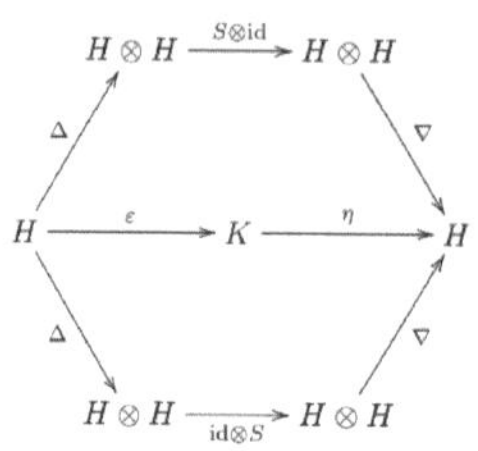

Figure 7. Commuting diagram of a Hopf bialgebra.

The role of a Hopf bialgebra in QM calculations over a lattice of quantum numbers emerges immediately when we consider that the "algebraic half" $H \times H \to H$ of the Hopf bialgebra applies when we must calculate, for instance, the energy E of a single particle; whereas the "coalgebraic half" $H \to H \times H$ applies when we have to calculate the total energy of two particles superposed in the same quantum state $E = E_1 + E_2$ (coproducts are effectively disjoint sums, as we know). In this case, the *commutativity* of the coproducts also makes sense because the total energy of the quantum state does not change by interchanging between themselves the two particles.

However, in dissipative QFT or, more generally, when we deal with open systems, the total energy E depends on the system and its thermal bath *balanced* contributions, which evidently cannot be treated on the same algebraic basis. This means that the commutativity of the coproduct terms for calculating the total energy cannot hold any longer. Indeed, the terms of the coproducts represent here, necessarily in a *non-interchangeable* and then in a *non-commutative* way, the system state energy value and the associated thermal bath energy value. In these cases, the proper algebraic tool is the *q-deformed Hopf coalgebras*, with *non-commutative coproducts*, strictly related to the Bogoliubov transform (Section 4.3), because the q deforming ("squeezing") parameter is a *thermal parameter* corresponding to the inverse of the θ "mixing angle" of the Bogoliubov transform [9,34].

Therefore, the non-commutative coproducts for a system state a and for its correspondent thermal-bath state $\tilde{a}$ are: $\Delta a_q = a \times q + q^{-1} \times a \equiv a \times q + q^{-1} \times \tilde{a}$. The q-deformed Hopf coalgebra mapping $H \to H \times \tilde{H}$ thus describes the *doubling* of the degrees of freedom system-thermal bath, i.e., $a \to \{a, \tilde{a}\}$, and for the phase space $\mathcal{F} \to \mathcal{F} \times \tilde{\mathcal{F}}$, with the operators a and $\tilde{a}$, respectively [9,34]. The "toy-model" of the quantum doubling system-thermal bath that is able to make it immediately intuitive is given by *two damped quantum harmonic oscillators* [37], which is the dissipative counterpart of the standard "two coupled quantum harmonic oscillators" for representing the quantum superposition in the standard non-dissipative case of QM (see Section 3.3). The dissipative model therefore plays an essential role in modeling Gerard 'tHooft's approach to quantum gravity that explains quantization in terms of "information dissipation" in a deterministic system [119]. What is worth emphasizing, anyway, is that by such a doubling, the composite system recovers the essential *unitarity condition* of the Hamiltonian: indeed, it is globally a *closed system*. Consequently, we can use the "doubling of the Hilbert spaces" $\mathcal{H} \to \mathcal{H} \times \tilde{\mathcal{H}}$ to recover the *canonical* or *Hamiltonian representation* of a dissipative quantum system.

We recall, indeed, that the *canonical* representation of whichever dynamic system, both in classical and quantum physics, is in terms of its *Hamiltonian function*. It gives the total energy of the dynamic system concerned because it is an extension of the *Lagrange equation* to the study of the evolution in time of a dynamic system. Since the Hamiltonian canonically represents the dynamic system as being energetically closed, this implies that if we want to write the Hamiltonian of a dissipative system, we must also include in it the thermal bath degrees of freedom to grant that it is "globally" energetically closed.

To sum up, from the *algebra doubling* intrinsic to the coalgebraic representation of the balanced state of an open system, we can naturally pass to the canonical, Hamiltonian, representation of the open system dynamics by extending the algebra doubling to its Hilbert space, i.e., $\mathcal{H} \to \mathcal{H} \times \widetilde{\mathcal{H}}$.

Therefore, because we are interested in the phenomena of QFT in condensed matter physics where the electromagnetic force field and hence photon condensates are involved, we can limit ourselves to the Bogoliubov transform for CCRs of bosons and not CARs of fermions (Section 4.3). In the CCR case, in the "doubled" Hilbert space, what are commuting are not, of course, the doubled (system/thermal bath) states but the associated operators $A(\theta), \widetilde{A}(\theta)$, according to the following relations [34]:

$$A(\theta) = A\cosh\theta - \widetilde{A}^{\dagger}\text{sinh}\theta \tag{5}$$

$$\widetilde{A}(\theta) = \widetilde{A}\cosh\theta - A^{\dagger}\text{sinh}\theta \tag{6}$$

Consequently, CCRs are:

$$\left[A(\theta), A(\theta)^{\dagger}\right] = 1,\ \left[\widetilde{A}(\theta), \widetilde{A}(\theta)^{\dagger}\right] = 1 \tag{7}$$

All other commutators are equal to zero. Equations (5) and (6) are nothing but the *Bogoliubov transformations* for the $\left\{A, \widetilde{A}\right\}$ couple, evidently applied in a *reversed* way between the system (5) and its thermal bath (6), to signify the *energy balance* for a given temperature $T > 0$ (= boundary condition) of the QV ground state, characterizing any *phase transition* of a dissipative QFT system. They, I repeat, are concerning phase coherence domains, including the system and its thermal bath states, i.e., they are "entangled" in only one phase coherence domain to constitute only one dissipative system.

Indeed, the "reversal of the arrows" has an immediate physical significance in the correspondent *reversal of the energy arrow* characterizing the energy balance in any dissipative system in *far-from-equilibrium conditions*. This allows us to define the powerful principle of the "QV foliation" as an *ordered family*—effectively, a "(sub)category" in CT formalization (see Section 4.8)—of "unitarily inequivalent representations" of QV $|0\rangle$ at its ground state, such as many pairs of doubled Hilbert spaces univocally indexed in θ (and/or in q, in the case of the correspondent category of the q-deformed Hopf coalgebras). Indeed, there exists a fundamental relationship between the q-deformation (or "squeezing") *thermal* parameter of Hopf coalgebras in the calculations over a specific set $\mathbf{k}$ of quantum numbers and the "θ-mixing angle" of the Bogoliubov transform, i.e., the θ-set dynamically *labeling* the different "vacua" (QV-foliation) is defined as: $\{\theta_{\mathbf{k}} = \ln q_{\mathbf{k}}, \forall\mathbf{k}\}$.

Quoting directly from [34], let us introduce the QV "splitting" $|0\rangle \equiv |0\rangle \times |0\rangle$ as denoting the vacuum "annihilated" by the Bogoliubov operators A and $\widetilde{A}: A\Big|0\rangle = 0 = \widetilde{A}\Big|0\Big\rangle$, respectively. This means that $|0$ is not annihilated by $A(\theta)$ and $\widetilde{A}(\theta)$ of Equations (5) and (6). On the contrary, the vacuum annihilated by the Bogoliubov operators $A, \hat{A}$ is:

$$|0(\theta)\rangle_{\mathcal{N}} = e^{i\sum_{\mathbf{k}}\theta_k G_k}|0\rangle = \prod_{\mathbf{k}} \frac{1}{\cosh\theta_k} \exp\left(\tanh\theta_k A^{\dagger}\widetilde{A}^{\dagger}\right)|0\rangle. \tag{8}$$

where the subscript $\mathbf{k}$ refers to a set of quantum numbers for quantum variables; $G_{\mathbf{k}} \equiv -i\left(A_{\mathbf{k}}^{\dagger}\widetilde{A}_{\mathbf{k}}^{\dagger} - A_{\mathbf{k}}\widetilde{A}_{\mathbf{k}}\right)$ is the generator of Bogoliubov transformations of Equations (5) and (6) for a whole category of *doubled Hilbert spaces*, effectively, a sub-category for the Bogoliubov functor of the category **Hilb**, as we see in Section 4.8; the subscript N refers to a given *unique* value (its "fingerprint"!) of the condensate of NG-bosons characterizing an annihilated QV state (i.e., in our case, a phase coherence of electromagnetic fields emerging from SSB of QV, corresponding to a dissipative quantum system); and θ denotes, as we know, the set $\{\theta_{\mathrm{K}}, \forall\mathbf{k}\}$, and ${}_{\mathcal{N}}\langle 0(\theta)|0(\theta)\rangle_{\mathcal{N}} = 1$.

Finally, it is possible to demonstrate that in the infinite volume limit $V \to \infty$, the phase space splits *into infinitely many inequivalent representations*, as the Haag theorem previews. In dissipative QFT, however, each of them is *dynamically labeled* by a specific θ—set $\theta_{\mathbf{k}} = \ln_{q\mathbf{k}}, \forall \mathbf{k}$. This is exactly what we intend by the notion of *QV-foliation* in QFT, as a powerful dynamic tool of "construction" used by nature of *complex* systems in physics and *memory* in biology and neuroscience, given that all biological systems are dissipative systems.

4.6. The QV-Foliation and the Principle of Doubling of the Degrees of Freedom as a Solution of the Complexity Issue in Fundamental Physics

The *QV-foliation* emphasizes the advantages of modeling QFT systems as dissipative systems, the main principles of which we illustrate in this section. This advantage can be synthesized in the following statement:

> *Because of the principle of the "doubling of the degrees of freedom" (DDF), the canonical Hamiltonian, i.e., the finite number of degrees of freedom for a faithful representation of the system, must not be supposed any longer such as in the SQ modeling of QFT (Section 3.3) but (thermo)dynamically justified in far-from-equilibrium conditions of many-body physics.*

Any dissipative system indeed must satisfy the energy balance condition corresponding to the *ground state* of the overall dynamics. This means that a ground state is a *0-sum total energy condition* E_{tot} between the system energy E_{sys} and the environment (thermal bath) energy E_{env}, i.e., $E_{tot} = E_{sys} - E_{env} = 0$. However, the same ground state $E_{tot} = 0$—and this is the deep significance of the Goldstone theorem—is compatible with different *orderings* of the concerned "order parameter", and then with different *structural conditions* of the balanced system, related to different *NG-boson condensates* N or *long-range correlations* for each N, including in the dissipative case both the system and its thermal bath, and corresponding in QFT to as many *degenerate states* of the QV "coexisting without interferences" at its ground state.

The notion of *QV-foliation* is therefore that "an ordered hierarchy" of different QV degenerate states indexed in N can coexist without interfering with each other in the same ground state of the balanced system. This is the basis at the level of QFT as the fundamental physics of condensed matter systems of the notion of *system complexity*, I. Prigogine's notion of *dissipative structures* is included [110,120]. A complex system can indeed be intuitively described as a system characterized by a hierarchy of different "emerging" levels of structural organization, each with its own "order parameter" [121]. How they are ordered and how they emerge using only one framework of reference at the level of fundamental physics remains unexplained *without referring to the notion of QV-foliation in QFT.*

In a word, as D. K. Morr stated on the *Science* journal [122] to introduce two further successes of QFT in explaining high-temperature superconductivity, and whose reports are published in the same issue of the journal [123,124], QFT of the dissipative systems is the only theoretical tool "for lifting the fog of complexity" both at level of fundamental physics and fundamental biology.

Moreover, because the principle of the "infinite QV-foliation" also implies per se an infinite foliation of the representational Hilbert spaces as previewed in the Haag theorem—and happens not casually for the infinite inclusion chains of NWF-set—in thermal QFT, it is possible "to exorcise", at least in a partial but significative way (see Section 4.8), such a "nightmare" for mathematicians, physicists, and computer scientists without any reference to an extrinsic "observer", i.e., the supposed knowledge of the system Hamiltonian.

In fact, the *minimum free energy* can be used here as a *dynamic choice criterion* of admissible states of the doubled Hilbert space, differently from quantum thermodynamics based on statistical mechanics [125], where the stability is studied at *equilibrium* in the so-called *asymptotic condition* of the perturbative techniques so that only open systems "near-to-equilibrium" can be studied by such a formalism [126].

On the contrary, the possibility in QFT of using the minimum free-energy function as a *dynamic selection criterion of physically admissible states* makes the notion of the *doubling of the degrees of freedom* (DDF) effective.

A fundamental consequence of the DDF principle is that, in finite temperature QFT, a fundamental statistical tool consists of computing the thermal average expectations of some observable, say $\mathcal{O}$. This requires the computation of "traces", with the trace of a square matrix **A**, $\langle n \times m \rangle$, which is denoted tr(**A**), being defined as the sum of elements on the main diagonal (from the upper left to the lower right) of **A**. Typically ([34], p. 45):

> one deals with matrix elements of the type $\mathcal{O}_{nm} = \langle n|\mathcal{O}|m\rangle$, with orthonormal states $\langle n|m\rangle = \delta_{nm}$ in the Fock space and $H|n\rangle = E_n$.

Here, $H|n\rangle$ is the statistical expectation value with respect to the state n and E_n is the associated energy value. In this way,

> the trace $\sum \mathcal{O}_{nm}$ is obtained by multiplying the matrix elements by δ_{nm} and summing over n and m.

Now, in the standard SQ approach to QFT, the supposition that we already know the Hamiltonian of the system (Section 3.3) consists in introducing δ_{nm}

> as an external (to the operator algebra) computational tool, which essentially amounts in picking up "by hand" the diagonal elements of the matrix and summing them. One may instead represent the in terms of the doubled tilde-states $|n\rangle$, $\langle\widetilde{n}|\widetilde{m}\rangle = \delta_{nm}$ with $\widetilde{H}\left|\widetilde{n}\right\rangle = E_n|\widetilde{n}\rangle$ and $\widetilde{H}\omega\widetilde{A}^{+}\widetilde{A}$ ([34], p. 45).

Then, using the notation $|n,\widetilde{n}\rangle = |n\rangle \times |\widetilde{n}\rangle$, since $\mathcal{O}$ operates only on the non-tilde states, we have:

$$\langle n,\widetilde{n}|\mathcal{O}|m,\widetilde{m}\rangle = \langle n|\mathcal{O}|m\rangle\langle\widetilde{n}|\widetilde{m}\rangle = \langle n|\mathcal{O}|m\rangle\delta_{nm} = \langle n|\mathcal{O}|n\rangle \tag{9}$$

In other terms, the DDF principle in QFT dissipative systems, requiring the doubling of the Fock state spaces, means that the introduction of δ_{nm} is *intrinsic* to the operator algebra because the determination of the orthonormal basis of the Fock space and then of the representation doubled Hilbert space is intrinsic to the dynamics of the system. Indeed, we are dealing with Pauli matrices, where

> in the states $|m,\widetilde{m},\rangle$ m is an integer number. We thus have a Kronecker delta in Equation (9) and not a Dirac delta ([34], p. 45).

Moreover, what is also relevant for our aims is that to each set of degrees of freedom $\{A\}$ and its "entangled doubled" $\left\{\widetilde{A}\right\}$ corresponds a *unique integer number* $\mathcal{N}$, i.e., $\mathcal{N}_A, \mathcal{N}_{\widetilde{A}}$, so that $|\mathcal{N}|$ *identifies univocally*, i.e., it *dynamically labels*, a given *phase coherence domain* of the quantum fields, including the system and its environment. This depends on the fact that in the QFT mathematical formalism, as we know, the number N is a numeric value expressing the NG-bosons condensate value, on which a phase coherence domain *directly depends* such as on its "control parameter". In an appropriate *set theoretic interpretation*, because for each "phase coherence domain" x, $|\mathcal{N}|$ effectively *identifies univocally* such a domain, it corresponds to an "identity function Id_x" that, in a "finitary" coalgebraic logical calculus, corresponds to the *predicate satisfied by such a domain because it univocally identifies it*. We discuss more extensively in Section 4.8 this fundamental *dynamic* indexing principle using the powerful operation of *colimit* as a comma category of indexed functors, which in our case is the comma category of the Bogoliubov functors for the generator G of all the Bogoliubov transform (see Equation (8), Section 4.8, and overall Appendix A).

Finally, the DDF principle justifies, as discussed elsewhere [34], the interpretation of the *maximal entropy* in a QFT "doubled" system as *a semantic measure of information*, i.e., as a statistical measure of *local truth* in the CT coalgebraic BAO logic, as we see in Section 5.2. This is strictly related to the notion of *doubled qubit* or *semantic qubit* characterizing the quantum computations in these classes of dissipative quantum systems. Indeed, in the QFT "composed Hilbert space", including the thermal bath degrees of freedom, $\widetilde{A}$, i.e., $\mathcal{H}_{A,\widetilde{A}} = \mathcal{H}_A \otimes \widetilde{\mathcal{H}}_{\widetilde{A}}$, to calculate the static and dynamic entropy associated with the time evolution generated by the free energy, i.e. $|\phi(t)\rangle, |\psi(t)\rangle$, of the qubit mixed states $|\phi\rangle, |\psi\rangle$, one needs to double the states by introducing the tilde states $|\widetilde{0}\rangle$ and $|\widetilde{1}\rangle$, relative to the thermal bath, i.e., $|0\rangle \to |0\rangle \otimes |\widetilde{0}\rangle$, and $|1\rangle \to |1\rangle \otimes |\widetilde{1}\rangle$. This means that such a QFT version of a qubit effectively implements the CNOT (controlled NOT) logical gate, which flips the state of the qubit, conditional on a *dynamic* control of an effective input matching [34,127,128]. Because it is demonstrated that it is possible to recursively calculate in such a computational architecture the *Fibonacci series,* this architecture is a particular type of (topological) quantum implementation of a "golden machine" (see [129] for a different model of this type of machine related to the "non-commutative anions" for the Hall-effect), i.e., it is a "universal computer" (see [34,129] for further details with the necessary bibliography).

4.7. The Cognitive Relevance of the QFT Modeling of Non-Linear Brain Dynamics

The main interest of Walter Freeman (1927–2016) during the many decades of active research with his group in the neuroscience lab of the University of California at Berkeley was the explanation of the neural basis of *intentional behavior* in humans and animals. More specifically, it was the explanation for the *dynamic integration* of the "mosaic" of the different brain modules of the cortex, each performing different sensory or motor functions. Effectively, this aim was pursued by Freeman using for the electroencephalogram (EEG) and electrocorticogram (ECoG) computational signal analysis, not the Fourier transform but the *Hilbert transform.* Indeed, while the former decomposes the signal into its frequency components, the Hilbert transform decomposes the signal into the analytical amplitude $A(t)$ and its analytic phase $\varphi(t)$. In this way, he discovered the empirical evidence of the massive presence of patterns of AM phase-locked oscillations in the *background activity* of the brain. This is registered by EEG, but often it is filtered as "noise" by neurophysiologists exclusively interested in studying the neuron (arrays) spike activity and their synaptic circuitry [130].

On the contrary, these *long-range correlation patterns* have their neural medium in the cortex *neuropil*, that is, "the dense felt-work of axons, dendrites, cell bodies, glia and capillaries forming a superficial continuum 1–3 mm in thickness over the entire extent of each cerebral hemisphere" in mammals ([39], p.95). These correlation patterns are intermittently present in resting and/or awake subjects, and the same subject actively engaged in cognitive tasks requiring a goal-directed interaction with the environment. These "wave packets" extend almost instantaneously over coherence domains covering much of the hemisphere in rabbits and cats, and regions of linear size of about 19 cm in human brains [131–134].

During the largest part of his research life, Freeman tried to find a suitable physical explanation of this amazing phenomenon, as I can personally testify during the many meetings I had with him during the 1990s because we were both interested at that time in studying the overall chaotic dynamics of natural (him) and artificial (myself) neural networking. I recall, therefore, when he enthusiastically announced to me, during one of our last personal meetings in 2008, that he solved the problem. Indeed, Vitiello demonstrated mathematically that the trajectories in the phase space between different phase coherence domains, along which a dissipative quantum system moves at the *microscopic* level, display a chaotic character at the *macroscopic* level [118]. Effectively, since 2002, a fruitful collaboration has started between Freeman and Vitiello, leading to a series of common publications till the year of Freeman's death in 2016.

The final result is that Freeman's discoveries constitute at the moment the more extended experimental confirmation at the level of condensed matter physics of the main notions of QFT for the "dissipative brain" model (see [39,40] for a synthesis). This excited my interest in searching for the most suitable *intensional* logic to associate with this modeling of the *intentional* behavior in brain dynamics conceived as a natural computational system (see Section 1.1). Additionally, this is surely the *coalgebraic semantics of Kripke relational logic* in the CT framework, given also the non-casual but essential role that coalgebras and coproducts generally play in QFT computations (see Section 4.5).

> To conclude, in this "holistic" framework of intentional behaviors, the concept of the boson carrier and the boson condensate does more; it enables an orderly and inclusive description of the phase transition that includes all levels of the macroscopic, mesoscopic, and microscopic organization of the cerebral patterns that mediate the integration of the animal with its environment, down to and including the electric dipoles of all the myriad proteins, amino acid transmitters, ions, and water molecules that comprise the quantum system. This hierarchical system extending from atoms to the whole brain and outwardly into the engagement of the subject with the environment in the action-perception cycle is the essential basis for the ontogenetic emergence and maintenance of meaning through successful interaction and its knowledge base within the brain. By repeated trial-and-error, each brain constructs within itself an understanding of its surroundings, which constitutes its knowledge of its own world that we describe as its *double*. It is an *active* mirror [24] because the environment impacts onto the self independently and reactively. The relations that the self and its surround construct by their interactions constitute the meanings of the flows of information that are exchanged during the interactions ([39], p. 108).

To intuitively understand what all this means and specifically the notion of brain–environment "entanglement" or "active mirroring" in QFT, as the cognitive counterpart of the DDF principle, it is noted, for example, that as the elementary evidence humans receive visual stimulations from the optical part of the electromagnetic spectrum while owls receive from the infrared part (see Figure 8). Intuitively, but consistently in the light of the DDF principle, this is because the cone and the rod cells of the human retina are oscillating in phase—so to constitute "only one phase coherence domain"—with the optical (visible light) part of the electromagnetic spectrum in the environment while the owl retina cells are in phase with the infrared part of the spectrum.

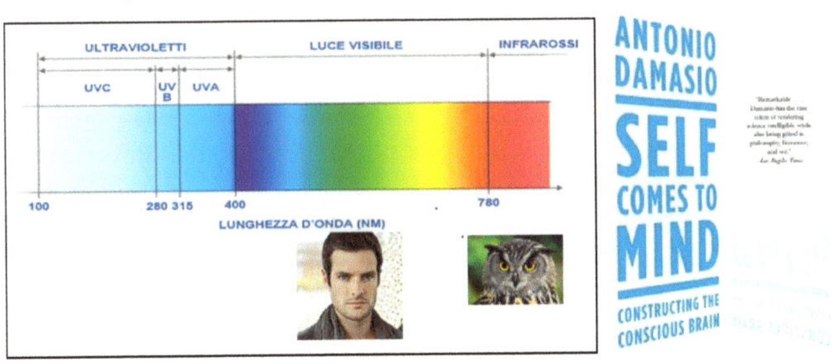

Figure 8. Intuitive representation of the "active mirroring" brain–environment in cognitive neuroscience, consistent with Damasio's biological foundation of intentionality on the physical homeostasis principle—as suggested metaphorically in the promotional picture of Damasio's book on the right—that holds in biology, from the epigenetic principle of cell specialization, till brains and beyond (see text).

All this is also consistent with Antonio Damasio's proposal of finding the biological foundation of *intentionality* in the *homeostatic non-linear mechanisms* characterizing all living systems as dissipative systems. Living systems are endowed, differently from non-living ones, with complex non-linear *self-regulation* processes [135]. The complex self-organizing "intentional" structure of living dissipative systems "entangled" with their environments thus ranges from the elementary level of the epigenetic mechanisms in cell specialization to the formation of tissues, organs, nervous systems, brains (see [120]), and even beyond to model the social behavior ("the society of brains") of animals [41,136]. The reader can usually refer to [137] to synthetically deepen the relationship between Damasio's and QFT approaches to the foundations of intentionality in biological and neural systems.

Finally, this physical foundation of the intentional behavior in biological and neural systems is consistent with the so-called interpretation of the *extended mind,* locating it not in the brain but in the physical (material and informational) intercourse between brains and their environments, given that a *living* brain is a *dissipative* brain [138,139]. This intercourse effectively ends only with the death of the animals when they finish being a dissipative system, stable in far-from-equilibrium conditions, and they are in equilibrium with their environment.

4.8. Categorical Definition of the QFT Systems as Initial (Terminal) Objects of Comma Categories (Universal Morphisms)

This final subsection is devoted to discussing the fruitfulness of modeling the dissipative QFT and its non-commutative coalgebraic formalism in the framework of CT. Indeed, in the light of what is discussed in this section, it is evident that the q-deformed Hopf coalgebras form a category for the Bogoliubov functor $\mathcal{B}$ ([34], pp.47–48), i.e., **qHCoalg($\mathcal{B}$).** Moreover, because of the existence of a common generator of all Bogoliubov transforms $G_{\mathbf{k}} \equiv -i\left(A_{\mathbf{k}}^{\dagger}\widetilde{A}_{\mathbf{k}}^{\dagger} - A_{\mathbf{k}}\widetilde{A}_{\mathbf{k}}\right)$ that are part of the Hamiltonian of the system, which is strictly related to the NG-boson condensate value $\mathcal{N}$ univocally indexing each phase coherence domain or degenerate state of QV $|0(\theta)\rangle_{\mathcal{N}}$ (see Equation (8)), the consequent principle of the QV-foliation can be properly formalized in CT by the powerful *colimit operation* (i.e., "direct limit operation" in mathematical analysis) applied to the category of the QFT coproducts for the Bogoliubov functor. Indeed, we have all the "ingredients" satisfying the categorical definition of QFT systems as objects that are *universal morphisms,* specifically, as initial (colimits) or terminal (limits) objects of *comma categories* (see Appendix A). Here, a comma category is a category $\mathcal{C}$ constituted by morphisms relating objects belonging to other two categories $\mathcal{A}$ and $\mathcal{B}$ having $\mathcal{C}$ as their common functorial target, i.e., $\mathcal{A} \xrightarrow{F} \mathcal{C} \xleftarrow{G} \mathcal{B}$ (see Definition A1).

It is worth emphasizing here the physical, mathematical, and then ontological relevance of this categorical construction, by which whichever object c in a category $\mathcal{C}$ is a functor, in turn, mapping from the category **1** with only one object $(*)$ (a "limit point") and its identity morphism 1_* to the category $\mathcal{C}$. In this way, c can always be represented as the initial/terminal object of a "triadic" comma category of morphisms (functors): $\mathbf{1} \xrightarrow{c} \mathcal{C} \xleftarrow{F} \mathcal{B}$. This diagrammatic construction leads to the universal formalization of the "constructive power" of nature of new systems by the general mechanism of SSB, where the new generated system corresponds to a "phase coherence domain" of the quantum fields at their ground state (=quantum vacuum), and then to the one-object category **1** of the comma category construction, i.e., mathematically, they are limit points of the *compact, Hausdorff, totally disconnected* topological spaces of the σ*-algebras* on which the probability spaces of quantum systems in physics and the "Stone spaces" of the RTBA in logic are defined (see Section 5.1.1 and Appendix B). The possibility of defining an ordered hierarchy of different phase coherence domains by the QFT notion of the "QV-foliation" can then obtain its proper diagrammatic universal formalization by the colimit operation using as the third category $\mathcal{B}$ of the comma category a particular small category of indices $j \in \mathcal{J}$ (see Appendix A), where in the QFT case, the set of indices corresponds to the set $\{\mathcal{N}\}$ of QFT coproducts, with each identifying a phase coherence domain of the QV, i.e., $\{j\} \equiv \{\mathcal{N}\}$, $\forall(\mathcal{N})$.

Indeed, the fundamental QFT construction of the "QV-foliation" as the fundamental dynamic tool used by nature for constructing "complex systems" in the physical reality and "memories" in the biological systems, and all the related QFT constructions discussed in this section, receive their proper "diagrammatic" *universal definition* in the CT metalanguage. Thus, we can obtain *the diagrammatic universal definition* of classes of QFT systems as initial (colimits) and terminal (limits) objects of indexed comma categories, which are, respectively, categories of "cocones" and "cones" of morphisms over a functor F (see Appendix A).

Indeed, following the classic Adamek's-Rossicky's book [140] in the characterization of *different classes of small categories* and then *their limits-colimits* according to the different cardinalities, $\aleph_0$, $\omega_1, \omega_2, \ldots$, of their objects (see Appendix A), one can usefully apply these distinctions to the QFT construction of the category of the q-deformed Hopf coalgebras for the Bogoliubov functor **qHCoalg($\mathcal{B}$)** and, by the principle of the "algebra doubling", to the category of the doubled Hilbert spaces **DHilb($\mathcal{B}$),** as a subcategory of the category **Hilb.** In this way,

1. We can formalize the general "particle-wave" duality principle in QFT of Equation (4) in Section 4.4, for which new objects (systems) "emerge" in nature as new phase coherence domains of the quantum fields at their ground state such as many SSBs of QV, using the categorical constructions of the initial/terminal objects in comma categories. Indeed, we can start from a given category $\mathcal{B}$ of quantum fields mapping its objects (quanta) and morphisms (fields) over a category $\mathcal{C}$ via the functor $F: B \rightarrow C$. In turn, the SSB mechanism of the phase coherence among quantum fields can be formalized as a functor from the category **1** with one object $(*)$ and its identity (reflexive) morphism 1_* as a functor c to C to obtain the structure of the comma category $1 \xrightarrow{c} \mathcal{C} \xleftarrow{F} \mathcal{B}$. In this way, we can define any new "emerging" object (system) in nature by the "universal morphism" (commuting diagram) from c to F as the *initial* objects (= colimits) in the comma category $(c \downarrow F)$ or dually as *terminal* objects (limits) in the comma category $(F \downarrow c)$ (see Definition A1 and Definition A2 with their explanations in Appendix A). Finally, it is worth emphasizing for the logical applications of this construction that set-theoretically this interpretation of objects as (self-containing) phase coherence domains of the quantum fields requires the *set self-membership* property characterizing the coalgebraic category of the NWF-sets (see Sections 2.5 and 5).
2. Because of the category **qHCoalg($\mathcal{B}$)**, and the existence of only one generator G of all the Bogoliubov transforms—strictly related to the NG-boson condensate value $\mathcal{N}$ indexing "dynamically" each phase coherence domain or degenerate state of QV $|0(\theta)\rangle_{\mathcal{N}}$ (see Equation (8))—we can categorically formalize the fundamental QFT construction of the "QV-foliation" by the colimit construction. Indeed, also in the QFT case, we can substitute the category $\mathcal{B}$ of Definition in Appendix A with the small category of indices $\mathcal{J}$, where the set of indices $\{j\} \in \mathcal{J}$ corresponds here to the set $\{\mathcal{N}\}$ of the NG-boson condensates univocally associated with the QFT non-commutative coproducts, i.e., $\{j\} \equiv \{\mathcal{N}\}$, $\forall \mathcal{N}$, with each $\mathcal{N}$ identifying a phase coherence domain (or "degenerate state") of the QV, i.e. $|0(\theta)\rangle_{\mathcal{N}}$. The "universal morphism" is therefore in terms of the initial objects of comma categories $(F \downarrow \Delta)$, which are *colimits* of "cocones of morphisms" according to Definition A3 in Appendix A.
3. This depends on the fact that in QFT, there also exists a constant diagram: $\Delta_c : \mathcal{J} \rightarrow \mathcal{C}$, mapping every object in $\mathcal{J}$ to c and every morphism in $\mathcal{J}$ to 1_c. We can therefore define the *diagonal functor* $\Delta : \mathcal{C} \rightarrow \mathcal{C}^{\mathcal{J}}$ as the functor assigning to each QFT object c of C the diagram (constant functor) Δ_c and to each morphism (Bogoliubov transform) $f : c \rightarrow c'$ in C the natural transformation $\Delta f = \Delta c \rightarrow \Delta c'$. Because $\Delta c, \Delta c'$ are constant functors, Δf is just the morphism $f : c \rightarrow c'$ for every object in $\mathcal{J}$.
4. Because the "diagonal functor" $\Delta : \mathcal{C} \rightarrow \mathcal{C}^{\mathcal{J}}$ is, in QFT, a categorical generalization of the *diagonal form* of the $n \times m$ Pauli's matrices of the dissipative QFT calculations on integer numbers (see Section 4.6 and Equation (9) with the relative comments), one could propose the following hypothesis requiring further studies and developments: namely, that the category **DHilb($\mathcal{B}$)** constitutes a *locally finitely presentable sub-category* of the category **Hilb**, with ordered objects (coproducts) characterized by infinitely many *countable $\aleph_0$-directed colimits* (see [140], pp. 2–3) because it satisfies the fundamental condition of having only one generator G of all the Bogoliubov transforms (see [37], p. 18).

Indeed, recalling here again the remarks of Adamek and Rosicki ([140], pp. 70; pp. 101–105), the category of Hilbert spaces **Hilb** is not "locally presentable", and even less it is "locally *finitely* presentable". **Hilb** is closed, indeed, under a particular class of colimits (the ω_1-*directed* colimits [25] in the category of Banach spaces **Ban**) [26]; however, therefore, it is not *cocomplete* (i.e., it does not satisfy the powerful "cocompleteness theorem", see Theorem A1 in Appendix A). Specifically, all the colimits are not in **Hilb** but in **Ban**, as the GNS-construction demonstrated (see Section 3.2 and Note 19 in it), so that **Hilb** is a full subcategory of **Ban**. A condition that the authors emphasize depends on the fact that the category **Hilb**, such as the strictly related category **Hopf** of Hopf bialgebras, differently from the category **Ban**, is made of objects that are *self-dual* [27].

In other terms, we know that in the operator algebra approach to QM and QFT, different families and then subcategories of the category **Hilb** exist. For instance, in the light of the GNS-construction, infinite subcategories exist, with each composed (by a family) of *finitely many unitarily equivalent* Hilbert spaces for the Gelfand functor, one for each *closed* quantum system, because it satisfies the Stone–Von Neumann theorem (see Section 3.1). In the light of our precedent discussion, however, the sub-category of the *doubled* Hilbert spaces *DHilb* for the Bogoliubov functor $\mathcal{B}$ also exists, i.e., **Hilb** $\hookrightarrow$ **DHilb($\mathcal{B}$).** It is composed of a *denumerable infinite number of inequivalent pairs* of *doubled* Hilbert spaces $\mathcal{H}, \widetilde{\mathcal{H}}$, with each pair being *dually unitarily equivalent* because it represents a *dissipative* quantum system in a balanced state.

Indeed, each pair, despite being defined on different non-commutative algebraic "footings", is *dually unitarily equivalent* for the contravariant application of the Bogoliubov transform, i.e., $\mathcal{H} \cong \widetilde{\mathcal{H}}$. Namely, despite each $\mathcal{H}$ of the pair at the level of its *inner structure* being, of course, self-dual and then belonging to the category **Hilb,** nevertheless, the two finite Hilbert spaces of each pair, generated by the contravariant application of the Bogoliubov transform, *share the same metrics* (effectively, they constitute one "σ-algebra" of the related probability space; see Section 5.1.1 and Appendix B) because they form only one dissipative system, and then they are *dually unitarily equivalent.* Therefore, the category **DHilb($\mathcal{B}$)** is constituted by infinitely many *inequivalent* (pairs of) finite Hilbert spaces, with *each dually equivalent pair* (i.e., satisfying the *coequalizer* condition for coproducts (see Definition A5 in Appendix A)) representing a different dissipative quantum system.

Moreover, because only one *generator* of all the Bogoliubov transforms characterizing the **DHilb($\mathcal{B}$)** indexed subcategory for the Bogoliubov functor exists in the dissipative QFT equations, it satisfies the fundamental condition for which this subcategory can be defined as a *locally finitely presentable category,* characterized by infinitely many *countable* $\aleph_0$-*directed colimits* (see [140], pp. 2–3). Therefore, the hypothesis, which, I repeat, needs to be properly mathematically developed and demonstrated, is that, differently from the category **Hilb,** its sub-category **DHilb($\mathcal{B}$)** is *cocomplete* (i.e., it satisfies the fundamental "cocompleteness theorem" (see Theorem A1 in Appendix A) because it is $\aleph_0$-*locally presentable.* Namely, despite it is constituted by infinitely many inequivalent pairs of doubled Hilbert spaces, whichever subset of them of any dimension is finitely denumerable, because a finite number of NG-bosons $|\mathcal{N}|$ is *always* sufficient for indexing each of these subsets (see Appendix A for further details).

The non-irrelevant consequence is that despite this subcategory being made of infinitely many inequivalent representations of CCRs and CARs modeled by the DDF principle for dissipative QFT systems, and then the pairs of *DHilb* spaces in the subcategory represent the outcomes of as many *phase transitions* of an overall non-linear dynamics in far from equilibrium conditions, nevertheless, the cocompleteness theorem grants the *compactness* of the underlying topological space. Namely, it contains all its limit points (colimits), with each representing a different phase coherent domain of the quantum field dynamics at their (balanced) ground state. Or, better, each represents a different class (cocone of morphisms) of (dissipative) quantum systems generated by nature through the universal mechanism of SSBs. For the very same reason, the trajectories in the phase space between different phase coherent domains are (macroscopically) *chaotic trajectories* because of the compactness of the space in which they are defined (see [34]).

To sum up, the QFT conundrum of the infinitely many unitarily inequivalent representations (Hilbert spaces) of CCRs (and CARs) related to the Haag theorem has an at least partial (?) solution in the hypothesis of the thermal interpretation of QFT, for which any quantum system is a dissipative system, so that the DDF principle holds because of the Bogoliubov natural transformation.

Finally, from the *logical* standpoint, another character (see [34], p. 44) of this subcategory of Hilbert spaces is that their topologies satisfy the condition of being *Chu spaces* [141], in which NWF-sets (see [34], p. 44; [142], and Section 2.5) and then "rooted trees" of Kripke structures (models) can be defined [73] (see Section 5.2.3). This opens the way to the core of the "relational natural realism" and then the NR-formal ontology based on the principle that the semantics of the descriptive propositional formulas of the ontological language, formalized in a modal Boolean algebra, are *validated directly by homomorphism* onto the (co)algebraic mathematical structure of the physical systems they describe and/or to which they refer. Of course, this also gives a categorical positive answer to Quine's logical issue about the same possibility of a naturalistic formal ontology, with which we started this work (see Section 1.2).

5. From Coalgebras in Physics (QFT) to Coalgebras in Kripke's Relational Semantics

To formalize all these "descriptive" statements of the NR-formal ontology, it is necessary to define therefore which is the proper algebraic (Boolean) modal logic of this ontology in a categorical setting, with evident consequences for a formalized philosophy of nature and science. The main steps of this construction, corresponding to as many subsections of this section, are the following:

1. An examination of the Boolean logic in the light of *Stone's momentous RTBA* and the consequent *Tarski's and Jónsson's construction of BAO's*, extending the operator algebra formalism from physics to logic. The core of both constructions is, indeed, the notion of *field of sets*, that is, the subalgebra of the power-set of a given set. Indeed, this subalgebra is a *σ-algebra*, that is, an algebra defined on a *measure space* and specifically on a *probability space* characterized by a finite number of degrees of freedom (the orthonormal basis of the associated Hilbert space, for instance) on which the statistical expectations are calculated. This opens the way for an extension of the operator algebra formalism over a topological complex algebra (algebra-subalgebras structures), from the statistical and quantum physics to the so-called "topological approach" to Boolean logic, properly to BAO logic.
2. A formal explanation of the algebraic formalization of the relational notion of *meaning function* in CT logic, in which the semantics of the Boolean propositional logic is validated by its homomorphism with a complex (co)algebra over a topological space.
3. Its extension to Goldblatt's and then to Kripke's *coalgebraic modal relational semantics* of BAOs in their categorical setting, which is the proper logic of the NR-formal ontology.

5.1. From Stone's Representational Theorem for Boolean Algebras to Tarski's Theory of the Boolean Algebras with Operators

5.1.1. Definition of "Field of Sets"

After Stone's RTBA [36], when we speak today in mathematical logic of an *algebra of sets*, we are speaking of a *Boolean algebra* (BA) and/or a *field of sets* [143]. Indeed, the more synthetic way for expressing the main statement of RTBA is: "RTBA states that every BA is *isomorphic* with a certain *field of sets*". Following [144], we can give the following definition of a "field of sets":

Definition 1. *(Definition of field of sets). A field of sets $(X, \mathcal{F})$ is a pair consisting of a set X and a collection (family) $\mathcal{F}$ of subsets F of X called "algebra over X", which contains the empty set as an* *element and is closed under the set-theoretic operations of taking complements, finite union, and finite intersection, i.e., such that:*

- $F \in \mathcal{F} \Rightarrow X \backslash A \in \mathcal{F}$ *(closed under complementation).*
- $F, G \in \mathcal{F} \Rightarrow F \cup G \in \mathcal{F}$ *(closed under union).*
- $F, G \in \mathcal{F} \Rightarrow F \cap G \in \mathcal{F}$ *(closed under intersection).*

In other words, $\mathcal{F}$ constitutes a *subalgebra* of the power-set atomic BA of X. Elements of X are called *points* while the elements of $\mathcal{F}$ are called *complexes* (i.e., subalgebras of the complex structure algebra-subalgebras we are considering here) and/or the *admissible subsets* of X.

Definition 1. *(Definition of a σ-field of sets). A particular field of sets* $(X, \mathcal{F})$ is called the *σ-field of sets—and then the algebra* $\mathcal{F}$ *is called a σ-algebra—if one or both the two equivalent further conditions are given:*

- $\cup_{i=1}^{\infty} F_i := F_1 \cup F_2 \cup \ldots \in \mathcal{F} \mid \forall F_i \in \mathcal{F}$ *(closure under countable unions).*
- $\cap_{i=1}^{\infty} F_i : +F_1 \cap F_2 \cap \ldots \in \mathcal{F} \mid \forall F_i \in \mathcal{F}$ *(closure under countable intersections).*

Because of its measurable character, a σ-field of sets is defined as "normal".

Therefore, given that with "representation theorem" we intend to prove that every abstract structure with certain properties is isomorphic with another structure, the precedent definitions help us to understand the core of Stone's RTBA, even though its formal proof is outside the limits of this work (see for this [36]).

5.1.2. Stone Representation Theory of Boolean Algebras

To understand this step, it is necessary for philosophers unacquainted with this notion to recall what is an *ultrafilter* in set-theory defined by the power-set of a given set, and its relevance for set-theoretic logic, and Boolean logic. A power-set always includes the empty set because it is given by the combinations of all the subsets of a given set, so that if the generic *finite* set X has cardinality n, its power-set will have cardinality 2^n because of the inclusion of both the empty set as its minimal element and the same set X as its maximal element so that it satisfies a *reflexive* relation (beside *transitivity* and *anti-symmetry*) and any power-set is a partially ordered set or *poset*.

We already know from the previous discussion that *abstractly*, a BA is isomorphic with the power-set of a given set and more precisely with a field of set. Now, when we think of some *concrete* realization of a BA, such as, for instance, in any logical application of a BA, all the subsets of the power-set must satisfy the *further condition* of representing some *property*, say p expressed by a propositional formula φ of the logical calculus. This means that we must include some *filtering* condition over the power-set, first, the condition of *excluding the empty set*, given that every subset must contain p and this condition cannot be evidently satisfied by the empty set.

In other terms, this means that we need an *atomic* BA, where the minimal element is an *atom a*. Whereas, in *order theory*, an element a of a poset with the least element 0 (e.g., a power-set) is an *atom* if $0 < a$ and there is no x, such that $0 < x < a$. Therefore, in Boolean logic, every Boolean term corresponds to a propositional formula φ of propositional logic expressing a given property p. In this translation between Boolean algebra and propositional logic, Boolean variables x, y... become propositional variables (or *atoms*) $p, q, \ldots$, Boolean terms such as $x \vee y$ become complex propositional formulas $p \vee q$, **0** becomes *false* or $\perp$, and **1** becomes *true* or $\top$. In this way, "an *atomic* BA is isomorphic to an *ultrafilter* of partially ordered sets, in our case a power-set" (see Figure 9).

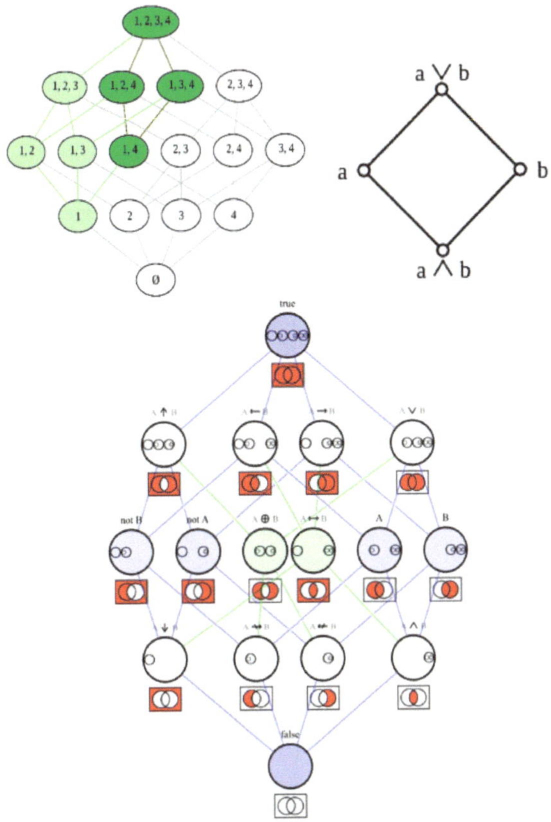

Figure 9. (**Top-left**): Representation of a *principal ultrafilter* (all the subsets colored in green), that is, the *maximal proper filter* of the power-set of the 4-element set S $\{1,2,3,4\}$ ordered by inclusion in a Hasse diagram. It has as the *principal element* or *upper set* ($\uparrow$) the *singleton* or 1-element set ($\uparrow \{1\}$). The subsets colored in dark green also constitute a *proper filter*, which is also a *principal filter*, because it has a principal element, i.e., ($\uparrow \{1,4\}$). However, it is not an ultrafilter because only by adding the subsets coloured in light green, can we obtain the maximal proper filter of this power-set with the principal element ($\uparrow \{1\}$). (**Top-right**): Schematic representation of the *meet* (*last lower bound*) ($a \wedge b$) and the *join* (*least upper bound*) ($a \vee b$) of a Boolean logical lattice, where a and b are two propositional constants. (**Bottom**): *Complete* Hasse and Venn diagrams of all the propositional connectives representable on a Boolean logical lattice, having as *meet* (bottom-element $\perp$) the *always false* proposition ($0 := a \wedge \neg a$) and as *join* (top-element $\top$) the *always true* proposition ($1 := a \vee \neg a$).

To understand this statement, let us explain briefly what an ultrafilter is in mathematical *order* theory. An ultrafilter is the *maximal filter* of a given set, that is, the maximal partially ordered subset of the power-set of a given set, with the empty set excluded.

More generally, if P is a set partially ordered by $\leq$ (and the power-set is always partially ordered), then:

- A subset $F \subseteq P$ is called a *proper filter* on P if:
    - F is nonempty (i.e., with the exclusion of the empty set $\varnothing$);
    - For every $x, y \in F$ $x \leq x$ and $x \leq y$ hold; and
    - For every $x \in F$ and $y \in P$, $x \leq y$ implies that y is in F too.
- A proper subset U of P is an *ultrafilter* on P if:

- ◦ U is a filter on P; and
- ◦ There is no proper filter F that properly extends U.

Filters and ultrafilters are called *principal* if they contain a *least* element, i.e., they are of the form $F_a = \{x : a \leq x\}$ for some but not all elements a of the poset P (a is an atom). In this case, a is called the *principal element* of the ultrafilter. Therefore, all ultrafilters that are not principal are called *free*.

For ultrafilters defined on the power-set of a given set S, $\mathcal{P}(S)$, a principal ultrafilter consists, therefore, of all subsets of S containing a given element $p \in S$. This is fundamental in Boolean logic for understanding why the semantics of a formula of the propositional calculus interpreted equationally on a Boolean equation logic is defined onto the principal ultrafilter of the power-set of some set.

For instance, if we take (see Figure 9 (top-left)) the lattice P representing the power-set of the set $S\ \{1,2,3,4\}$, ordered by inclusion in a Hasse diagram $(\mathcal{P}(S), \subseteq)$, the colored subsets represent the maximal filter starting from the *principal element* p (= *least element* or the "meet" of the corresponding Boolean lattice), which in our case is $\{1\}$. In this way, we can say that all the elements of the filter are *upward closed*, with respect to p: $\uparrow p$, i.e., they are *upsets* because they satisfy the condition: $(\uparrow p := \{\forall x \text{ in } F \mid x \geq p\})$, i.e., they are opsets (finite *intersections* or set products) closed by other opsets (finite intersections).

Moreover, if we observe the subsets of S colored in dark green in Figure 9 (top-left), we see immediately that these also constitute a *principal filter* with the principal element $(\uparrow \{1,4\})$. However, it is not the *maximal proper filter* on S without the addition of the other light-green-colored subsets, which therefore is an *ultrafilter* with a principal element $(\uparrow \{1\})$.

Finally, in order theory, the *dual* of a filter F is called the *ideal I*, and the dual of an ultrafilter U is the *prime* or *maximal ideal*. Generally, we can obtain an ideal I from a filter F and a prime ideal from un ultrafilter U simply by *inverting the ordering relations* in F (U), i.e., $x \leq y$ with $x \geq y$, and then, in the corresponding Boolean logic, by substituting products (finite intersections) $\wedge$ with disjunctions (finite unions) $\vee$. In this way, we are satisfying the duality ($\wedge$, $\vee$) of the fundamental *De Morgan logical laws* [28], i.e., we are satisfying in Boolean logic the so-called *De Morgan duality*. In the case of a principal ideal I, the principal element p is the *last element*, in our case $\{1,2,3,4\}$. In this way, again dually, as to filters, we can say that all the elements of the ideal are *downward closed* with respect to p: $\downarrow p$, i.e., they are *downsets* because they all satisfy the condition: $(\downarrow p : +\{\forall x \text{ in } I \mid x \leq p\})$, i.e., they are downsets (*disjoint unions* or set coproducts) closed by other downsets (unions).

However, to *properly* pass to posets that are BAs, the ultrafilters must be characterized by containing exactly its Boolean complement $\neg a$ for each element a (atom) of BA to satisfy the BA signature of the proper operations defining BA, i.e., ($\wedge$,$\vee$,$\neg$) (see Figure 9 (top left and right)). Therefore, if P is BA and F is a proper filter, the following statements are equivalent ([145], p.186):

1. $(F \uparrow)$ is an ultrafilter on P.
2. $(F \downarrow)$ is a prime ideal on P.
3. For each a $\in$ P, either a $\in$ P, or $\neg$ a $\in$ P.

In the light of this discussion, it is possible to understand at least intuitively the notion of *Stone representation*, which is the core of his RTBA. Indeed, every *finite* BA can be represented as a power-set of its *atoms*, with each element of BA corresponding to the set of atoms below it, i.e., the join of which is the element. This power-set representation can be constructed for any *complete* atomic BA, where "complete BA" means BA in which every subset has a "supremum" or *least upper bound* [29].

In the light of this discussion and that in Section 5.1.1, Stone's RTBA stating the *isomorphism* and then the *Stone duality* between a BA B and a topological field of sets also becomes easily understandable. Specifically, this is a field of sets constituting the basis of the *Stone space* associated to BA B, i.e., $S(B)$. The *points* in $S(B)$ are indeed the *ultrafilters* on

B, or equivalently the *homomorphisms* from B to the two-element BA. Conversely, given any topological space $(X, \mathcal{T})$, its subsets that are clopen form BA.

Therefore, on the basis of this and of Appendix B, we can understand why a simpler version of Stone's RTBA states that every BA B is isomorphic to the algebra of clopen subsets of its Stone space $S(B)$. The isomorphism indeed sends an element $b \in B$ to the set of all ultrafilters that contain b, as the previous illustration of the notion of the topological field of sets showed.

In the framework of CT, the Stone RTBA states that a *dual equivalence* for the contravariant application of the Stone functor $\mathcal{S}$ between the category of BAs **Bool** and the category of Stone spaces **Stone** exists, i.e., $\mathbf{Bool}(\mathcal{S}) \simeq \mathbf{Stone}(\mathcal{S})^{\mathbf{op}}$. This depends on the fact that in addition to the (invertible) homomorphisms between BA B and its Stone space $S(B)$ making them isomorphic, i.e., $B \cong S(B)$ (= "Stone duality"), each homomorphism (monotone function) from a BA A and a BA B in the category **Bool** corresponds naturally (by a natural transformation) to an homomorphism (continuous function) going in the opposite direction from the Stone space $S(B)$ to the Stone space $S(A)$. This theorem is thus a special case of the more general *Stone duality* between topological spaces and partially ordered sets.

For our ontological aims, what is relevant is indeed the extension of Stone RTBA to *measure spaces* that are the topological spaces used in physics. As we have just said, Stone RTBA by the construction of the fields of sets and the ultrafilters over them holds for arbitrary set unions and intersections, i.e., it does not require *complete* atomic Bas, and this gives the theorem its full and powerful generality. However, if an algebra over a set is closed over countable unions and intersections, it is a *σ-algebra* (see Section 5.1.1) and the corresponding field of sets is a *measurable space,* and its complexes are *measurable sets* [30]. Now, a theorem demonstrated independently by L. H. Loomis [146] and R. Sikorski [143], the *Loomis–Sikorski theorem,* grants the Stone-duality between *σ-complete* BAs, $\mathbb{BA}^{\sigma}$ (or "abstract σ-algebras") and *measurable spaces.*

We recall, indeed, that in applied mathematical sciences, and physics, a *measure space* $(X, \mathcal{F}, \mu)$ is a measurable space and μ is a measure defined on it. If μ is a *probability measure,* we have a *probability space,* and its underlying measurable space is a *sample space.* The points of this space are *samples* representing the outcomes of measuring operations while the measurable sets (complexes) are called *events* and represent the outcomes of physical processes to which we want to assign probabilities. In physics and specifically in quantum physics, we work on measure and probability spaces derived from significant algebraic structures such as the inner products of Hilbert spaces and the topological groups with the associated topological spaces and complexes. This is confirmed by the fact that the topological spaces of Stone's RTBA and Hilbert space complexes (Hilbert spaces with their C*-subalgebras) are the same (see [45], pp. 805–833, and Appendix B).

5.1.3. Jónsson–Tarski Theory of the "Boolean Algebras with Operators"

In this framework, the Jónsson–Tarski representation theorems for the *Boolean algebras with operators* (BAO), effectively extending Stone's RTBA to the operator algebra approach [4,5], acquire a fundamental relevance. To understand the theoretical core of these theorems, we must introduce the notion of representation of *interior algebras* by field of sets *preorders* [31]. Indeed, a *preorder field* is a triple $(X, \leq, \mathcal{F})$, where $(X, \leq)$ is a preordered set and $(X, \mathcal{F})$ is a field of sets.

The preorder fields play an important role in the representation theory of interior algebras, given that an *algebraic* field of sets can be defined only on a (topological) *complex algebra* (algebra-subalgebras structure) or *algebra of complexes* [32] $\mathbb{A}^{+}$, so that $x \leq y$ if $S \in \mathbb{A}^{+}$, $(y \in \mathbb{A}^{+} \Rightarrow x \in \mathbb{A}^{+})$ for every complex. To pass to BAOs, we must consider structures $(X, (R_i)_I, \mathcal{F})$, where $(X, (R_i)_I)$ is a *relational structure* $\mathbb{S}$, i.e., a set with an indexed family of relations defined on it, and $(X, \mathcal{F})$ is a field of sets. Therefore, the *complex algebra* $\mathbb{A}^{+}$ determined by the field of sets $\mathbf{X} = (X, (R_i)_I, \mathcal{F})$ on a relational structure is BAO:

$$\mathbb{BA}(\mathbf{X}) := (\mathcal{F}, X, \cap, \cup, \top, \bot, (f_i)_I)$$

where for all $i \in I$, if R_i is a relation of arity $n+1$, then f_i is an operator of arity n and $\forall(S_1, \ldots, S_n \in \mathcal{F})\, f_i(S_1, \ldots, S_n)\, \{x \in X : \exists(x_1 \in S_1, \ldots, x_n \in S_n), R_i(x_1, \ldots, x_n, x)\}$.

Such a construction can be generalized to fields of sets on arbitrary algebraic structures, where both operators and relations as operators can be viewed as a special case of relations. If $\mathcal{F}$ is the whole power set of X, i.e., 2^X, then $\mathbb{BA}(\mathbf{X})$ is called a *full complex algebra* or *power set algebra* $\mathbb{P}(S)$. Therefore, Jónsson and Tarski [4,5] demonstrated that every BAO can be represented as a field of sets on a *relational structure* $\mathbb{S}$ in the sense that it is *isomorphic* to the complex algebra corresponding to the field (see [147,148] for such a reconstruction). Particularly, in such a way, Jónsson and Tarski extended Stone's RTBA to a particular type of Boolean algebras, the "σ-complete Boolean algebras" $\mathbb{BA}^\sigma$, and then, in the light of the Loomis–Sikorski theorem introduced in Section 0, they extended the RTBA to the *measurable* topological spaces [148]. This formally exemplifies the extension in CT logic of the (topological) "operator algebra" formalism from physics to logic and vice versa via the category of BAOs, **BAO.**

Finally, an important extension of this representation theory for the category **BAO** that is fundamental in its application to *Kripke relational semantics* (see Section 5.2) consists in the extension of this representation theory at the level of *morphisms*. Specifically, an *algebraic homomorphism* between two Boolean algebras $\mathbb{BA}_1 \rightarrow \mathbb{BA}_2$ *dually* induces a certain type of structure-preserving *reversed* mapping between the correspondent topological structures $\mathbf{X}_{\mathbb{BA}_2} \rightarrow \mathbf{X}_{\mathbb{BA}_1}$ called *bounded morphism*. This means the extension of the categorical dual equivalence between Stone spaces and BAs via the contravariant application of the Stone functor $\mathcal{S}$, $\mathbf{Stone}(\mathcal{S}) \simeq \mathbf{Bool}(\mathcal{S})^{\mathbf{op}}$, to the categorical duality between BAOs and the relational structures on Stone spaces, i.e., $\mathbf{Stone}^{+}(\boldsymbol{\Omega}) \simeq \mathbf{BAO}(\boldsymbol{\Omega})^{*}$. In both cases, indeed, these dual equivalences depend on the respective dual functors, generally rewritten as $(\cdot)^{*} = \Omega = (\cdot)$ [147].

5.1.4. The Meaning Function in Relational Semantics

As discussed, in their seminal papers on BAOs, Jónsson and Tarski introduced the notion of *relational semantics* and then the notion of an *algebraic interpretation* of the set-theoretic notion of *meaning function* that we discuss here in its categorical formalization for the category **BAO**, essentially following Yde Venema's exposition in ([6], pp. 331–426).

The meaning function is the arrow-theoretic version of the set-theoretic semantics of the propositional calculus, for which a propositional function (e.g., $(p \wedge q)$) is evaluated as true/false on the operations over correspondent sets, P, Q, that is, on the propositional function *extension*. So, in our example, $(p \wedge q)$ is *true* if and only if the intersection of the correspondent sets holds, i.e.,$(P \cap Q)$.

Correspondingly, the *meaning function* $[\cdot]$ maps a propositional formula ϕ to its extension $[\varphi]$; *that makes* ϕ *true*. One can then impose an *algebraic structure* $\mathbb{S}$ on the formulas of the propositional calculus, i.e., the so-called *τ-formula algebra* $\mathbb{F}\,\phi_\tau$, where $\{\tau\}$ is the set of propositional connectives or propositional predicates («and», «or», «not», etc.), by which substitutions are *completely determined by their values on the variables*. Namely, for any function σ assigning a formula to a variable, the substitution by σ is the unique extension $\widetilde{\sigma}$ of σ to an *endomorphism* on $\mathbb{F}\,\phi_\tau$. Therefore, given an arbitrary algebra $\mathbb{A}$ of type $Bool_\tau$, any assignment, mapping variables to elements of the carrier-set of $\mathbb{A}$, has a unique extension $\widetilde{\sigma}$, which is a *unique homomorphism* from $\mathbb{F}\,\phi_\tau$ to $\mathbb{A}$: $\mathbb{F}\,\phi_\tau \rightarrow \mathbb{A}$.

For instance, in the well-known case that A is the «two-valued Boolean algebra», 2BA, its carrier is given as the set 2 = {0,1} while the classical *truth tables* give the interpretation (semantics) of the Boolean symbols/functions. Namely, given a *valuation* V: $X \rightarrow 2$ of truth values to propositional variables, we can simply arithmetically *compute* the truth value $\widetilde{V}(\varphi)$ of any complex propositional formula $\phi\ (p_1, \ldots, p_n)$, using the unique homomorphism $\widetilde{V}: \mathbb{F}\phi_\tau \rightarrow 2\mathbb{BA}$ *extending* the assignment V. This formalizes the usual statement of Boolean logic, according to which we can "extend" the valuation functions of propositional calculus to the homomorphic binary "arithmetical" operations of a Boolean algebra.

To generalize the precedent example to whichever relational semantics of propositional logic [6], pp. 337–342, to complete the transition from a set-theoretic to an arrow-theoretic relational semantics, we refer to the already introduced notion of *complex algebras* $\mathbb{A}^+$. We can therefore redefine this notion, descriptively introduced in the precedent subsection, in the CT logic formalism [6], p. 339:

Definition 1. *(Complex algebra of a τ-frame). Given an n + 1-ary relation R on a set S, define the n-ary map* $\langle R \rangle$ *on the power set of S by:*

$$\langle R \rangle (a_1, \ldots, a_n) := \{s \in S | R s_1 \ldots s_n \text{ for some } s_1 \ldots s_n \text{ with } s_1 \in a_1, \forall i\}$$

The complex algebra $\mathbb{S}^+$ *of a τ-frame* $\mathbb{S}$ *is obtained by expanding the power set algebra* $\mathbb{P}(S)$ *with operations* $\langle R_\nabla \rangle$ *for each connective*∇*. Specifically:*

$$\mathbb{S}^+ := \langle \mathcal{P}(S), S, \varnothing, \approx_S \cap, \cup \{\langle R_\nabla \rangle | \nabla \in \tau\} \rangle. \tag{10}$$

where $\approx_S$*. denotes all the equivalences in S. All this means that, from the perspective of complex algebra, a valuation is nothing but an assignment of variables of* $\mathbb{S}^+$*. Much more significantly, all this means that given a valuation V on a frame* $\mathbb{S}$ *, it is possible to prove by induction that:*

$$\mathbb{S}, V, s \Vdash \varphi \Leftrightarrow s \in \widetilde{V}(\varphi) \tag{11}$$

where $\widetilde{V} : \mathbb{F}ma_\tau \to \mathbb{S}^+$ *is the unique homomorphism extending V.*

Consequently, as Venema emphasizes, "what equivalence (11) reveals is that, in a slogan, *meaning is homomorphism*" ([6] p. 338). In other words:

Definition 1. *(Meaning function). Let V be some valuation on a τ-frame* $\mathbb{S}$*. Then, the meaning function is the unique homomorphism* $\widetilde{V} : \mathbb{F}ma_\tau \to \mathbb{S}^+$ *that extends V to* $\widetilde{V}$*.*

From this, the following theorem holds:

Theorem 1. *Because of* Definition 13.*, let* $\phi^{\approx}$ *denote the equation* $\phi \approx \top$*, where* $\approx$ *is the algebraic equality symbol and* $\top$ *stays for "true" (e.g., "1" in a two-valued Boolean logic). Then, we have that for any τ-frame* $\mathbb{S}$*, and any τ-formula* $\phi, \psi, \ldots$*, the following validations hold:*

$$\mathbb{S} \Vdash \varphi \Leftrightarrow \mathbb{S}^+ \vDash \varphi^{\approx}; \text{ and } \mathbb{S} \Vdash \varphi \leftrightarrow \psi \Leftrightarrow \mathbb{S}^+ \vDash \varphi \approx \psi. \tag{12}$$

Specifically, the validity of a formula in the frame $\mathbb{S}$ *corresponds to that of an equation in the complex algebra* $\mathbb{S}^+$*of* $\mathbb{S}$ *, and vice versa.*

In this case, the metalogical biconditionals $\Leftrightarrow$ in (12) express the core of the notion of *algebraization of a propositional logic*, that is, of its translation into an *equational logic* by which substitutions are *completely determined by their numerical values on variables*.

Finally, it is worth emphasizing that the *validation direction* of the homomorphism constituting the algebraic construction of the meaning function is from the formula structure $\mathbb{F}$ to its extension $\mathbb{F}'$, i.e., $(\mathbb{F} \vDash \varphi) \to (\mathbb{F}' \vDash \varphi)$. However, and this is very interesting for our aims, when *a selection criterion of admissible sets* is present, e.g., an *ultrafilter Uf*, the validation direction is obviously *reversed*, i.e., $(\mathbb{F} \vDash \varphi) \leftarrow \left(Uf(\mathbb{F}')\right) \vDash \varphi)$. This is the case of the partially ordered sets of Stone's RTBA. However, the usage of ultrafilters in standard set theory is limited by their *non-finitary* character because of the "ultrafilter lemma" in **ZF(C)** (see [149], pp. 57–68), as we discuss (Section 5.2.1) [33], which indeed implies *second-order* semantics. This limitation can be "locally" avoided if we define the (coalgebraic) topologies of the Stone theorem over NWF-sets such as in Kripke's relational semantics, as we see immediately.

5.2. The Coalgebraic Relational Semantics of Kripke Models in Modal Logic

5.2.1. From Goldblatt's to Kripke's Development of a Modal Relational Semantics

To properly understand Kripke's modal relational semantics, we must previously understand the extension of RTBA categorical duality to the category of *modal Boolean algebras* because of the fundamental "Goldblatt–Thomason theorem". It is not possible nor necessary to illustrate here the formal details of this demonstration for which we refer to ([6], pp. 354–356). For us, it is sufficient to recall that this theorem demonstrated the dual equivalence between the category of the *descriptive general frames DGFs*, which are the frames of Definition 12. for a set $\{\tau\}$ of *modal* propositional connectives, and the category of *modal* BAOs, i.e., $\mathbf{DGF}_\tau$ **(Ω)** $\simeq$ $\mathbf{BAO}_\tau$ **(Ω*)** In this way, we are effectively extending to the modal logic ML_τ the Stone categorical dual equivalence **Stone(Ω)** $\simeq$ **Bool(Ω*)** for the Stone functor $\mathcal{S}^* := (\cdot)_* \circ \Omega \circ (\cdot)^*$, rewritten as the dual equivalence between the category of *descriptive fields of sets*, which are the equivalent in set-theory of the Stone spaces in topology, and the category of Boolean algebras [147,148].

The consequent construction of the meaning function can be extended from a *Goldblatt general frame* $\mathfrak{G}$ to a *Kripke frame* $\mathfrak{F} = W, R$, where W is a set of "possible worlds" and R is an "accessibility relation" between pairs of them (see Section 5.2.3), and its extension F′= ⟨W′,R′⟩. This, indeed, is effectively a *general frame* $\mathfrak{G} = \langle \mathfrak{F}, \mathbb{W} \rangle$ extending F with a τ-algebra of the W power-set $\mathcal{P}(W)$, closed under Boolean operations and modal operators $\{R_\alpha\}_{\alpha \in \tau}$. W thus plays the same role as F of a complex algebra S^+ as S in Definition 12., so we can denote it with G^+. Therefore, also in the case of Kripke frames F, the truth preservation direction is normally from a frame F to its extension $F' = G^+$, that is, $(F \models \phi) \rightarrow (F' \models \phi)$.

To pass the Goldblatt–Thomasson theorem, we must consider the ultrafilter extensions of $\mathfrak{G}$, $Uf(\mathfrak{G})$, denoted as *descriptive general frames*, with the fundamental difference as to the simple general frames $\mathfrak{G}$, for which, in this case, just as in the Stone duality of the Stone RTBA, the truth preservation direction is reversed, that is, it goes from $\mathfrak{F}'$ to $\mathfrak{F}$, that is, $(Uf(\mathfrak{F}) \models \varphi) \rightarrow (\mathfrak{F} \models \varphi)$. This reversal of the validation direction depends on the fact that ultrafilters act, here as everywhere in logic and mathematics, as a *second-order selection criterion of admissible sets on the power-set of a given set.* This depends on the so-called "ultrafilter lemma" for which any proper filter is the intersection of *all* ultrafilters containing it, requiring that free ultrafilters for existence must be defined on *infinite* sets (see [149], pp. 57–68). Therefore, the ultrafilters lemma supposes in **ZF** the "axiom of choice" (hence **ZFC**), or the "Zorn's lemma", as was the case of Stone's demonstration of RTBA.

What is typical of Kripke's modal relational semantics is that it also allows a *first-order selection criterion of admissible sets* related to the possibility of defining the truth valuation function $V(p)$ of Kripke models/structures on some *restriction* W' over the whole set of possible worlds $\{W\}$, i.e., $W' = \upharpoonright \{W\} := (W' \subseteq W)$, that is, exclusively for all the worlds accessible by a given world.

If *ontologically* these restrictions over the possible states of the world give us back *the logic of the physical causality principle* (light-cone) (see Section 1.2 and Figure 1) this formally immediately leads us to Kripke structures defined onto a coalgebra (disjoint sums) of NWF-sets, for which only set *partial orderings* are allowed, defined on Stone spaces. This is, indeed, the original intuition underlying the seminal work of S. Abramsky in 1988 [150], who proposed for the first time the possibility of using the so-called *Vietoris construction* on a coalgebra of NWF-sets over Stone spaces. Indeed, this allows us to use Aczel's powerful construction of the *final coalgebra theorem,* justifying the duality between a final coalgebra and an initial Boolean algebra, using the *Vietoris functor* V/V* as *dual functors.* The core of this extension is the *isomorphism* between the category of descriptive general frames **DGF** of Goldblatt's theorem and the category of coalgebras on Stone spaces or *Stone coalgebras, SCoalg*, using the Vietoris functor V, **SCoalg(V)** as its endofunctor, that is, $\mathbf{DGF}_\tau \cong$ **SCoalg(V)**. In this way, from Goldblatt's dual equivalence $\mathbf{GDF}_\tau$ **(Ω)** $\simeq$ $\mathbf{BAO}_\tau$ **(Ω*)**, we can derive immediately the other one: **SCoalg(Ω)** $\simeq$ **MBA(Ω*)** for the Vietoris functor: $\mathcal{V}^* := (\cdot)_* \circ \Omega \circ (\cdot)^*$, between the categories of the Stone coalgebras and the modal Boolean algebras [6]. We return in Section 5.2.3 to the logical and ontological relevance of such a construction.

5.2.2. Bisimulation and Co-Induction in Coalgebraic Logic

As a further step, let us introduce the notion of *bisimulation,* as an arrow-theoretic counterpart of the notion of algebraic *congruence* effectively defined on coalgebras. This notion was defined for the first time by Aczel with respect to NWF-sets in [82], even though in TCS, this notion today has a wider application for defining the *behavioral equivalence* between computational systems (automata) in any algebraic fashion (see [6], p. 388):

Definition 1. *(Bisimulation (symbol: ⇆)). Let* $\mathbb{S} = \langle S, \sigma \rangle$ *and* $\mathbb{S} = \langle S', \sigma' \rangle$ *be two systems for the set functor* Ω*. A relation* $B \subseteq S \times S'$ *is called bisimulation between* $\mathbb{S}$ *and* $\mathbb{S}'$ *if we can endow it with a coalgebra map* $\beta = B \to \Omega B$ *in such a way that the two projections* π*:* $B \to S$ *and* π'*:* $B' \to S'$ *are homomorphisms from* $\langle B, \beta \rangle$ *to* $\mathbb{S}$ *and* $\mathbb{S}'$*, respectively:*

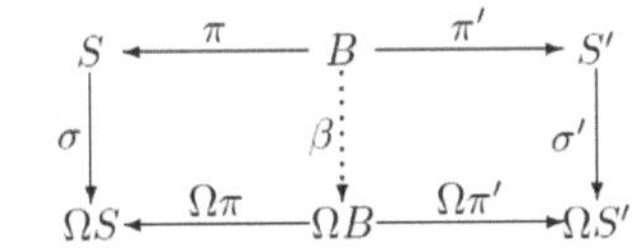

If a bisimulation B with states (s, s') ä B exists, we say that s and s' are bisimilar, i.e., sÆ s'.

Now, let us suppose two deterministic systems formally defined as pointed $2 \times \mathcal{I}^{\mathcal{C}}$-coalgebras, $\mathbb{S}0$, $\mathbb{S}1$, where $\mathcal{I}$ is the set of the identity relations Id_s univocally indexing each s state of the system. In case the functor Ω admits a final coalgebra $\mathcal{Z}$, we can formalize the notion of *observational equivalence*, with an evident relevance for quantum physics, by expressing the equivalence between the state s_0 in coalgebra S_0 and the state s_1 in coalgebra S_1, as $!_{\mathbb{S}_0}(s_0) = !_{\mathbb{S}_1}(s_1)$. More generally (see [6], p.389):

Definition 1. *(Definition of observational equivalence). Let* $\mathbb{S} = \langle \mathbb{S}, \sigma \rangle$ and $\mathbb{S}' = \langle \mathbb{S}', \sigma' \rangle$ *be two systems for the set functor* Ω*. The states* $s \in S$ *and* $s' \in S'$ *are observationally equivalent, i.e., S, s* $\equiv_\Omega S', s'$*, if an* Ω*-system* $\mathbb{X}\langle X, \xi \rangle$ *exists and homomorphisms f:* $S \to X$ *and f′:* $S' \to X'$*, such that f (s) = f′ (s′). In case* Ω *admits a final coalgebra Z, then S, s* $\equiv_\Omega S', s'$ *iff* $!_S(s) = !_{S'}(s')$*.*

This is precisely the case of the NWF-sets [11] for which the powerful "final coalgebra theorem" holds [82] (see Section 2.5).

However, this is also the case of the physical category of QFT coalgebras related to the category of the non-commutative coproducts of Hopf algebras for the Bogoliubov functor **qHCoalg($\mathcal{B}$)**, which also satisfy the final coalgebra theorem, and a categorical indexing by a diagonal functor $\mathcal{C}^{\mathcal{J}}$, as we explained at length in Section 4.5 and Appendix A. This conversely means that NWF-sets are those on which the mathematical formalism of dissipative QFT can be naturally defined! Not casually, indeed, in NWF-sets, the set self-membership and then infinite chains of set inclusions are allowed, just as in dissipative QFT systems, the phase coherences of quantum fields are related to the infinitely many SSBs of QV (see also the discussion of the cocomplete category of **DHilb($\mathcal{B}$)** and its coalgebraic internal structure in Section 4.8).

Afterward, Lawrence C. Paulson demonstrated, in the style of Peter Aczel for NWF-sets, a final coalgebra theorem for **ZF** set theory, even though it was significantly limited to *denumerable sets* [151], showing the generality for both standard and non-standard set theories of the categorical duality between initial algebras and final coalgebras, and their usefulness in TCS [152], in quantum physics and then in quantum computing [34].

The theoretical relevance of these CT notions for a *finitary* constructive mathematics and for functional programming in TCS but also for a *dynamic* approach to the evolutionary cosmology in fundamental physics is that, in addition to the (algebraic) *inductive* and *recursive* methods for the set definition and proof, we now have the (coalgebraic) *co-inductive* and *co-recursive* methods for the set definition and proof. That is, for the "set unfolding" from a common root in NWF-set trees of *different posets* along different and reciprocally irreducible branching (edges of the set tree) from the common root (see Section 5.2.3). Intuitively, "the crucial feature here is that processes need not be *bottom-up*, inductive, but it can instead be *top-down* co-inductive streams of events". More precisely (see [152,153], p. 46)—and waiting for a more explicit formalization of this duality between induction/coinduction using Kripke's relational logic, given in Section 5.2.3—we have:

Definition 1. *(Sets inductively/co-inductively defined by F). For a complete lattice L whose points are sets, and for an endofunction F, we have the sets:*

$$F_{ind} := \cap\{x|F(x) \leq x\}$$
$$F_{coind} := \cup\{x|x \leq F(x)\}$$

Specifically, the meet of the pre-fixed points and the join of the post-fixed points, i.e., the least and the greatest fixed-points if F is monotone, are, respectively, the sets *inductively* defined by F and the sets *co-inductively* defined by F.

Therefore, the following rules hold:

Definition 1. *(Induction and co-induction as proof principles). In the hypothesis of Definition 16., we have:*

$$if\ F(x) \leq x\ \ then\ F_{ind} \leq x\ \text{(induction as a method of proof)}$$
$$if\ x \leq F(x)\ \ then\ \ x \leq F_{coind}\ \text{(co} - \text{induction as a method of proof)}$$

In the light of the two precedent definitions, it is easy to grasp the main idea underlying the usage in TCS of coalgebra structures of NWF-sets to model *concurrent computations* [34], which was evident since the preface of Aczel's book (see [11], p. xiv), and as it was finally synthesized in J. M. Rutten's construction of the *Universal Coalgebra* as a "general theory of systems" (see [81], especially pp. 69–70). In this framework, the semantics of BAOs, either in functional programming, or in propositional logic, is given directly *by the physical states* of the system interpreted as a "labeled state-transition system" (LTS) and *coalgebraically* modeled [81]. This means conceiving the coalgebraic co-recursive computations (co-inductively defined: ↓), and the algebraic recursive computations (inductively defined: ↑) as two *concurrent computations* (see Note 34), respectively, for a final coalgebra (defined on a Stone space) and for an initial Boolean algebra to give a logical/computational counterpart in TCS of the categorical duality **SCoalg(Ω)** $\simeq$ **MBAO(Ω*)**. When the two computations "match" with each other, we have a *Boolean algebra with operators* (BAO), that is, a Boolean algebra whose signature ($^-$, $\wedge$, $\vee$, $\top$, $\bot$) is constituted by *operators* acting on an algebra of complexes [4,5] and whose "top" and "bottom" ($\top/\bot$) operators indicate that computations are coinductively and inductively lower and upper bounded on a *finitary* basis.

This depends on the powerful notion of the *functorial bounded morphism* between Kripke models $\mathfrak{M} \overset{\rightleftarrows}{\leftarrow} \mathfrak{M}'$, as we discuss in Section 5.2.3 (see also the related powerful construction of the "infinite-state black-box machine"$\mathbb{M}$ in TCS and its application to the problem of "data streaming" in AI *machine learning* in [81,154]).

To provide an intuitive illustration of the notion of the concurrent coalgebraic/algebraic computations, see Figure 10 in which the coalgebra (up) "functorially mirrors" its structure onto the algebraic structure (down) to give the latter a *selection criterion* of admissible sets, in a word, its *finitary semantics*, as we discuss.

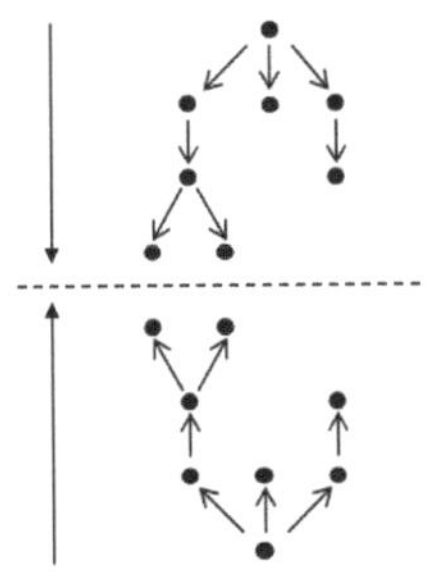

Figure 10. Intuitive representation of the principle of the *coalgebraic/coinductive* (final: upper part of figure) *algebraic/inductive* (initial: lower part of figure) concurrent computations, where the coalgebra functorially *mirrors* its structure over the algebra, and for which, in both directions, "the *finitary* is the limit of the successions of finites" (Abramsky).

However, we have already discussed the relevance of the coalgebraic construction of coinduction not only in TCS and logic but primarily in mathematical analysis and mathematical physics because of the strict categorical connection between the notion of "direct limits" in mathematical analysis and colimits in CT (see [71,72], and Section 4.8 with Appendix A).

5.2.3. The Relevance of Kripke's Modal Relational Semantics for an Ontology of Quantum Physics

Referring to [6,43] for a more complete formal discussion about Kripke's modal relational semantics, let us sketch its main elements and its relevance for a formal ontology and then for a *formal philosophy of nature* based on QFT as the fundamental physics.

As a starting point, let us recall some fundamental axioms of ML in Lewis' axiomatization that we introduced in Section 1.1 while for a complete treatment of this axiomatic approach to ML, we refer to classic handbooks such as [14]. On this regard, we recall here the so-called *normal modal system* **K**, based on the *necessitation axiom* **N**:

$$((X \vdash \alpha) \rightarrow (\Box X \vdash \Box\alpha))$$

where X is a set of propositions, α is a propositional meta-symbol, and $\Box$ is the necessity operator of ML. This axiom is sufficient for validating in modal propositional logic, ML, all the logical laws (tautologies) in their modal form, starting from the modal *modus ponens* for propositional formulas:

$$\Box(\vdash \varphi \wedge \vdash (\varphi \rightarrow \psi) \rightarrow (\vdash \psi))$$

Now, this looks like a standard second-order semantics of first-order formulas, where $\vdash \varphi$ means "φ is provable" and $\Box$ behaves like $\forall$.

Therefore, on the basis of Definition 13. and Theorem 1. about the "algebraic" meaning function for BAOs and its extension to modal BAOs through the Goldblatt–Thomasson theorem (Section 5.2.1), we can state this other fundamental theorem, necessary for defining the *minimal* modal algebra for the normal modal system **K** (for the theorem complete statement and relative proof (see [6], pp. 340):

Theorem 2. *(Minimal modal algebra for the normal modal system* **K***). Let Γ be a set of τ-modal formulas. Then, the class of atomic Boolean algebras with operators BAO_τ (Γ), which validates the set of equations* $\Gamma^{\approx} := \{\gamma \approx \top \mid \gamma \in \Gamma\}$, algebraizes $\mathbf{K}_\tau\, \Gamma$. In particular, the class of modal algebras MA(Γ) algebraizes $\mathbf{K}\,\Gamma$.

Let us apply the previous definitions and theorems to Kripke's relational semantics by reducing the polyadic frames $\mathbb{S}$ to the dyadic *Kripke frames* $\mathfrak{F} = \langle W, R \rangle$, constituted by "a universe" (and/or "a world") W, which is a set of "possible worlds" (and/or of "possible states of a world") $\{w_1, w_2, \ldots, w_n\}$, and by R as a dyadic "accessibility" relation between a pair of w defined on the Cartesian product of all possible worlds: $W \times W$, i.e., $(R \subseteq W \times W)$ (see Figure 11).

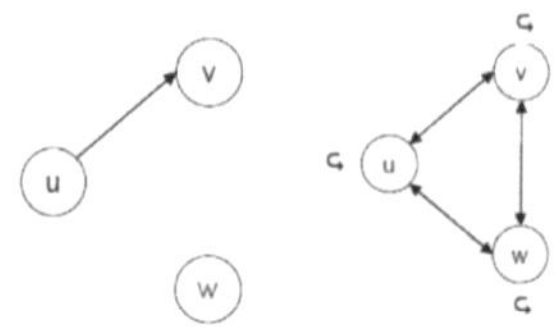

Figure 11. Representation of the Kripke $\mathfrak{F}$ with $W = \{u, v, w\}$ and with (**Left**) $(R \subseteq u, v)$, where v is accessed by u and not vice versa while w is irrelated. (**Right**) with $(R \subseteq W \times W)$ to define an equivalence class of possible worlds, or *universe* because all satisfy the reflexive, symmetric, and transitive accessibility relations among them. In parentheses, the right graph is a representation of a **S5 (KT45)** modal system, as we see.

Therefore, if we take as the basic semantic entity of relational semantics a *Kripke model* $\mathfrak{W} = \left(W, \{R_\nabla\}_{\nabla \in \{\tau\}}, V(p)\right)$, where ∇ is a propositional connective of the connective set $\{\tau\}$ of the propositional calculus, and $V(p)$ is an evaluation function of a proposition p defined on W. The model $\mathfrak{M}$ is effectively a *relational structure*, denoted as a *Kripke structure*. We have, indeed, a domain of quantification W, a collection of dyadic "accessibility" relations R over this domain, and a collection of unary (1/0) valuation relations $V(p)$ for each propositional symbol $p \in$ PROP. This means that properly, it is not necessary to speak about Kripke models using modal languages: "they provide us with everything needed to interpret classical languages too" [42], p. 10. In other terms, for our aims, we can also apply Kripke's model theory to the *mathematical languages of physics* and not only to the modal languages of philosophy (see [148]).

Moreover, in philosophical logic and formal philosophy, Kripke's relational semantics was developed for formalizing modal logics (ML) because of the "stipulatory character" of Kripke's "possible worlds" notion and their accessibility relations. Namely, in this sense, they are "possible", as far as satisfying the logical rules of a given (modal) language. This implies that the notion of "possible worlds" are applicable to whichever intensional logic interpretation (either alethic, or epistemic, or deontic (see Section 1.1) of the MC), and not only to the ontological (alethic) one, differently, for instance, from Leibniz' "possible world" semantics, whose meaning is essentially cosmological and then "alethic". Of course, in our ontological and then alethic application, the dyadic accessibility relations R's of the Kripke structures are interpreted as *causal relations* defined on the physical causal light-cone (see Figure 1), given that the causal event satisfies the notion of *terminal object* in the (sub-)category of its effects (see Section 1.2 and Definition 7.). This, again, emphasizes the coalgebraic nature of the causal light-cone in Kripke's relational semantics.

Finally, what makes Kripke semantics so attractive in logic is the possibility of satisfying, because of the Van Benthem's "correspondence theorem" [155,156], both a "second-order relational semantics" on Kripke *frames*, in which we quantify over all the *valuation functions* $(\forall V)$ for the proposition validity over all the possible worlds, and a "first-order relational semantics" on Kripke *structures*. In it, the validity holds for a given state of the world, and for all the other ones accessible from this state. In this way, we can always quantify at the *first-order*, either over *all* states $(\forall x,y)$ of a given universe W, e.g., when we state something about our universe as "generated" from the big-bang causal event at its origins, or over *some* states of the universe causally accessible from another state $(\exists x,y)$, that is, over a partition of the universe $(W' \subseteq W)$. In both cases, however (universal and particular quantification), we are referring to the first-order notion of *local* truth of the type: *it is locally the case* (F. W. Lawvere) (see Chapter 14 of [7], pp. 359–437). For instance, an example is when we speak about the radiating electromagnetic force that is true in our expanding universe only after the stage in which "cosmological microwave background radiation" (CMBR) originated, about 300,000 years after the big bang. That is, to speak about the electromagnetic radiation, and then of Maxwell's electromagnetism laws, is true only for a *partition* of possible universe states, from a stage of the universe evolution onward. "Cosmogony is the legislator of physics", and Kripke's relational semantics is its logic!

This logical evidence is the core of Patrick Van Benthem's celebrated *correspondence theory* between modal relational semantics and a variable-free fragment of first-order logic [155–157]. Indeed, coming back to Lewis' axioms in ML (see Section 1.1), in addition to the "necessitation axiom" **N** defining the "normal modal system" **K,** all the other axioms of modal systems (**T, D, 4, 5, ...**) used for axiomatically extending **K** can be defined via *first-order formulas.*

In this light, the result of Theorem 2. associated with Definition 13. allowed logicians and computer scientists to think at axiomatic extensions of **K** (e.g., **KT, KT4, KT5,** ... modal systems) not as giving rise to new systems of the modal calculus but as different *theories* over the minimal system **K,** "just as a first-order theory (e.g., of linear orders) is constructed over first-order validities" [42], p. 35.

This means that we can answer the fundamental question about the types of Kripke frames $\mathfrak{F} = \langle W, R \rangle$ that are able *to validate* the different modal axioms extending **K**. Because of Definition 13., whereas we interpret the generic frame S of the CT relational semantics (see Section 5.1.4) as a Kripke frame $\mathfrak{F}$ simply by limiting R to only dyadic (accessibility) relations, if the property defining a particular class of Kripke frames is *a first-order relation over sets,* it is possible to answer the precedent question. In fact, it is enough to interpret the set **Γ** of *modal formulas* extending **K** as the set of *modal axiomatic formulas extending* **K.** In a word, in Kripke modal semantics, we are faced with a particular solution of the famous "Löwenheim -Skolem paradox" [158] in set-theoretic semantics.

According to this paradox, indeed, despite all the axioms of **ZF** set-theoretic semantics being expressible through first-order formulas, nevertheless, their justifications need a higher-order logic. In modal semantics, however, because a modal formula evaluation typically distinguishes between "actual" and "possible" states (worlds) satisfying it, we can have different semantic levels because of the two double dichotomies: 1) *Kripke frames*: $\mathfrak{F} = \langle W, R \rangle$, versus *Kripke structures (models)*: $\mathfrak{M} = \mathfrak{F}, V$; and 2) *total,* versus *local* truths that we introduce now and justify as follows (see [43], p. 252).

Indeed, given that the *basic* semantic notion in Kripke modal logic is the *truth of a formula at a world-state* $\langle w \rangle$ *in a Kripke structure (model),* this notion is *local* and of a *first-order* nature. Therefore, the passage from structure semantics to frame semantics and then from *local* to *total* truth depends on looking or not at *all the valuations over a frame.* Namely:

1. By an abstraction through a universal *second-order* quantification over all *valuations* on *propositional formulas*: $\forall V(V(\phi))$.
2. Or, on the contrary, as we anticipated some paragraphs before but we now discuss in a more formal way, we can pass in Kripke's model theory to a first-order semantics of a propositional formula $\boldsymbol{\phi}$ in a *total* and/or in a *local* way if we do not quantify over all the valuation functions (second-order), but if we quantify at the *first-order* over *all* (respectively *some*) possible worlds. Namely:
   a. Either, over *all* the possible world-states w of the universe W for which a given formula φ is true, i.e., $\forall w \in W(V(\varphi_w))$: *total* truth.
   b. Or, over a *restriction of possible world states* for which a given formula ϕ is true: *local* truth. Specifically, $\forall w' \in W' | \, W' \subseteq W(V(\varphi_{w'})$. Namely, if $R \subseteq W \times W$ is any binary relation over W, and $W' \subseteq W$, we write $R¸W'$ for the *restriction* of R to W', i.e., $R¸W' = R$ Ç $(W' \times W')$. Similarly, for a valuation V on W', $V \,|\, ¸W'$ stands for the restriction of the evaluation function to the formula defined on the partition W'.

This means that in Kripke modal relational semantics, we can validate a formula ϕ over a given world-state w and on the sub-set of all the possible world-states accessed by w, which is evidently the logic of the light-cone causality in fundamental physics.

For the sake of brevity, we cannot develop a full formal treatment of the CT notion of meaning function here, applied to Kripke's modal semantics for which, among a wide literature, we essentially refer to two fundamental chapters of the monumental *Handbook of Modal Logic* (see [6], and especially [43]). We can then synthesize the main theoretical passages of their treatment as follows.

The first step is to extend **K,** by introducing the notion of Kripke frames *validity* that in turn is a *modal* extension of the frame validity in relational semantics (see Definition 13.). Namely [43], p. 253:

Definition 18. *(Kripke frame validity). Let φ $(p_1, \ldots, p_n)$ be a modal formula consisting of the atomic proposition symbols $p_1, \ldots, p_n$, with the associated monadic predicates for each atomic proposition, $P_1, \ldots, P_n$. φ is locally valid on a frame at a state (point, node, world) $w \in W$, if for each valuation V of its proposition symbols, φ is satisfied in the resulting model at w, i.e.,* M, *$w \models \varphi$, for each* M *over* F, *so that we write F, $w \models \phi$. Specifically, φ is valid in the pointed frame* (F, *w). Consequently, we say that φ is valid in* F, *denoted as* F $\models \varphi$, *if* F,w *φ for every $w \in W$. Finally, φ is valid, denoted as $\models \varphi$, if* F $\models \varphi$ *for every frame* F.

To sum up, we can say that a first-order modal formula φ defines *a class of frames* F if it is valid on every frame in F and falsified on any frame that is not in F. Then, we can define several classes of frames, as far as satisfying basic modal formulas, in the classical axiomatic approach to ML [14]. There are effectively as many axiom schemes for modal systems, extending the normal system **K.**

In this introductory exposition, for sake of clarity, for each (Lewis') axiom and the correspondent FO relation in a Kripke's frame, we make its significance for the philosophical logic explicit, according to Van Benthem's *correspondence theory* ([42], pp. 36). Of course, this makes explicit the subdivision into "three ages" (axiomatic, Kripkean, algebraic) of the modern history of ML during the XX cent., as we anticipated in Section 1.1, quoting [28]. In the present expositions, we limit the application of the correspondence theory only to some modal axioms: **T, D, 4, 5(E)** and their definitions that are more relevant for an ontology of physics.

Proposition 1. *(Definition of some classes of Kripke frames validating Lewis' modal axioms).*

- $\Box p \rightarrow p$ ($\equiv_{\text{def}}$ **T**) *defines (is validated by) the class of frames, which consists of isolated reflexive points/worlds such that $\forall x,y\ (Rxy \leftrightarrow x = y)$.*
    - *[The meaning of axiom* **T** *(from "truth") is evident: it is the axiom scheme of all alethic logics in modal formal logics and ontologies. It says, indeed, that if a proposition p is true in all possible worlds, it is evidently true also in the actual one]. For example, if the Galilean law of falling bodies is true in all possible physical worlds, it is also true in ours.]*
- $\Box p \rightarrow \Diamond p$ ($\equiv_{\text{def}}$ **D**) *defines the class of frames where the frame relation R is "serial": $\forall x \exists y\ (Rxy)$.*
    - *[The meaning of axiom* **D** *(from "deontic") is evident too. It says that if p is necessary, it is possible as a necessary condition. It is therefore the axiom scheme of all "deontic logics". Nobody, indeed, can be morally or legally obliged to something that is impossible for him/her: the possibility of satisfying a moral oughtness is a necessary condition of the validity of a moral obligation ("impossibilia nemo tenetur", in Latin). In the difference between axiom* **T** *and axiom* **D**, *the core of the famous "Hume principle" of not confusing alethic and deontic necessity is hidden, the "world of facts" and the "world of values"].*
- $\Box p \rightarrow \Box\Box p$ ($\equiv_{\text{def}}$ **4**) *defines the class of frames, where the frame relation R is "transitive": $\forall x,y,z\ ((Rxy \wedge Ryz) \rightarrow Rxz)$.*
    - *[The meaning of the axiom* **4** *is, indeed, the "transitivity of necessitation's". It is typically the axiom of the modal formalization of the "scientific necessity" according to*

distinct levels of necessitation (ordered natural laws) against any naïve reductionism]. For example, the laws of physics are necessary also in chemistry, even though they are not sufficient for justifying all the chemical phenomena, etc.]

- $\Diamond p \rightarrow \Box\Diamond p$ ($\equiv_{\text{def}}$ **5** *or* **E**) *defines the class of frames, where the frame relation R is "Euclidean", sometime denoted as a "weak transitivity": ∀x,y,z ((Rxy ∧ Rxz → Ryz)).*
  - *[The axiom* **5** *(in Lewis' enumeration) or* **E** *(from "Euclidean") is in some sense the axiom of the formalization of "metaphysics", because it states that if something is possible, it is "necessarily possible". In this sense, it formalizes the notion of "faculty" as a "power" that necessarily pertains to something/somebody because it characterizes its/her/his "nature" or "essence"]. For example, think of the faculty or the "necessary possibility" of thinking or of freely deciding as characterizing each human person, as an irreducible subject of rights and duties in society.]*
- *(...)*

As we know, by properly combining modal axioms, we obtain different systems of the modal calculus, each constituting the common "syntax" of different philosophical theories (i.e., of different "semantics"). Therefore, we can say, for instance, that Lewis' axiomatic modal system **S4**, i.e., **KT4,** is defined over a frame-class simultaneously satisfying the reflexive and transitive accessibility relations. The axiomatic system **S5**, i.e., **KT45,** is a **KT4** defined over the frame class that also satisfies the Euclidean relation characterizing the axiom **5**. On the contrary, with respect to **S5**, the **deontic S5** system, i.e., **KD45** that substitutes the axiom **T** with **D** as is necessary in deontic contexts, is defined over Kripke frames that satisfy the serial relation instead of the reflexive one, and so on.

Let us discuss some exemplifying applications in philosophical logic (modal semantics) of two modal systems: the **KT45 (S5)** and **KD45**, which is also defined as a **generated S5** system, for the reason we explain below, which has applications not only in formal ethics (deontic logic: **deontic S5**) but also in formal epistemology and formal ontology. Indeed, the **S5** modal system is the more powerful one because in a proper sense, it includes all the other ones [14].

Not casually, therefore, **S5** is universally recognized as the "modal syntax" of whichever *metaphysical theory* because it effectively represents a complete *universe of possible worlds W* as constituting only one equivalence class of entities that all "might exist" according to the very same defining axioms of a given metaphysical theory. Let us think, for instance, of a physical theory satisfying a TOE that would reveal in this way its "metaphysical" nature. In Figure 11, the right graph is a representation of the **S5** system for an oversimplified universe of only three worlds.

Now, if we observe Figure 12, the left graph represents a **KD45** system for a simplified four-world universe, justifying its definition as a **generated S5** system, with the inaccessible world *u* as the "generator" of an equivalence class of the other three worlds because all are accessible to *u* but not vice versa. The right part of the same figure represents for the sake of simplicity the calculus of relations generating an equivalence sub-class of two worlds for an oversimplified **KD45** system of a three-world universe, with the world *u* as the generator, according to the following relational steps: (∀*u,v,w*) ((*u*R*v* ∧ *u*R*w*) → *v*R*w*) > [transitive rule: **4**]; (∀*u,v,w*) ((*u*R*v* ∧ *u*R*w*) → (*v*R*w* ∨ *w*R*v*)) [Euclidean rule: **5**]; (∀*u,v,w*) ((*u*R*v* ∧ *u*R*w*) → (*v*R*v* ∨ *w*R*w*)) [serial rule: **D**]. To sum up, we *can generate a transitive-symmetric-reflexive (equivalence)* relation for the world sub-class $\{v, w\}$ starting from the asymmetric accessibility relation of $\{u\}$ with $\{v, w\}$ by a suitable "composition of morphisms" (accessibility relations *R*) satisfying the **KD45** axioms, i.e., (∀*u,v,w*): ((*u*R*v* ∧ *u*R*w*)) → (*v*R*w* ∧ *w*R*v* ∧ *v*R*v* ∧ *w*R*w*) (see Figure 12 (Right)).

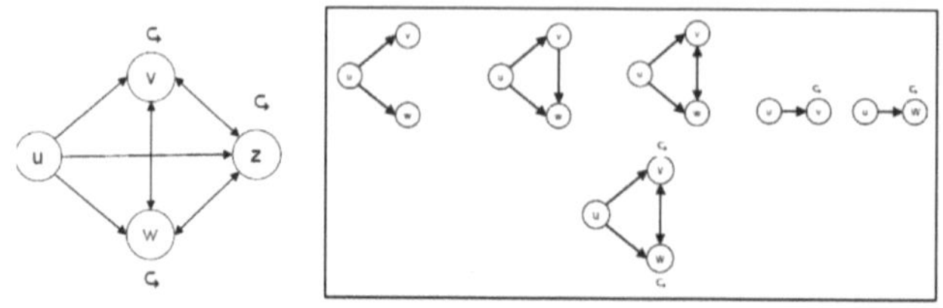

Figure 12. (**Left**): Representation of a KD45 modal system for a four-state world W, where the frame sub-class $\{u, w, z\}$ constitutes an equivalence class generated by only one inaccessible world-state $\{u\}$. KD45 constitutes a "secondary" or **generated S5** system. (**Right**): Oversimplified calculus of relations underlying a three-state world W for a **generated S5** system, where the accessibility relations from an inaccessible world-state $\{u\}$ generates the equivalence sub-class $\{v, w\}$ (see text).

When we integrate this simple relation calculus with the powerful construction of the *generated rooted trees* of Kripke structures over a coalgebra of NWF-sets, we can understand why a *nested structure* of **KD45** systems is the proper logic of the NR-formal ontology as I summarize in the general conclusions (Section 6), and as I anticipated in a semiformal way in [30], and, finally, as proposed for the first time by Francesco Panizzoli in [159].

As a further step, what about the *extensions* of Kripke frame definability and Boolean validity in the framework of the above Definition 13. and Theorem 1. concerning the "meaning function" $[\![\phi]\!]$? For this, let us extend the notion introduced in Section 5.1.3 of "bounded morphism" for a topology of *Kripke structures (models)* $\mathfrak{M}$. Specifically, these wre Kripke frames $\mathfrak{F}$, endowed with *valuation functions V*, over the atomic propositional symbols p of molecular modal formulas ϕ associated at each state/point w ä W (see [43], pp. 258–259). Let us, therefore, define the fundamental notion of "bounded morphism" for Kripke models that we introduced in discussing in the Section 5.1.3. Jónsson–Tarski Theory of the "Boolean Algebras with Operators".

Definition 20. *(Bounded morphism between Kripke structures).*

Let $\mathfrak{W} = \left(W, \{R_{\nabla}\}_{\nabla \in \tau}, V(\varphi)\right)$ *and* $\mathfrak{W}' = \left(W', \{R'_{\nabla}\}_{\nabla \in \tau}, V'(\varphi)\right)$ *be two Kripke structures, where* ∇ *is a modal connective of the set* τ *of modal connectives, and* φ *is the meta-symbol for propositional variables p. A morphism* $\rho : W \to W'$ *is a bounded morphism from* $\mathfrak{M}$ to $\mathfrak{M}'$ *if its graph is a bisimulation (see Definition 14.) from* $\mathfrak{M}$ *to* $\mathfrak{M}'$*, denoted as* $\mathfrak{W} \overset{\rightleftarrows}{\rightarrow} \mathfrak{W}'$*. Bounded morphisms between frames are similarly defined.*

Remark 2. *It is important to emphasize the "truth preservation" property of the bounded morphisms. Specifically, if* $\rho : \mathfrak{W} \overset{\rightleftarrows}{\rightarrow} \mathfrak{W}'$ *is a bounded morphism and* $\varphi \in ML_{\tau}$*, then* $\forall u \in$ *dom(F):* $\mathfrak{M}$, u $\models \varphi$ if $\mathfrak{M}'$, ρ *(u)* $\models \varphi$*. Therefore, if* $\mathfrak{F}$, $u \models \varphi$*, then* $\mathfrak{F}$ *p (u)* $\models \varphi$*.*

Remark 3. *If* ρ *is defined onto Kripke structures, then* $\mathfrak{M}'$ *is a "bounded morphic image" of* $\mathfrak{M}$ *(the same holds for frames). Therefore, for each u* $\in$ *W, a bounded morphism* ρ *"uniquely singles out" a bisimilar state* ρ *(w) in W'.*

Remark 4. *Each model* $\mathfrak{M}' = \mathfrak{F}', V'$ *over frame* $\mathfrak{F}'$ *can be "pulled back" to give a model* $\mathfrak{M} = \mathfrak{F}, V$ *over the frame* $\mathfrak{F}$ *via* $V(p) := \rho^{-1}[V'(p)] = \{\, w \in \mathrm{dom}(\mathfrak{F}) \mid \rho(w) \in V'(p) \,\}$*. This turns* ρ *into a "reversed bounded morphism" from* $\mathfrak{M}'$ *to* $\mathfrak{M}$*, i.e.,* $\mathfrak{W} \overset{\rightleftarrows}{\leftarrow} \mathfrak{W}'$*. Nevertheless, not every model* $\mathfrak{M}$ *over* $\mathfrak{F}$ *can be obtained by such a construction.*

Indeed, it holds only for Kripke models (structures) $\mathfrak{M}'$ endowed with a *selection criterion* of the admissible sets on which the semantics of $\mathfrak{M}$ can be defined, as we see immediately. Now, what about the truth preservation in the case that $\mathfrak{F}'$ is interpreted as an *extension* of $\mathfrak{F}$ in the sense of S and S' of Definition 13.? Effectively, this application of the relation semantics notion of "bounded morphism" (already introduced as BAOs in

Section 5.1.3) to Kripke models/structure is based, as the same used symbol ($\overrightarrow{\leftarrow}$) emphasizes very well, on the notion of "bisimulation" of Definition 14. (symbol: $\leftrightarrows$), where the "long arrow" in $\overrightarrow{\leftarrow}$ stays for an endofunctor in the category of Kripke models.

Indeed, we know that the first usage of the bisimulation notion was by Aczel to justify the "set unfolding" of subsets from the common root of NWF-sets that acquires all its logical relevance in the light of the functorial dual equivalence **SCoalg(Ω)** $\simeq$ **MBA(Ω>*)**, and all its *ontological relevance* when we recall that BAOs are defined on topological measurable (probability) spaces (σ-algebras: see Note 30) such as the Hilbert spaces. In this way, for what we discussed in Section 4.8, the same dual equivalence holds if we substitute the category **SCoalg(V)** for the Vietoris endofunctor, with the category of **qHCoalg($\mathcal{B}$)** for the Bogoliubov endofunctor at the coalgebraic footing of the category of the "doubled Hilbert spaces" **DHilb($\mathcal{B}$)** and its cocompleteness property (see Section 4.8 and Appendix A). Indeed, because the algebraic footing of the "doubled Hilbert spaces" consists in the non-commutative q-deformed Hopf coproducts, as explained in Section 4.5 and Section 4.6, this means that the dual equivalence **SCoalg(Ω)** $\simeq$ **MBA(Ω*)** for the Vietoris functor $\mathcal{V}^* := (\cdot)_* \circ \Omega \circ (\cdot)^*$ can be rewritten as **qHCoalg(Ω)** $\simeq$ **MBA(Ω*)** for the Bogoliubov functor: $\mathcal{B}^* := (\cdot)_* \circ \Omega \circ (\cdot)^*$.

Indeed, as the Vietoris functor $\mathcal{V}$ in ML endows the system with a first-order criterion of the admissible sets, making the $\mathcal{V}^*BA$ dually associated with ("induced by") its $\mathcal{V}SCoalg$ naturally "modal" as explained below, the same also holds for the Bogoliubov functor $\mathcal{B}$ with respect to the category of the q-deformed Hopf coalgebras in QFT. Indeed, the category of **qHCoalg($\mathcal{B}$)** is also endowed with a (dynamic) selection criterion of the admissible sets naturally making $\mathcal{B}^*BA$ dually associated with ("induced by") its $\mathcal{B}qHCoalg$ "modal". A further advantage is that the q-deformed Hopf coproducts and then the associated q-deformed Hopf Coalgebras are dynamically indexed via the construction of the $\aleph_0$-directed (i.e., denumerable locally presentable) comma category of the associated colimits, as explained in Section 4.8 and Appendix A.

The indexing value $\mathcal{N}$ defined on positive integers and univocally denoting a given condensate of NG-bosons, characterizing (or dynamically labeling) a given phase-coherence domain of a degenerate QV state ("QV-foliation") in a dissipative QFT system "balanced" with its environment, therefore confirms itself, also from the logical and computational standpoints, as a *finitary computable* "dynamic memory address" for biological systems and for natural and artificial neural systems (see Section 4.7). On this regard, see [127], and more recently [154] for an application addressed to the "data streaming" problem in machine learning. Finally, it is worth emphasizing that the non-commutative character of the q-deformed Hopf coproducts implies the non-commutative character of the associated "skew" Boolean algebras (see Note 6 and [160]), enhancing the "discriminant" semantic power of this computational architecture based on the "double qubit" computational principle (see above Section 4.6).

Therefore, coming back to Kripke's model theory, it is not casual that in logic and TCS, one of the most important constructions related to the bounded morphism equivalence between Kripke models, and with the pullback of evaluation functions in them, is the *unfolding* or *tree unravelling* of a Kripke structure modeled on a coalgebra of NWF-sets according to the categorical duality **SCoalg(Ω)** $\simeq$ **MBA(Ω*)**. To understand this, see the powerful notion of *generated and rooted sub-structures (sub-frames)* of a Kripke structure (frame) ([43] pp. 259–261).

For the moment, and for the sake of simplicity, we can say that the rooted graphs of Figure 10 can also be interpreted as representing such a construction of *generated rooted trees* of point-set Kripke structures, where each point represents a world-state $w \in W$. Now, the core of the dual equivalence **SCoalg(Ω)** $\simeq$ **MBA(Ω*)** for the contravariant application of the Vietoris functor $\mathcal{V}^* := (\cdot)_* \circ \Omega \circ (\cdot)^*$ is that the Vietoris endofunctor in the category of the coalgebras of NWF-sets **SCoalg($\mathcal{V}$)** acts as an equivalent of the power-set functor (see [6], p. 418). Therefore, when we apply the Vietoris construction to the rooted trees of Kripke structures defined on NWF-sets, this allows us to define the relation of *converse* *membership* (*comembership*) $\ni$ in the category **SCoalg($\mathcal{V}$)** as functorially dual to the relation of *membership* $\in$ in the equivalent category **MBA($\mathcal{V}$*)** (see [6], p. 393 for a formal exposition). In other terms, the Vietoris construction in this case acts as a first-order (*local*) selection criterion of admissible sets for the Boolean logical calculations.

For our aims, apart from the formalisms, it is very important to understand the logical meaning of this construction from a philosophical standpoint. As we anticipated many times, what characterizes the set "unfolding" of subsets from the root of an NWF-set tree is that because of the "anti-foundation axiom" (see Section 2.5), the set-subsets relationship along the edges of the set trees can be justified along reciprocally irreducible, infinitely many *arbitrary* "paths". This means that we cannot use the usual set-subsets inclusion relationship $\subseteq$ such as in **ZF** that supposes the set total-ordering to signify the membership to the root-set of the "disjoint unions (coproducts $\cup$))" of subsets along different unfolding paths or of the "intersections (products $\cap$)" of unfolded subsets along different paths but the modal notions, respectively, of "possible comembership"$\ni$ and "necessary comembership"$[\ni]$ ([6], p. 393). Here, the angular and square parentheses stay for the modal operators $\Diamond$ and $\Box$, respectively, but *relativized* to some partitions of the whole set of possible world states, and so justify FOL of *local* truths.

In this way, we can express this modal comembership of subsets to a set by saying that in the coalgebraic Kripke rooted structures on NWF-sets—constituting the extensions validating in the reversed direction $(\mathbb{F} \vDash \varphi) \leftarrow (f\mathbb{F}') \vDash \varphi)$ the correspondent modal Boolean propositional formulas because the extensions are endowed with a first-order *principal filter condition f* (see Section 5.1 and above)—the superset (i.e., the common root of the NWF-set tree) *admits* or better *generates* (not includes) its subsets. Just as we are acquainted to say, when we speak about "natural kinds" in a naturalistic ontology, that a *genus* (e.g., "mammalians") *admits* (not includes!) its several different *species* ("horses", "dolphins", "elephants", etc.), given that the genus–species relationship cannot satisfy any total ordering condition. Which ordering relation $\leq$, indeed, could be defined between different species and between subsets belonging to different branches of the unfolding process of NWF-set trees (see Figure 8)?

On the other hand, from the ontological standpoint, the notion of *generated* rooted trees of Kripke structures rigorously formalizes the notion of the *local* truth of a formula, evaluated at a current (actual) state of the world, and preserved (and carried) along the edges of accessibility (=causal) relations to other states. This justifies the usage of a "relativized" (indexed) universal quantifier and the related necessity modal operators. Both characters make this coalgebraic modal logic a "guarded fragment of FOL" (see [43], p. 323), particularly suitable for modeling an evolutionary cosmology (and biology) based on QFT as fundamental physics, where the "cosmogony is the legislator of nature". Therefore, let us discuss briefly, to conclude this section, the logical and ontological relevance of these constructions for a formalized philosophy of nature.

In the light of what we have discussed till now, it is easily understandable why, since its first appearance during the 1960s [15,16], Kripke's possible world semantics seemed more adequate for logically representing the contemporary evolutionary cosmology than the classical Tarski model theory, whose truth condition is for *only one* state of affairs, i.e., for only one "actual world" [17]. Particularly, the relational semantics of possible (states of) world(s), where a formula is evaluated at the *current (=actual) state* of the world and preserved along the edges of the accessibility relations to other possible states, seemed immediately consistent with the causality principle of the *light-cone* of special relativity theory [20].

Based on what we demonstrated in this contribution, we therefore have to interpret the accessibility relations of the rooted trees of Kripke models/structures on a coalgebra of NWF-sets as *causal relations* according to the light-cone causality principle in fundamental physics.

In this way, the modal distinction suggested by Kripke in *Naming and Necessity*—and acclaimed as one of the most relevant contributions to the analytic philosophy movement of the XX cent. (see Section 1.1)—between "*epistemic* (logic) necessity" and "*ontic* (causal) necessity" and then between *logical classes* (class-subclasses logical structures) and *natural kinds* (genus-species, intended as causally generated structures), where the latter ones are validating the former ones, has its proper formalization in the "reversed validation" between Kripke coalgebraic and algebraic models. This holds in the framework of the categorical dual equivalence **qHCoalg(Ω)** $\simeq$ **MBA(Ω*)**, which is the counterpart in the ontology of fundamental physics of the dual equivalence in modal logic **SCoalg(Ω)** $\simeq$ **MBA(Ω*)**.

This "local" validation of the (Kripke) models of the modal descriptive philosophical language over the (Kripke) models of the mathematical language of physics and then of natural sciences is therefore the semantic core of the *NR formal ontology* we propose in this contribution as a formalized philosophy of nature. This ontology is, at the same time, our categorical version of the approach of the so-called *ontic structural realism* in the philosophy (ontology and epistemology) of quantum physics while at the same time solving the debate between its "causal" and "statistical" justification [1,2,161]. The DDF principle in QFT, indeed, constitutes a dynamic (causal) selection criterion of the admissible sets, validating both the statistical and the logical representations of the QFT system at hand.

Historically, this ontology, as we discuss in Section 6.2, is a re-proposal, by the functorial *dual equivalence* coalgebra/algebra, of the Aristotelian categorical duality between the *ontic necessity*, i.e., the *cause–effect entailment* ("an effect without a cause is impossible") that means the converse implication between a cause α and effect β: $\Box((\alpha \Leftarrow \beta)) := (\neg \diamond (\beta \wedge \neg\alpha))$, where the cause α is the "necessary condition", and the dually correspondent *logical necessity*, the *premise-conclusion entailment* that means the direct implication between premise α^* and consequence β^* ("it is impossible that a premise is true, and the consequence is false", that is $\Box((\alpha^* \Rightarrow \beta^*)) := (\neg \diamond (\alpha^* \wedge \neg\beta^*))$, where the cause α^* is represented as a "sufficient" condition made "sound" by the dually equivalent causal entailment.

In a word, coming back to our precedent biological example of the *logical truth* of the statement: "horses are mammals", the logical membership ($\in$) between the sub-class of horses and the class of mammals is functorially induced in the NR formal ontology by the *ontic truth* of the dual statement: "the genus of mammals causally admits ($\ni$) the species of horses" from a given step $m \leq n$ onward of the universe evolution. This happens for other species of mammals but following different generation paths from the common root of a shared progenitor. In synthesis:

$$\Box_{\forall n \geq m}\left(\underbrace{horse^* \in mammmalian^*}_{\text{Modal Boolean Algebra}(\Omega^*)} \underbrace{\xleftarrow[\Omega^*/\Omega]{\rightleftarrows}}_{\text{Bounded Morphism}} \underbrace{horse \ni mammalian}_{\text{Stone Co-Algebra}(\Omega)}\right) \tag{13}$$

where the symbol $\overset{\rightleftarrows}{\leftarrow}$ stays for a functorially induced "*onto/logical* dual equivalence" between modal statements, respectively, in a coalgebraic (*ontic, causal*) and algebraic (*logical, representational*) formalization. Effectively, this is a *reversed bounded morphism* $\mathfrak{W} \overset{\rightleftarrows}{\leftarrow} \mathfrak{W}'$ between a physical coalgebraic Kripke model $\mathfrak{M}' = \langle W', R', V' \rangle$ and its logical algebraic dual homomorphic image $\mathfrak{M} = \langle W, R, V \rangle$, with *local* evaluations V', defined on a world-state $w'_m \in W'$ and preserved along the states $w'_{n \geq m}$ *causally* accessible to it. This exemplifies the fundamental notion of *local truth* in modal relational semantics illustrated above, which therefore has its proper syntax in *nested structures/sub-structures* of **KD45** systems (see [43] for an extensive formal treatment of this logic of nested trees of Kripke models/structures).

6. Some Final Remarks from a Historical Perspective

6.1. The Logic of NR-Formal Ontology as a Formalized and Ecological Philosophy of Nature

I showed in this work the usefulness of using CT as the proper metalanguage of formal philosophy, and specifically of formal ontology and formal epistemology, and therefore for a formalized philosophy of nature and philosophy of science, respectively. Specifically, this is the formal ontology and epistemology of the "natural realism" (NR).

This justifies why I dedicated Section 2 of this work to make philosophers more acquainted with the main notions of CT, in a direct comparison with ST, more known by philosophers of the analytic tradition, with special attention given to the axiomatic modal logic as the proper logic of the philosophical languages.

The main results that I tried to show using this formalism are the possibility of modeling and validating the descriptive statements of an ontology and an epistemology of the natural sciences directly on their mathematical models at the level of their fundamental physics (QFT), in the common framework of the operator algebra formalism extended from physics to logic.

The *physical core* of the NR formal ontology (see Sections 3 and 4) is summarized in the possibility of modeling the powerful construction of the "QV-foliation" for classes of dissipative QFT complex systems sharing the same dynamic (causal) structure, in terms of the categorical *universal* construction of the functorial "comma category" (see Appendix A). In this categorical construction, each class of complex quantum systems is modeled as the one-object category **1**, and then as a "colimit" or the "initial object" of the (cocone of morphisms of the) correspondent comma category $\mathcal{C} := 1 \xrightarrow{c} \mathcal{C} \xleftarrow{F} \mathcal{J}$, so that its complex structure is indexed by the correspondent "diagonal functor" $\mathcal{C}^{\mathcal{J}}$ (see Section 4.8 and Appendix A). In this way, it is possible to also give a physical foundation to the mathematical methodology of the "memory evolutive systems" (MESs) proposed by Ehresmann and Vanbremeersch in another paper of this journal issue as the core of a renewed philosophy of nature to formalize the fundamental notions of *emergence, complexity,* and—in the case of biological and neural systems—*memory* [38]. This methodology is extended in our interpretation from an evolutionary biology to an evolutionary cosmology, as required by a formalized philosophy of nature.

Moreover, the theoretical core of the extension of QFT to dissipative quantum systems, consisting of the DDF principle between the system and its thermal bath, means the proposal of a fundamental physics that is *intrinsically ecological.* Indeed, it is compliant with a generalized vision of physical systems that can no longer be conceived as *closed* systems. Additionally, this can be performed without renouncing the mathematical apparatus of statistical mechanics, as it is synthesized in a nustshell by the possibility of recovering the Hamiltonian canonical representation of dynamic systems simply by inserting the environment (thermal bath) degrees of freedom. Simply, we must consider from a new local perspective the perturbative methods of statistical mechanics as a useful abstract tool for modeling dynamic classical and quantum systems that does not capture, however, the intrinsically dissipative character of the "many-body physics".

On the other hand—without charging the physical research and practice of excessive social and cultural responsibilities—it is straightforwardly evident that a mathematical physical formalism conceiving in a generalized way physical systems as closed systems has contributed significantly to the development of a modern technology that is disrespectful of the environment. In this framework, indeed, the environment is naturally considered as an "irrelevant boundary condition", only good for being indefinitely energetically despoiled. This approach, indeed, is compliant with a modern culture that is no longer aware of the intrinsic *relational nature* of whichever physical non-living and living entity, the human individuals included. For example, to pass to social contexts the ideologies of the *individualist liberalism* or the *collectivist communism* characterizing the Modern Age, both forget the essential character of the human individual as a *person,* i.e., as an *individual-in-relation* with her natural and social environment, starting from her biological and cognitive faculties. Both ideologies, indeed, forget at the level of the social, economic, and political modern sciences the fundamental ethical consequence of a personalistic anthropology. Namely, the "common good" that society ought to pursue is the *personal flourishing* of human individuals and groups (see [162]).

Of course, such an "ecological" approach to fundamental physics and its ontology is perfectly compliant with the NR formal ontology as a categorical version of the *relational natural realism* (see Section 1.4 and Figure 2), here proposed as a formalized philosophy of nature and sciences.

Indeed, the *mathematical and logical core* of the NR formal ontology is the possibility, in the framework of the CT formalism, of using Kripke's coalgebraic "many world" relational semantics, both on the *physical* or *ontic side,* and on the *logical side.* On the ontic side, this is because it is compliant with the causal light-cone in fundamental physics for the mathematical modeling of QFT systems. On the *logical side* of the modal logic—effectively a modal BAO—is the descriptive language of ontology. In both cases, indeed, Kripke's model theory can express all its mathematical and logical representational power if models are defined onto a coalgebra of generated rooted trees of NWF-sets, validating the correspondent (modal Boolean) algebraic models of the descriptive language of an ontology. This depends on the fact that the former ones are endowed with a selection criterion of the admissible sets, the DDF principle in QFT, on which the semantics of the ontology can be truthfully defined. The NWF-set theory, indeed, via its "anti-foundation" axiom, makes a set capable of "containing" itself $\{\{\cdot\}\}$, so violating **ZF** set theory "total ordering", and then allowing an indefinite number of *arbitrary* "partial orderings", along different unfolding paths sharing the same root.

This is the deep reason for which the "generated rooted-trees" of Kripke structures and models defined onto a (topological) coalgebra of NWF-sets (see Section 5.2.3 and [43]) can constitute the proper representation in mathematical logic of the QFT process of the "QV-foliation", indexed by the comma-category of colimits applied to the (non-commutative) Hopf coproducts of QFT, such as many "unitarily inequivalent representations" of QV. The related q-deformed Hopf coproducts (coalgebras) constitute, indeed, only one category for the Bogoliubov endofunctor, as discussed in Sections 4.8 and 5.2.3 of this paper. For this reason, I showed that by the powerful CT logic construction of the (functorially reversed) "bounded morphism" between Kripke's coalgebraic models in the mathematical physics, and Kripke's algebraic models in the modal languages of the natural philosophy, the statements of the latter are validated onto the statements of the former.

All this holds in the logical framework of the dual equivalence **SCoalg(Ω)** $\simeq$ **MBA(Ω*)** between the category of coalgebras on NWF-sets defined onto Stone spaces and the category of modal BAOs for the contravariant application of the Vietoris functor $\mathcal{V}^* := (\cdot)_* \circ \Omega \circ (\cdot)^*$ (see [6]). In this contribution, we extended this logic to the *non-commutative case* of the dual equivalence, **qHCoalg(Ω)** $\simeq$ **MBA(Ω*),** between the category of the q-deformed Hopf coalgebras on NWF-sets defined onto the Chu spaces of QFT and the category of modal BAOs of the ontological descriptive languages for the contravariant application of the Bogoliubov functor $\mathcal{B}^* := (\cdot)_* \circ \circ (\cdot)^*$.

Finally, this *onto*-logic semantics of the NR-formal ontology is a significant realization of what we anticipated as being the core of the CT logic. Namely, the fact that a statement α over a category $\mathcal{C}$ is true if and only if the opposite statement α^{op} in the dually equivalent category $\mathcal{D}^{op}$ is true (see Section 2.4).

6.2. The NR-Formal Ontology from a Historical Perspective

From a historical standpoint, this particular version of the *relational natural realism* (see Section 1.4) is consistent with the anti-Platonic solution of the long-standing issue of the *realism of universals* in the history of logic and epistemology, having its ancestor in the Aristotelian naturalism. We emphasized in Section 2.1 that the categorical diagrammatic formalization of the Aristotelian syllogism, as Peirce first emphasized in the Modern Age [163], is more adequate than the Leibnizian extensional interpretation that, as such, is unable to formalize the modal syllogistic forms [164]. The intrinsic "relational character" of the Aristotelian logic is confirmed by the *dual character* of the epistemology and ontology of the Aristotelian modal naturalism. Indeed, the dual core of the Aristotelian naturalism is synthesized in the famous Aristotelian motto according to which, epistemologically, "what is *last* in being (the cause known starting from its effects), is *first* in reasoning (it becomes the *sound* premise of the demonstrative reasoning) and vice versa" [35].

Of this dual relationship, during the Middle Age, Thomas Aquinas gave a justification in terms of the modal propositional logic, well known by him [36], in his commentary on a passage of the Aristotelian Second Book of *Physics* (II, 199b,34ss.). In this passage, Aristotle used this reversal of the arrows between *real (causal)* and *logical* relations to distinguish between the *logical* necessity of the mathematical demonstrations, i.e., the *logical entailment* "premise $\rightarrow$ conclusion"), and the *physical* necessity of a causal process, the *causal entailment* "effect $\rightarrow$ cause", since from the same cause several different effects can derive.

In his commentary on this passage, Aquinas, after having correctly explained the two fundamental logical laws (tautologies) in the hypothetical reasoning of the *modus ponens* and the *modus tollens*, and the relative "paradox of the consequent", for which the truth of the consequence cannot grant the truth of the premise, since "sometimes true consequences can be derived by false premises", to logically justify the Aristotelian "reversal of the arrows" between the logical and the causal entailments, Aquinas adds:

> However, in those things that happen *because of something* (*propter aliquid*), either by technology (*secundum artem*), or by nature (*secundum naturam*) [37], this reversal holds, because if some final state [effect] is or will be, *it is necessary that something before this final state, or will have been or is* [cause]. (...) Therefore, the similarity is from both sides, even though with *an inversion of the relation between the two ones* (*quamvis e converso se videatur habere*). ([86] *In Libros Physicorum*, II, lect.15, n.5).

What Aquinas in this passage is implicitly affirming is the categorical duality, i.e., the reversal of the similarity (of the homomorphism) between the modal *logical entailment* "it is impossible that the premise is true, and the consequence is false": $\neg \diamond (\alpha^* \wedge \neg\beta^*)$, and the modal *causal (ontic) entailment* "it is impossible the effect without the cause" according to the physical causality principle, i.e., $\neg \diamond (\beta \wedge \neg\alpha)$. Synthetically, using the (functorially reversed) bounded morphism symbolism of CT for this universally valid *local* truth:

$$\Box_n(\alpha* \rightarrow \beta*) \xleftarrow[\Omega^*/\Omega]{\rightleftarrows} (\alpha \leftarrow \beta) \tag{14}$$

In this way, the well-known Aristotelian logical duality between the *causal syllogism* or *"quia"* (literally, "because") *syllogism* (from effect to cause, on the right side of the functorial dual equivalence above) can be easily understood. It gives the *demonstrative* or *"propter quid"* (literally, "because of which") *syllogism* (from premise to consequence on the left side) its *sound* (true) premise. It eliminates the possibility of the premise being false and the consequence true such as in the hypothetical syllogism without a causal entailment semantically validating it. In this way, which can be consistently formalized *only* in the functorial framework of CT logic, Aristotle and Aquinas obtained what C. I. Lewis searched for in vain with his theory of the purely logical "modal implication", as the correct metaphysical logic, as Quine's criticism of Lewis emphasized very well (see Section 1.2). For this reason, Aquinas, commenting on the Aristotelian teaching in the First Book of his *Posterior Analytics*, defined the demonstrative syllogism, based on *causally sound* premises, as the "because-of-which" (*propter quid*)" syllogism (see [86], *Expositio libri Posteriorum Analyticorum*, I, lect. 23, nn. 6–7; see also [87], *Summa Theologiae*, I, 2, 2).

In this sense, indeed, we must also interpret Aristotle's and Aquinas' epistemological theory of the *ontological truth* expressed in the motto of "the adequacy of the intellect and the thing" (*adaequatio intellectus et rei*) in the sense of a *universally valid local truth* in propositional logic (see the relativized modal operators (and/or quantifiers) in the bounded morphism of the modal formula (14)) [38]. This is based, indeed, on the *conformity* (*conformitas*)—in CT language, the *homomorphism up to isomorphism*—between the causal structure genus-species (or species-individual) of a natural kind, on the physical or *ontic* side, and the reversed logical structure or logical composition subject-predicate (the membership relation subclass/class) of a truthful sentence referring to the former, on the *logical side*.

The "circular conformity" (isomorphism via a natural transformation) of the adequation relationship is indeed, a homomorphism from the ontic side to the logical side (intellect $\leftarrow$ thing) in the causal constitution ("formal causality" as a "homomorphic mapping") of a *sound* predicative sentence as far as dually "mirroring" a causal generative process (natural kind) to which it refers. Afterward, it goes in the opposite direction from the logical side to the physical side (intellect → thing) to justify the *predictive* power of a sound premise in logic and mathematics of natural sciences, and so complete the semantic reference relationship (see [86], *Quaestiones De Veritate*, I, 1–4, for such a reconstruction).

This double reversed homomorphism—or isomorphism, effectively a natural transformation between the functors of the correspondent categories justifying a reversed *bounded morphism* between Kripke models in the CT modal coalgebraic semantics—therefore constitutes the *composite circular* notion of truth as *adequation*. This is epistemologically interpreted by Aquinas as an *intentional cognitive notion* (see the reference to the "appetitive faculty" or "emotional will" in the next quotation), straightforwardly synthesized by him in the following passage:

> The movement of the cognitive faculty terminates into the mind: it is therefore necessary that the known be in the knowing according to the knowing modality; on the other hand, the movement of the appetitive faculty terminates into the thing. Therefore, this is the sort of circle in the acts of mind that the Philosopher affirms in his III Book of *De Anima*. According to it, the thing that is outside mind moves the intellect, then the intellectualized thing (*res intellecta*) moves the appetitive faculty, and this directs itself toward the thing for reaching that from which the cognitive movement started (see [86], *Quaestiones Disputatae De Veritate*, I, 2co.).

Walter Freeman was therefore right in vindicating the continuity between his "action-perception cycle" in the QFT foundation of intentionality in cognitive neuroscience (see above Section 4.7) and Aquinas' theory of intentionality [165].

Unfortunately, this Aristotelian logical and epistemological theory of the ontological truth and its ML version given by Aquinas strictly depended on Aristotle's physics being abandoned, starting from the XIV-XVI centuries. Together with it, both the notion of *natural kinds* (genus-species) of the causal generative processes in nature and, more generally, the same modal logic were abandoned during all the Modern Age until their re-proposal in XX cent.

During the XVI cent. the duality between the causal and the logical entailments transformed itself—particularly in the work of the more known "Aristotelian" logician at that time, Jacopo Zabarella (1533–1589) —into the confused and confusing relationship between the *resolution* (inductive) and the *composition* (deductive) methods in logic, by which the duality between the "causal" *quia* (from the effect to the cause) and the demonstrative "logical" *propter quid* syllogisms, just recalled, is inconsistently interpreted without having the capability of distinguishing between the *real* and *logical* categorical modalities. Specifically, Zabarella interpreted the former as the *induction* of the premise (cause) from its consequence (effect) and the latter as the *deduction* of a consequence (effect) from its premise (cause), confusing the logical and the causal entailments that belong to different modal logic categories.

Now, Zabarella's fatal confusion, emphasizing how severe the abandonment of the modal logic in the XVI cent. was, deeply influenced the same debate between Galilei and the Inquisition at the beginning of the Modern Age. Indeed, as recently noted by Enrico Berti, one of the more recognized contemporary historians of the Aristotelian philosophy:

> It is (...) interesting, though not always remembered, the fact that Galileo also adhered to this (resolution-composition) method, which he learned from his visits as a young man to the Jesuits of the Roman College, who were profoundly influenced by Zabarella. In fact, Galileo also believed that physics, in particular astronomy, was structured like mathematics, that is, that it proceeded first with the resolutive and then with the compositive method, and that way was able to provide "necessary demonstrations", that is demonstrations endowed with necessity, not only from causes to effects, but also from effects to causes. Furthermore,

> in logic he always considered himself, as we know, totally Aristotelian, referring to the Aristotelianism of his time, that is, above all, of Zabarella. The novelty that Galileo introduced in *regressus* were the experiments, the "sensible experiences", that is, the so-called experimental method, aimed at assuring the truth of the effects, which is the truth of the conclusions. However, he did not doubt that, once the truth of the conclusions was determined, they would be enough to guarantee the truth of the hypotheses from which they sprung, transforming them in unmitigated principles (...). As we know, Galilei claimed to have found the argument that proved in an absolute necessary way the truth of Copernican theory, and he pinpointed it in the phenomenon of tides, which he explained as a consequence of the earth's movement ([166], pp. 289–290).

In a word, Galilei pretended to follow Zabarella to justify the soundness of the Copernican heliocentric hypothesis by his astronomical observations through his telescope. The inconsistency of the Zabarella method applied to the Galilean epistemology of modern science led Leibniz—much less naïf in logic than the great Italian physicist—to give the only possible consistent interpretation of the "reversal of the arrows" in the extensional logic he used to justify the Aristotelian syllogism. In extensional logic, indeed, the reversal of the arrows premise-consequence is formally consistent in the *tautologies* of the double-implication (i.e., of the logical equivalence, $\leftrightarrow$).

This logical elementary evidence led Leibniz to distinguish between the meaningless tautologies of the *a priori analytic judgements*, and the meaningful but contingent empirical *a posteriori synthetic judgements*. In turn, this distinction led Kant to define their synthesis by his notion of the *a priori synthetic judgments,* whose logical inconsistency was at last demonstrated by Quine, as we know [167]. All these "contortions" of the modern ontological conceptualism are an evident consequence of the abandonment of the modal non-extensional logic in the academy that started in the late XV cent. and endured until the actual recovery during XX and XXI centuries.

Effectively, indeed, from Gottfried Leibniz on, the causality relation was interpreted as the *sufficient condition* of the premise-conclusion deductive inference, therefore reducing the causal relation to the logical one, as Kant explicitly demonstrated, making "causality" (differently from Aristotle [39]) one of the logical (predicative) categories of his table in the *Critique of the Pure Reason*. Leibniz even made the "Sufficient Reason" or "Reason-Consequent Principle" together with the "Contradiction Principle" the two pillars of his *Théodicée* (Section 44) and his metaphysics (*Monadologie*, Sections 81–82).

What is worse for modern ontology and metaphysics is that this confusion between the causal and logical entailments led to the acritical generalized extension of the Leibnizian interpretation of "cause" as "sufficient reason" in a demonstrative procedure, also to Aristotle's and Aquinas' metaphysics [40]. In the Aristotelian logic and ontology, on the contrary, the causality is not a categorical (simple) predicate such as in Leibniz and Kant, but it is the result of the composition of three predicative categories of the Aristotelian table: *relation, action, and passion*. Therefore, when we speak about a "primary cause" in the Aristotelian cosmology, we are referring not only to the "heavenly spheres" (=acting principle) but necessarily also to the "primary matter" (=passive principle) of the heartily physics on which the heavenly bodies act.

Moreover, as we have seen before, in the *modal* causal entailment, the cause plays the role of the *necessary condition* (it is categorically on the left side of the reversed causal arrow $\leftarrow$ "from the effect to the cause"), thus justifying, evidently, the resolution into some *primary cause* for closing "upwardly" the causal chain, either on the physical plan (Aristotle) or on the metaphysical plan (Aquinas) [30–32]. To say all this more synthetically with the words of Michael Heller in a paper where he proposed the CT logic as the proper logic of formal metaphysics and theology for the construction of an updated "theology of nature", what is lacking in Leibniz's logic and metaphysics is the notion of "categorical duality" [168]. Indeed, in the modal coalgebraic logic of CT, as we have seen, the reversal of the arrows of the Aristotelian logic and ontology acquires its full intelligibility and soundness, i.e., its categorical, diagrammatic *universality*.

Finally, it is evident that this re-proposal of the categorical duality between the causal and logical entailments can suggest a solution not only to the *ontological* but also to the *epistemological* conundrum from Galilei to Popper concerning the justification of the soundness of the mathematical hypotheses in modeling physical processes and events, according to the hypothetical-deductive method of modern natural sciences, as I anticipated in the introduction of this paper.

Funding: This research received no external funding.

Informed Consent Statement: Not applicable.

Data Availability Statement: Not applicable.

Conflicts of Interest: The Author declares no conflict of interest.

Abbreviations

BA	Boolean Algebra
BAO	Boolean Algebra with Operators
CCR	Canonical Commutation Relation (in QM and QFT)
CDM	Cold Dark Matter (Model in GR)
CMB	Cosmic Microwave Background (Radiation)
CSM	Cosmological Standard Model (in GR)
CT	Category Theory
DDF	Doubling of the Degrees of Freedom
ESR	Epistemic Structural Realism
FOL	First Order Logic
GR	General Relativity Theory
ML	Modal Logic
NBG	Von Neumann-Bernays-Gödel (Set Theory of)
NG	Nambu-Goldstone (bosons) in QFT
NR	Natural Realism (formal ontology)
NWF	Non-well-founded sets (Theory of)
OSR	Ontic Structural Realism
PC	Predicate Calculus
QFT	Quantum Field Theory
QM	Quantum Mechanics
QV	Quantum Vacuum
RTBA	(Stone's) Representation Theorem for Boolean Algebras
SM	Standard Model (in QFT)
SQ	Second Quantization (interpretation of QFT)
SR	Special Relativity Theory
SSB	Spontaneous Symmetry Breaking (of the QV)
ST	Set Theory
TCS	Theoretical Computer Science
TOE	Theory Of Everything (in Fundamental Physics)
Z	Zermelo (Set Theory of)
ZF	Zermelo-Fraenkel (Set Theory of)
ZFC	ZF with Choice Axiom (Set Theory of)

Appendix A. The Categorical Operations of "Limits" and "Colimits"

Landsman, in his short but significant account about the relevance of the CT meta-language for formalizing the main notions of operator algebra in quantum physics and quantum logic (see [45], pp. 805–833), indicates in the categorical operation of *(co)limits* another fundamental contribution of clarification for mathematical and functional analysis in physics (see [45], p. 805). We give in this appendix some fundamental definitions of the *limits* and *colimits* operations, mainly referring to a recent paper by Kairui Wang, aimed at using limits and colimits operationally [169], even though we also refer to more classical works such as [170], pp. 22–23, [171], pp. 16–18].

The first step is to introduce the notion of *comma category*. Effectively, this is a category whose objects are morphisms, given that "limits" and "colimits", interpreted in CT as "vertices" of "cones" and "cocones" of morphisms that share, respectively, the same "domain" or the same "codomain" of morphisms, are *specific types of comma categories* [41], as we see below.

To approach the notion of the comma category, it is fundamental to understand which problem we want to solve by introducing it. Wang thus explains the *diagrammatic universality* problem that the comma category solves in CT logic and mathematics.

> One type of problem in category theory is the universal mapping problem. Informally, these problems look for a morphism (called the "universal morphism") *that satisfies some desired property*, such that any other morphism satisfies the property "factors through" it in the sense that it is the same universal morphism composed of some other morphism ([169], p. 3).

From this characterization, the relationship between limits and colimits as comma categories clearly emerges, and our problem of a categorical formalization (diagrammatic universality) in formal ontology of the dynamic process of *factorizing through* (limits) or *constructing from* (colimits), respectively, but *universally* some properties/functions/predicates both in physics and logic is identified, as we have seen in the rest of this paper. What is highly significant for our aims is that this is true in formal ontology because it is true in logic and mathematics that *colimits* give a diagrammatic universalization to the fundamental operations of *direct limits* $X = \varinjlim X_i$ in mathematical and functional analysis, and then in physics and logic [42]. Therefore, we can refer to the definition of *comma category* as a "category whose objects are morphisms", following Wang himself ([169], p. 3):

Definition A1. *(Comma category). Let A, B and C be three categories, and $F : \mathcal{A} \to \mathcal{C}$ and $G : \mathcal{B} \to \mathcal{C}$ be two functors with the same target category (codomain), i.e., $\mathcal{A} \xrightarrow{F} \mathcal{C} \xleftarrow{G} \mathcal{B}$. The "comma category" $(F \downarrow G)$ has objects that are triples (α, β, f), where α is an object in A, β is an object in B, and $f : F\alpha \to G\beta$ is a morphism in C that thus is constituted by morphisms relating objects belonging to the other two categories.*

This allows us to apply to our construction of comma category the notion of "final objects", that is, *initial* and *terminal objects* (see Definition 7.). For this, it is necessary to consider comma categories where one functor of the two considered has as its source the category **1** with only one object $(*)$ and its relative identity morphism 1_*. This functor simply maps to an object *c* ä Ob(C) and therefore we call this functor *c*. We can then easily understand the notion of *universal morphism* as the initial (respectively, the terminal) object in each of the *dual* comma categories $(c \downarrow F)$ and $(F \downarrow c)$ ([169], pp. 4–5):

Definition A2. *(Universal morphism as the initial (terminal) object of a comma category). Let c be an object in a category C and let F: B $\to$ C be a functor. If we consider c as a functor from category **1** with one object $(*)$ and its identity morphism 1_* to C, we define the "universal morphism" (natural transformation) from c to F as the initial object in the comma category $(c \downarrow F)$, and dually, the "universal morphism" from F to c as the terminal object in the comma category $(F \downarrow c)$.*

To pass from this last definition of a comma category $\mathbf{1} \xrightarrow{c} \mathcal{C} \xleftarrow{F} \mathcal{B}$ to the definition of *colimits* as categorical universalization of the notion of *direct limits* —and dually of *limits* as categorical universalization of the notion of *inverse limit*—it is sufficient to recall that direct and inverse limits ultimately consist of a *family of sets indexed by a fixed set* or, equivalently, by a function from the indexing set to a class of sets (see note 42). In the CT generalization, a colimit (and a limit) interpreted as a diagram is therefore a collection of objects and morphisms, labeled by a fixed "small" *category of indices* $j \in \mathcal{J}$; or, equivalently, it is a *functor* from a fixed small *index category* $\mathcal{J}$ to an arbitrary category [43].

Indeed, one can interpret the category of *J*-shaped diagrams in C as the *functor category* $\mathcal{C}^{\mathcal{J}}$. It means that, for each object *a* in C, a constant diagram exists: $\Delta_a : \mathcal{J} \to \mathcal{C}$, mapping every object *j* in $\mathcal{J}$ to *a* and every morphism in $\mathcal{J}$ to 1_a. One can therefore define the *diagonal functor* $\Delta : \mathcal{C} \to \mathcal{C}^{\mathcal{J}}$ as the functor assigning the diagram (*constant functor*) Δ_a to each object *a* of C and $f : a \to b$ in C to each morphism the natural transformation $\Delta f = \Delta_a \to \Delta_b$ in $\mathcal{C}^{\mathcal{J}}$. Moreover, because Δ_a, Δ_b are constant diagrams, this construction implies a correspondent natural transformation ι between the functors Δ and *F*, therefore involving the morphisms between objects, *i,j* in $\mathcal{J}$.

At this point, we have all the necessary components for understanding the definitions of colimits (and limits) in CT as a diagrammatic generalization of the notions of direct (and inverse) limits in mathematical and functional analysis (see [169], pp. 5–7).

Definition A3. *(Categorical notion of colimit). We can connote a functor F from a small category $\mathcal{J}$ to a category C, $F : \mathcal{J} \to \mathcal{C}$, as a diagram Δ over $\mathcal{J}$ in C. It is evident that F is an object in the functor category $\mathcal{C}^{\mathcal{J}}$ for the diagonal functor $\Delta : \mathcal{C} \to \mathcal{C}^{\mathcal{J}}$. The colimit of a diagram F as an object in C and denoted as Colim F is therefore the universal morphism from Δ to F.*

Specifically, Colim F is the initial object in the comma category $(F \downarrow \Delta)$. It is a natural transformation $\iota : F \to \Delta_{\text{Colim } F}$, where Colim *F* is an object in C. Because $\Delta_{\text{Colim } F}$ is a constant functor, the naturality of ι produces the following commutative diagram for every morphism f: $i \to j$ in $\mathcal{J}$:

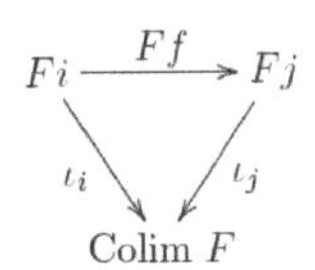

From this derives the further categorical universalization of Colimits as a *cocone of morphisms* (see note 41):

Definition A4. *(Categorical notion of a cocone of morphisms). A natural transformation from a diagram $F : \mathcal{J} \to \mathcal{C}$ to a constant functor $c \in Ob(\mathcal{C})$ is denoted as a "cocone" over F.*

Remark A1. *From the definition of cocones, the interpretation of the comma category $(F \downarrow \Delta)$ as the category of cocones over diagram F immediately derives. In it, the colimit of F is the initial object. This comma category is therefore denoted as $Cocone_F$.*

The universality problem solved by the comma category $Cocone_F$ is the following [169], p. 6.

Given a diagram $F : \mathcal{J} \to \mathcal{C}$, there is an object Colim *F* with associated morphisms from each F_j, where $j \in Ob(\mathcal{J})$ to Colim *F* such that Colim *F* and its associated morphisms ι_i, ι_j commute with all morphisms *Ff*, where *f* is a morphism in $\mathcal{J}$. This construction is universal in that if an object *Y* and its associated morphisms v_i, v_j from each F_i, F_j to *Y* also commute with all morphisms *Ff*, then a unique (universal) morphism *h* exists: Colim $F \to Y$, such that for any f: $i \to j$ in $\mathcal{J}$, the following diagram commutes:

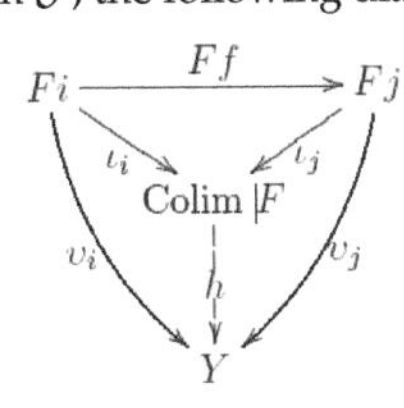

An analogue *dual* construction concerns the categorical operation of *limits* as the comma category $(\Delta \downarrow F)$ and its diagrammatic universalization as the **Cone**$_F$ category of the *inverse limit* operation for different categories of objects and morphisms (see Note 42).

Moreover, to immediately realize the relevance of these very abstract constructions, it is appropriate to recall here that, in the framework of the comma category construction, the notion of *diagonal functor* Δ is a generalization granting diagrammatic universality in CT to the *diagonal form* of a $n \times m$ matrix. It is thus a notion whose relevance is difficult to exaggerate in mathematics and physics!

Indeed, following the synthesis offered in [140], because all the *small* categories are categories defined not on *proper classes* with $Card \geq V$ (*large* categories) but on *sets* with $Card < V$, we can distinguish among *different classes of small categories* and then *their limits-colimits* according to the different cardinalities $< V$, $\aleph_0$, ω_1, ω_2, ... of their objects (sets). We can therefore have *locally finitely presentable categories* with $\aleph_0$*-directed colimits*, or *locally not-finitely* ω_1*-presentable categories* with ω_1*-directed colimits*, or *locally not-finitely* ω_2*-presentable categories* with ω_2*-directed colimits*, etc., where the attribute "local" means that the cardinality at issue *does not concern the category* but the objects in it. For instance, "**Set** is locally finitely presentable. In fact (i) every set is a directed colimit of the diagram of all of its finite subsets (ordered by inclusion), and (ii) there exists, up to isomorphism, only a (countable) set of finite sets" while the category of finite sets is not $\aleph_0$-countable and then it is not locally *finitely* presentable (see [140], p. 17). The categories **Pos, Grp, Aut, ...** are also locally finitely presentable while the categories **CPO** (*complete posets* in which every directed set has a join) and **Top** are not finitely presentable.

In Section 4.8, always following [140], we briefly introduced two examples of ω_1*-presentable categories* characterized by ω_1*-directed colimits* that are, respectively, either *locally* presentable (the ω_1- *locally presentable* category **Ban** of Banach spaces) or *not-locally* presentable (the category **Hilb** of Hilbert spaces that is a full sub-category of **Ban** because **Hilb** is not "cocomplete" since its colimits are in **Ban** not in **Hilb**). To them, in the light of the precedent discussion, we added the subcategory of the doubled Hilbert spaces for the Bogoliubov functor, i.e., **Hilb** $\hookrightarrow$ **DHilb(**$\mathcal{B}$**)**, which splits into an infinite number of pairs of *finite* Hilbert spaces *pairs* $\mathcal{H}, \widetilde{\mathcal{H}}$, dually equivalent between them. Regarding this subcategory, we proposed the hypothesis, which needs to be further developed and rigorously mathematically justified for its novelty and relevance, that this subcategory of infinitely many doubled Hilbert spaces for the Bogoliubov functor **DHilb(**$\mathcal{B}$**)** is *cocomplete* because its colimits are within it, and it is an $\aleph_0$*-locally finitely presentable* category, because the indexing small category $\mathcal{J}$ of its colimit construction is constituted by the denumerable sets $\{\mathcal{N}\}$ of NG-boson condensates, i.e., each finite subset—no matter how large it is—of the infinitely many pairs of the doubled Hilbert spaces is univocally indexed by a finite number of NG-bosons, i.e., it is $\aleph_0$*-countable*. Therefore, to understand the sense of our hypothesis, it is necessary to understand at least the statement of the fundamental *cocompleteness theorem* that the category **DHilb(**$\mathcal{B}$**)** satisfies while the demonstration of this theorem—and of its dual *completeness* theorem—is outside the limit of the present paper and can be found in [169], pp. 7–9.

To understand the statement of the categorical cocompleteness theorem, only the definition of the categorical notion of the *coequalizer* for pairs of morphisms between coproducts is necessary, which applies in our case to each pair of dually equivalent doubled Hilbert spaces $\left(\mathcal{H}, \widetilde{\mathcal{H}}\right)$ as far as both are indexed by the same finite value of NG-boson condensates $|\mathcal{N}| := \left(\mathcal{N} \equiv \widetilde{\mathcal{N}}\right)$.

Definition A5. *(Categorical notion of coequalizer). Given two objects A, B in a category C with two morphisms f, g:* $A \to B$ *the "coequalizer" of f and g is an object denoted* $Coeq(f,g)$*, and a morphism* $p : B \to Coeq(f,g)$*, such that* $fp = gp$*, and it is universal in that if Y is an object with morphism* $v : B \to Y$ *such that* $fv = gv$*, then a unique morphism* $h : Coeq(f,g) \to Y$ exists such that the following diagram commutes:

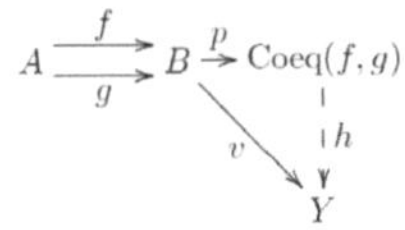

It is not difficult to see that $Coeq(f, g)$ is nothing but the colimit of F, that is, a diagram F in the precedent definition of the comma category of $Cocone_F$.

The notion of *coequalizer* allows us to understand the notion of *cocomplete category,* that is, a category where colimits over diagrams with a small source category *exist.* Specifically, it satisfies the cocompleteness theorem (see [169], p. 9):

Theorem A1. *(Cocompleteness Theorem). A category C is cocomplete if the coproduct of any set of objects in C exists and the coequalizer between any two morphisms with the same source and target exists.*

The statement of the theorem (and much more of its proof in [169], p. 9) emphasizes, in our application of the theorem to the category **DHilb($\mathcal{B}$),** its strict formal relationship with the category of coproducts **qHCoalg($\mathcal{B}$)** that we justified from a physical standpoint in Sections 4.6 and 4.8. Additionally, this indirectly supports the consistency of our hypothesis. To understand the physical and mathematical relevance of this result, see [71], in which it was recently demonstrated that the internal coalgebras of a locally presentable cocomplete category satisfy the *compactness* of the topological space on which they are defined because they contain all their limit points (colimits). This is not an irrelevant result in our case, given that by the **DHilb($\mathcal{B}$)** category, we model the *phase transitions* among quantum fields in non-linear (even chaotic) systems in far-from-equilibrium conditions.

Finally, the discussion in this Appendix and this whole paper reinforces, in a proper sense, what V. Pratt and D. Pavlović said about the relevance of the category of coalgebras for mathematical analysis and the same notion of continuum in mathematics as far as it is coinductively defined and, consequently, for logic in its duality with the category of algebras (see [72]):

> It is reasonable to ask why the continuum should be defined coinductively rather than inductively. It seems to us that the coinductive nature of the continuum is a consequence of our computing not with the reals themselves but inductively with rationals as approximations to reals. The computational status of the reals is not as *elements* but as *predicates* on the rationals, specifically the Dedekind cuts in the rationals. But elements and predicates are dual notions: whereas elements of a set X can be understood as functions *to* X, predicates on X are functions *from* X ([72], p. 119).

Effectively, this synthetic statement summarizes what I said in this paper, vindicating the ante-predicative nature of the CT logical analysis as far as it does not consider the membership relation as a primitive such as ST. Indeed, what is the notion of colimit as a cocone of morphisms just illustrated if not a sort of "structural" *relational anatomy* of the constitution of a class of elements as the "domain" of a given predicate?

Appendix B. The Core of Stone's Representation Theorem for Boolean Algebras

In the light of what we have already said in Sections 5.1.1 and 5.1.2 and pictorially represented in Figure 9, we know that every finite BA can be represented as the power-set of its set of atoms. Each element of BA corresponds to the set if the atoms below it, that is, the join of which is an element. As we know, this power-set representation can be constructed in general for any *complete* atomic *BA.*

The core of *Stone representation* is that we can generalize the precedent construction for *non-complete and atomic* BAs using fields of sets instead of power-sets (see Section 5.1).

Indeed, the atoms of a finite BA correspond to its ultrafilters and an atom is below a given element of BA if and only if it is included in the ultrafilter corresponding to that atom. This leads to the possibility of constructing a representation of BA by taking the set of its ultrafilters and defining complexes by associating each element of BA with the set of ultrafilters containing that element.

Equivalently, as we know, we can consider the set of *homomorphisms* onto the two-element BA and constructing complexes (and then fields of sets) by associating each element of the BA with the set of such homomorphisms mapping it to the top element. All this leads us to the so-called *Stone representation of BA as a certain field of sets,* that is, the fields of sets just illustrated, satisfying some topological requirements we discuss immediately as follows.

Indeed, the final step in understanding Stone's RTBA stating the *duality* between a given BA A and its corresponding *Boolean topological space* or, more synthetically, its *Stone space* $S(A)$, consists of extending the previous construction to *topological* fields of sets. In fact, we can consider the complexes of a field of a set representing BA as a *basis* for generating a given topology $\mathcal{T}$. Indeed, in mathematics, the basis of a topology $\mathcal{T}$ for a *topological space* $(X, \mathcal{T})$ is a family B of open subsets of X such that every open set of the topology can be considered as the *union* of some members of B (see [172], p. 30). Now, the fields of sets, the complexes of which can be the basis of a Stone topological space associated (isomorphic) with a given BA B, i.e., $S(B)$, must satisfy the following two conditions (see [173], pp. 69–76):

1. The field of sets must be *separative,* i.e., for every pair of distinct points, there is a complex containing one point and not the other.
2. The field of set must be *compact*, i.e., for every proper filter over X, the intersection of all the complexes contained in the filter is non-empty.

Therefore, by denoting with $\mathbf{X} = (X, \mathcal{F})$ the fields of sets whose complexes form a basis for a topology, and with $T(\mathbf{X})$ the corresponding topological space $(X, \mathcal{T})$, whose topology $\mathcal{T}$ is formed by taking arbitrary unions (and intersections) of complexes, then:

1. $T(\mathbf{X})$ is always a *zero-dimensional space* (i.e., graphically representable as a "point").
2. $T(\mathbf{X})$ is a *Hausdorff space* (i.e., whose points have disjoint neighborhoods) if $\mathbf{X}$ is separative.
3. $T(\mathbf{X})$ is a *compact space* with compact open sets $\mathcal{F}$ if $\mathbf{X}$ is compact.
4. $T(\mathbf{X})$ is a *Boolean space* or a *Stone space* with clopen sets $\mathcal{F}$ if $\mathbf{X}$ is both separative and compact.

From this reconstruction, the statement that a Stone space S is a *compact, Hausdorff, totally disconnected* [44] topological space becomes more understandable for philosophers. A topological space that has the same properties of the topological spaces on which the C*-subalgebras of Hilbert spaces in quantum physics formalism are defined, so that we can say that Stone RTBA is at the basis of any architecture of topological quantum computers, as we know.

In the light of what we said in this Appendix B and Section 5.1.1, Stone's RTBA stating the *isomorphism* and then the *Stone duality* between a BA B and a topological field of sets and the functorial dual equivalence between the correspondent categories also becomes easily understandable. Specifically, this is a field of sets constituting the basis of the *Stone space* associated with BA B, i.e., $S(B)$. The points in $S(B)$ are indeed the ultrafilters on B or, equivalently, the *homomorphisms* from B to the two-element BA. Conversely, given any topological space $(X, \mathcal{T})$, its subsets that are clopen form BA.

Notes

1 Following W. V. O. Quine's reconstruction ([29], pp. 133–136), according to this principle formulated by Russell to justify his solution of Cantor's and Frege's antinomies by a "ramified type theory", each object in logic, either individual or collection (set, class) can exist only as a *member* or *element* of the domain of a given predicate (i.e., of an "higher type" class), and finally, because it is so satisfying a *self-identity* relation, as a member of the *universal class* V, intended as the domain of the meta-predicate "being true". As Quine synthesizes, "V (stays) for '$\hat{x}(x = x)$' V is, by definition, the class of all those elements which are self-identical, i.e., since everything is self-identical (...), V is simply the class of all elements" ([174], p. 144). It is worth emphasizing that all this is equivalent to affirm that in standard ST no set *self-membership* is allowed, i.e., no set can be an element of itself. A condition that in ZF, for example, is granted by the "foundation" and the "pairing" axioms, which, because of the consequent Zermelo's "well-ordering theorem", at the same time grant that every set has an "ordinal rank", according to Von Neumann's "ordinal cumulative hierarchy" construction (see [66] for a synthesis).

2 For instance, from the truth of the propositions: "Julius Caesar wrote the *De Bello Gallico*" and "Julius Caesar fought in Gallia", by applying the connective "and", we can deduce the truth of the composed proposition "Julius Caesar wrote the *De Bello Gallico* and fought in Gallia". On the contrary, we cannot deduce the truth of the composed proposition "Julius Cesar wrote the De Bello Gallico while he was fighting in Gallia", typical of the *tense-logic* that is one of the possible *alethic* interpretations—historically the first one since Aristotle—of the MC. For the truthfulness of tense-logic propositions it is necessary, indeed, in ML to consider the relationships between the *present* or "actual" state of the world, and other *past* and/or *future* "possible" states of the world.

3 The distinction between "optimality" and "maximality" conditions for the ethical constraints —where "optimal" stays for "good in all the possible worlds", and then for *all* the human groups/cultures, and "maximal" stays for "good in some possible worlds", and then for *some* human groups/cultures—where introduced in the contemporary debate by the 1998 Nobel Prize in Economy Amartya Sen. This distinction is the core of his theory of the *Comparative Distributive Justice,* based on the notion of *equity (fairness)*, instead of the abstract (and false) "equality" in social sciences, of which he proposed also a *formal version* [162]. This was done in the framework of the newborn discipline of the *social choice theory,* he contributed to create, and to which he significantly dedicated his Nobel Lecture [175]. Today the "social choice theory" is a branch of the *formal philosophy*, the branch concerning the "decision theory and social philosophy" (see [176], pp. 611–725).

4 Effectively, in a formalized deontic logic in our global society and economy, the *optimal choice* (absolute) criterion must be substituted by a more effective and fair *maximal choice* criterion for *different social/economical situations* and *value systems,* relative to different groups in the society, according to a *comparative theory of distributive justice as fairness,* having in the *personal flourishing* of human individuals and group the "common good" to be pursued. This social theory was developed by the Nobel Prize in Economics Amartya Sen into the so-called *social choice theory,* conceived as a formal version of the political and social philosophy [162].

5 In the case of "balanced" open systems, the summands of the coproducts cannot commute with each other because representing the system and the thermal bath energy contributions in the calculation of the total energy of the quantum state.

6 Effectively, a difference occurs between them. The coproducts of the Bogoliubov construction in QFT for dissipative systems are *non-commutative,* so that the corresponding Boolean algebras are *non-commutative or skew Boolean algebras* that satisfy the same axioms of the general Boolean algebras except for the commutativity between the $\wedge$ and $\vee$ operators (see [160] for an extended examination).

7 For a connection with the actual use of "formal ontology" in computer science, it is sufficient to recall that "transcendental subject" in philosophy does *never* refer to a human *individual,* but to the common way of thinking and believing shared by a group of individuals, in the limit, by all the human individuals, as *conscious*—and then *intentional*—agents.

8 It is significant that for the conceptualist ontology there is *per se* no representative in the Ancient and Middle Ages, because it is typical of the Modern Age. It starts indeed with Descarte's foundation of the logical truth on the mental *evidence* and then on *consciousness,* and not on the "conformity" (homomorphism) of the structures of language with the structures of reality like in the Middle Age Platonic (logicism) and Aristotelian (naturalism) philosophies. Descarte's and modern conceptualist positions can therefore be synthesized with the slogan: "a statement is true because it is evident, and not it is evident because it is true", as it is in the logical and natural realisms.

9 Please, note that because $\in$ is not a primitive in CT, objects for existing in CT must not satisfy a self-identity relationship and then their membership to V like in ST, where they must satisfy Russell's set-elementhood principle for being consistently defined/demonstrated as existing in the theory (see Note 1).

10 For understanding immediately, the relevance of an arrow-theoretic way of thinking as to the set-theoretic one, let us think at the oldest proof method in the Western logic, which is the Aristotelian deductive (categorical) syllogism in its more fundamental form, the so-called *In Barbara* form. For instance: "If all humans (B) are mortal (A), and all the Greeks (C) are humans, then all the Greeks are mortal". Now, in the *extensional* interpretation that Leibniz (followed by Euler and Venn) gave of the Aristotelian scheme: "*AB* & *BC*.·. *AC*", this corresponds to stating predicatively: $((B \in A) \wedge (C \in B)) \rightarrow (C \in A)$. Such a predicative formula has according to Leibniz its extensional proof in the transitive inclusions of the respective classes $((\mathbf{B} \subseteq \mathbf{A}) \wedge (\mathbf{C} \subseteq \mathbf{B})) \rightarrow (\mathbf{C} \subseteq \mathbf{A})$. In CT where the set-elementhood is not a primitive, the universality of this demonstration takes the form of the commutative triangular diagram, *ABC* we discussed before, whose objects are *categories*—which in the Aristotelian syllogisms are always

"natural kinds"—and the morphisms are *functors* (see Definition 3. and [177] for this functorial interpretation of the syllogism "triadic" structure). Significantly, this categorical formalization of the syllogism can justify also the Aristotelian non-extensional (modal) syllogisms that, on the contrary, Leibniz's extensional interpretation cannot do, as J. Łukasiewicz first noticed [164].

11 On this regard, it is significant the fundamental work of R. Maddux [178] who demonstrated the strict relationship of Peirce's naïve triadic algebra of relations [179] with its axiomatic development into a calculus of relations by Tarski [69]. Not casually, indeed, the last book published by A. Tarski with S. Givant [180] (see also [181]) concerns precisely the demonstration of two fundamental results. (1) Before all, the demonstration that an *irreducible triadic algebra of relations* is sufficient for expressing faithfully any *first-order logic* (FOL) formula up to logical equivalence. That is, any FOL formula of the predicate and propositional calculi can be expressed faithfully in an *equation logic* (having arithmetic operators as connectives and numbers as their arguments) on a triadic basis. This fragment of FOL and the corresponding *variety* of relation algebras (**RA**)—i.e., the class of relation algebras defined by purely equational postulates—are therefore sufficient for expressing not only the *Peano arithmetic,* but also practically all *axiomatic set theories* ever proposed. Secondly, (2) just because of this expressive power, **RA** suffers in logic the same limitations imposed by Gödel's incompleteness theorems. I.e., the logic based on **RA** is incomplete and undecidable. However, and this is the second fundamental result, the Boolean FOL fragment of **RA** results to be *complete* and *decidable,* since its semantics is defined over *partially ordered* sets. All this means that we can express algebraically almost all mathematics in terms of a triadic **RA**, and more significantly we can express FOL without using quantifiers ($\forall, \exists$), connectives ($\wedge, \vee$), and turnstiles ($\vdash, \Vdash$), but essentially the *equation logic* of a Boolean algebra. If all this explains the odd title of Tarski's and Givant's book "A formalization of set theory without variables" [180], this algebraic construction of logic and mathematics is completed by the possibility in CT "arrow-theoretic" logic of demonstrating the natural number construction by *primitive recursion* without any (impredicative) reference to numbers as predicative numerals like in **ZF** (see [182–184] and [68], p. 285).

12 Effectively, the *present-time* event in the causal light-cone (see Figure 1) can be categorically interpreted as the *final* object F sharing with all the other events A belonging to its past/future light-cones a *dual* causal morphism in the sense of Definition 7. Indeed, in the case of the set of events $\{A\}$ belonging to the *past* light-cone, F plays the role of an initial object I defined by the unique morphism ι_A pointing to the set $\{A\}$ of its causes, i.e., $F = I := \iota_A : I \to A$. In the case of the set of events $\{A\}$ belonging to the *future* light-cone, F plays the role of a terminal object T defined by the unique morphism τ_A since the set $\{A\}$ of its effects are pointing to F as to its shared cause, i.e., $F = T := \tau_A : A \to T$.

13 Effectively, as Rieger rightly recalls, the foundation axiom is not *per se* necessary for avoiding Cantor's antinomies in the transfinite induction construction (see also [76]). The other axioms of **ZF**, first the "separation" and the "power-set" axioms, are sufficient for avoiding them. Effectively, the foundation axiom was introduced by Zermelo essentially for granting his well-ordering theorem.

14 All Aquinas' works are here quoted using their Latin title, according to the online edition of Aquinas' *Collected Works* in [86]. The translations into English of the different passages are mine. Effectively, in Aquinas' ontology the *physical* objects in nature or "substances", either "individuals" ("*primary* substances") or "species" ("*secondary* substances", for existing must satisfy a simple *reflexive/identity* relation (*reditio ad semetipsum,* "return onto itself") like objects in CT, and not a *double-reflexive/self-identity* relation like objects in ST. The self-identity relation, indeed, for Aquinas characterized the *logical* objects in mind, as far as they are *abstract* objects.

15 $\exists x Px$ stays in standard set-theoretic predicate logic for $x \in \mathbf{P}$, where $\mathbf{P}$ is the class connoted by the predicate P denoting the identity Id_x shared by all the elements x of the class, and where, therefore, the class $\mathbf{P}$ must be of a higher ordinal rank with respect to its elements x, i.e., belonging to the domain of the predicate P, if we must avoid the "Russell antinomy" in Frege's theory of classes (see Note 1).

16 Roughly speaking, this means that the state (classical mechanics) and/or the phase (statistical mechanics) space representing the system dynamics is *invariant* by exchanging each other the two canonical variables onto the orthogonal axes of their graphic vectorial representation.

17 Effectively it is a sort of "resonance" or "constructive interference" among statistical wave functions "oscillating coherently" with the same phase, as the famous "double-slit" experiment exemplifies very well also for the wave functions of only one particle (pure states).

18 In physics an "observable" is a physical magnitude that we can measure, for instance, the position and the momentum. In Classical Mechanics, an observable is a real-valued function on the set of all possible states. In quantum physics (QM and QFT), it is an "operator" because the properties of the quantum state, that is, the probability distributions for the outcomes of any possible measurement performed over it, can be determined only by some sequence of operations, e.g., by submitting the systems to the action of several electromagnetic fields, and then reading the resulting different values.

19 In functional analysis, a C*-algebra is a *Banach algebra*—that is, an associative algebra over the fields of real or complex numbers that is also a "Banach space", i.e., a space with a defined norm $||\cdot||$ complete in the metrics induced by the norm – together with an involution (a reversal of the morphisms like between $f(x)$ and $f^{-1}(x)$) satisfying the property of *adjointness.* In the specific case of quantum formalism, it is an *algebra B* over the complex number field $\mathbb{C}$ of *continuous linear operators* on a *complex Hilbert space.* In this case, the adjointness condition is strictly related to the Hermitian one (denoted by the symbol *), of the "inner products" $\langle \cdot, \cdot \rangle$ characterizing generally the Hilbert spaces. That is, without going deeper in the technicalities, given a linear operator $A : H_1 \to H_2$ between Hilbert spaces, the *adjoint* (dual) operator $A^* : H_2 \to H_1$ fulfills the condition between the relative inner

products: $\langle Ah_1, h_2 \rangle_{H_2} = \langle h_1, A^* h_2 \rangle_{H_2}$. In the case that the Hilbert spaces concerned are identical, A is an endomorphism on the same Hilbert space satisfying therefore a *self-adjointness* property and then the duality between a Hilbert space $\mathcal{H}$ and its operator space $\mathcal{H}^*$. For this reason, Hilbert spaces are *self-dual*. In the case of C*-algebras, this adjointness condition is extended to the operators acting on Banach spaces $A : D \rightarrow E$, with corresponding norms $||\cdot||_D, ||\cdot||_E$. Its adjoint operator is $A^* : E^* \rightarrow D^*$. In this case, the Banach algebra B satisfies two other properties: (1) it is *topologically closed* in the norm topology of the operators; (2) it is closed under the operation of taking adjoints of the *operators*. In the CT formalization, this means that the category of Hilbert spaces **Hilb** is effectively a *full subcategory* of the category of Banach spaces **Ban**, and then that **Hilb** is not *cocomplete* because all its colimits are in **Ban** not in **Hilb** (see [140] and below Section 4.8). Finally, one can extend the C*-algebra construction also to non-Hilbert C*-algebras. This class includes the algebras of the *continuous functions* $C_0(X)$, i.e., vanishing in the infinite limit. This justifies Landsman's reading of all classical and quantum physics in this framework of the algebra of operators, *per se* born, as we have seen, in the framework of quantum physics formalization.

20 The algebraic *tensor product* $V \otimes W$ between two vector spaces V and W over the same numerical field, is itself a vector space, endowed with the operation of *bilinear composition* denoted by $\otimes$ from ordered pairs in the Cartesian product $V \times W$ to $V \otimes W$, so to generalize to tensors the matrix *outer product*. Where: the "bilinear map" is a function combining elements of two vector spaces to yield elements of a third vector space, and it is linear in both of its arguments, while the "outer product" of two vectors of dimensions n and m is a $n \times m$ matrix. In the case of two tensors, the outer product is another tensor. A fundamental property of tensor products between finite dimensional vector spaces is that the resulting vector space has dimensions equal to the product of the dimensions of the two factors: $\dim(V \otimes W) = \dim V \times \dim W$. This distinguishes the tensor product from the *direct sum* vector space, whose dimension is the sum of the dimensions of the two summands: $\dim(V \otimes W) = \dim V + \dim W$. Just as – for giving another example of the direct sum operation in algebra well known by everybody –, the direct sum $\mathbb{R} \otimes \mathbb{R}$—where $\mathbb{R}$ is a coordinate space defined on real numbers—is the bidimensional *Cartesian plane.*

21 Effectively, the "quantum entanglement" acquires, in the light of the long-range correlations among quantum fields in QFT related to the Goldstone Theorem, an immediate intelligibility, showing that it does not imply any absurd "causal interaction" among quantum particles violating c (the light velocity). That is, a physical signal propagating at a superluminal velocity, which is the deep reason for which Einstein refused the quantum "non-locality" (entanglement) in his famous discussion with Niels Bohr, during the 30's of the last century. For showing this, it is sufficient to recall the notion of "phase velocity" V_P in the vacuum of SR that holds also in QFT. Now, $V_P = \frac{E}{p}$, where P is the field phase, E is the total energy, and p is the momentum of a given physical signal. Therefore, in SR, $V_P = \frac{E}{p} = \frac{\gamma mc^2}{\gamma mv} = \frac{c^2}{v}$, where γ is the Lorenz constant, m is the mass, and v is the velocity of the physical signal that is always less or also much less than c. This means that the phase propagation (or the propagation of correlation waves among quantum fields) in microphysics (QFT) is practically instantaneous without violating c.

22 Think at the everyday experience of the boiling water, exemplifying the continuously changing correlation-length among the water molecules, and then the continuously changing "dynamic boundary" of the vapor-liquid phase transition of water.

23 The dynamic mechanism according to which the water molecules, beyond a given density threshold, can condense into coherence domains (CDs) among their electric dipoles fields is today well known (see [120,185] for a more recent synthesis with several bibliographic references). The core of such a mechanism is that in each water CD the molecules oscillate *coherently* between two configurations of their electronic clouds, so to produce an electromagnetic field oscillating with the same frequency. The water CD can, therefore, attract by resonance a small number of "guests" molecules different from water, which share thus the energy stored in the CD. In this way, we have a much more efficient way than the random "diffusion process" introduced by the last work of A. M. Turing as the fundamental method of *morphogenesis* in biological matter [186], to make possible that *selective chemical reactions occur,* given that the chemical forces propagate only at short distances. For instance, this is the dynamic core of "cell specialization" in *epigenetics,* where only some sequences of the DNA that is the same for all the cells of a given organisms are activated/de-activated, because of the presence/absence of the proper molecules in the cell environment. In short, "the interplay between chemistry and electromagnetic field produces a collective oscillation of all the CDs that, according to the general theorem of quantum electro-dynamic coherence, gives rise to an extended coherence, where the CDs of water and "guest" molecules become the components of much more extended 'super-domains' which could just be the various organs" ([120], p. 37), at different level of the biological matter self-organization. Another well-studied phenomenon strictly related to the dipole CDs is the formation, propagation, and the reciprocal synchronization of *solitons,* that is a self-reinforcing solitary wave (a wave packet or pulse) that maintains its shape while it travels at constant speed. Solitons are caused by a cancellation of the nonlinear and dispersive effects in the medium. In macroscopic fluid dynamics, the formation of a "tsunami-wave" in the sea is a terrible example of "sea water soliton"! In biological matter electro-dynamics, the soliton presence is well established both in DNA and in protein dynamics, displaying a fundamental role for the efficiency of the cell metabolism through the cell microtubules, whose relevance for a quantum foundation of biology is today well recognized [120].

24 This notion, as Freeman and Vitiello explain elsewhere [40], is a critical reference to the much more famous theory of the "mirror neurons" by Giacomo Rizzolatti and his group [187]. The criticism consists in the fact that—apart from the fact that Rizzolatti's mirror neurons are limited to the brain interaction with the social environment—the measurements concerning the mirror system in the ape and in the human brains concern essentially the *passive* answer of neuron arrays of the motor neurons of one animal to stimulations deriving by the motor neurons of another animal, without explaining the underlying *dynamic mechanism* that, on

the contrary can have an elegant explanation at the fundamental physical level in the DDF principle of QFT as a result of the system-environment entanglement.

25 We recall that, following Von Neumann's construction of "cumulative hierarchy of ordinal number ranks" for justifying consistently the "transfinite induction" for infinite sets in **ZF**, ω_1 is the limit ordinal number of the set of transfinite numbers with cardinality immediately successive to $\aleph_0$, i.e., to the "cardinality of the denumerable sets", that is, of all the infinite sets with the cardinality of $\mathbb{N}$.

26 We recall that a *Banach space* is a *Hilbert space if* it satisfies the "parallelogram law" characterizing the "inner structure" (inner product) of a Hilbert space, and for which it is self-dual.

27 In fact, *each* Hilbert space, precisely for its self-dual character, is *complete*, i.e., it contains all the *limits* necessary and sufficient for its computations in functional analysis. However, the category **Hilb** it is not *cocomplete* (i.e., it does not satisfy the fundamental "Cocompleteness theorem" (see Theorem A1 and the relative comments in Appendix A), because the category does not contain in itself the colimits for indexing the infinitely many inequivalent Hilbert spaces (see below).

28 For the reader convenience, I recall the statements of the two De Morgan laws of propositional logic: $(\neg(P \vee Q) \Leftrightarrow (\neg P \wedge \neg Q))$; and $(\neg(P \wedge Q) \Leftrightarrow (\neg P \vee \neg Q))$.

29 In parenthesis, is highly significant that complete BAs are a necessary ingredient for constructing Boolean-valued models of set theory using P. Cohen's *forcing* notion [188].

30 The connection of σ-algebras and measurable spaces in (statistical and quantum) physics is much more evident when we recall that any "probability space" in statistics is a probability triple $(\Omega, \mathcal{F}, P)$, where Ω is the set of all possible outcomes; $\mathcal{F}$ is an event space, which is a set of events $\mathcal{F}$, an event being a set of outcomes in the sample space; P is a probability function, which assigns each event in the event space a probability that is a number between 0 and 1. Now, generally $\mathcal{F}$ is a σ-algebra, $\mathcal{F}$ consisting in the collection of all the events we would like to consider according to the type of the statistical analysis we want to perform on a given probability space. More generally, a σ-algebra or σ-field on a set X is a collection Σ of subsets of X, closed under complement, under countable unions, and under countable intersections so to constitute a *measurable space*.

31 We recall that a "set preorder" means that sets satisfy "reflexive" and "transitive" order relations $\leq$. So that, a set preorder is a "set partial order" if sets satisfy also "antisymmetric" order relations, while a set preorder is a "set equivalence class" if they satisfy also "symmetric" order relations.

32 This terminology originates with G. Fobenius in 1880s, who refers to a collection of elements of a group as a "complex".

33 Effectively, Stone in the demonstration of his theorem, does not use the Choice Axiom like ZFC but the Zorn Lemma that is an equivalent of this axiom for this type of application.

34 Generally, in TCS "concurrent computations" stay for two computational processes developing themselves *in parallel*—during overlapping time periods—instead of *sequentially*, with one completing itself before the other, and where the final state of the former gives the initial state to the latter according to a "circular" overall procedure.

35 For this historical reconstruction I am referring mainly to the final part of [32].

36 To Aquinas, indeed, is also attributed a short treatise about the modal propositional logic *De Propositionibus Modalibus* (available online in [86]), in which he offers an original interpretation of the *de re* and the *de dicto* modalities. Indeed, the propositional logic, unknown to Aristotle, given that it was defined and developed by his disciples, the Stoics, was well known in the Latin Scholasticism, because of the logical teaching of Anicius Manlius Severinus Boëthius (shortly, "Boethius": 477–524 A.D.). He not only translated into Latin Porphyry's *Isagoge* but wrote two treatises about the categorical (Aristotelian) syllogism and the hypothetical (Stoic) syllogism, by which the two logical calculi—the predicate and the propositional calculus, respectively—were introduced separately into the Medieval and then into the Modern logic, till their unification in Frege's formal calculus of classes.

37 Where, *by nature* means: those things that happen "or always or frequently" and then "not randomly" (*a casu*), i.e., by deterministic or statistical physical processes, as Aquinas explained before (see [86], *In Libros Physicorum*, II, lect. 13, n. 2).

38 That Aquinas is here referring to the notion of local truth (with relativized quantifiers) is evident from the following two quotations always from the *De Veritate* ("About Truth") book: «If therefore we take truth in the proper way, according to which the things are said true secondarily (i.e., relatively to an intellect), there are of several true objects (*plurium verorum*) several truths, and of a true object many truths in different intellects» (see [86], *Quaestiones Disputatae De Veritate*, I, 4 co.). «On the other hand, the truth that is in the human intellect is not related to things like an extrinsic and common measure to measured things [against Sophists, evidently], but like a measured to a measuring, (…) and therefore it must vary according to the variety of things» (see [86], *Quaestiones Disputatae De Veritate*, I, 4, ad 2).

39 For Aristotle, indeed, the predicate "being cause of" is not a category because resulting from the composition of three categories: "relation", "action", "passion" in his Table of Categories.

40 Particularly, it was the Scottish philosopher and logician Sir William Hamilton (1788–1856) who applied systematically, in his monumental work published posthumous *Lectures on Metaphysics and Logic*, the principle of sufficient reason for the construction of the whole metaphysical building, using a logic refusing explicitly, not only any symbolism, but also any *modality*, either real (*de re*) or logical (*de dicto*).

41 The terms "cone" and "cocone" helps to understand intuitively the arrow-theoretic notions of "limits" and "colimits" as "terminal" and "initial" objects, respectively. Indeed, if we take a cone, its vertex with respect to its basis can be connoted, either as the unique "terminal object" (common target/codomain) of all the arrows having in each point of the basis their own sources/domains (i.e., "cones of morphisms"), or *dually* as the unique "initial object" (source/domain) of all the arrows having in each point of the basis their own targets/domains (i.e., "cocones of morphisms").

42 We recall that generally in mathematics a *direct limit* is a way for constructing a *larger* object from many *smaller* objects "put together" in a specific way, generally by referring to an higher rank "class" of objects in the predicative set-theoretic approach in logic and mathematics. In CT these objects are from any category (e.g., **Set, Grp, Vect**$_k$, **Top**, ...) and the way for putting together the smaller objects is specified by the *homomorphisms* (or more generally, the morphisms) typical of the category concerned. For instance, in the case of sets, let $\{A_i : i \in I\}$ be a family of sets indexed by I, and $f_{ij} : A_i \rightarrow A_j$ be a homomorphism for all $i \leq j$. Then, the pair A_i, f_{ij} is called a *direct system* over I, the *direct limit* of the direct system is denoted by $\underrightarrow{\lim} A_i$, and its underlying set is constituted by the disjoint union (coproduct) of A_i's, "modulo" a given equivalence relation $\sim$. I.e., $\underrightarrow{\lim} A_i = \underset{i}{\cup} A_i / \sim$. That is, if $x_i \in A_i$ and $x_j \in A_j$, then $x_i \sim x_j$ if there is some $k \in I$ with $i, j \leq k$ such that $f_{ik}(x_i) = f_{jk}\left(x_j\right)$. From this definition, we derive the other definition of *canonical function* in terms of the homomorphism $\varphi_i : A_i \rightarrow \underrightarrow{\lim} A_i$ sending each element to its equivalence class. Dually, we can define the notion of *inverse limit* $\underleftarrow{\lim} A_i$ affirming that an element is equivalent to all its images under the maps of the direct system, i.e., $x_i \sim f_{ij}(x_i)$ for all $i \leq j$. The duality between direct and inverse limits can be expressed as the following relation: $\mathrm{Hom}\left(\underrightarrow{\lim} X_i, Y\right) = \underleftarrow{\lim} \mathrm{Hom}(X_i, Y)$.

43 We recall that a *small* category is a category whose objects are *sets* with *Card* < *V*, while a *large* category is a category whose objects are (Von Neumann's) *proper classes* with *Card* = *V* or even larger if we accept Gödel's generalized CH in **NBG**.

44 In a zero-dimensional topological space, indeed, only one-point sets (i.e., the empty set and the unitary sets) are connected.

References

1. Ladyman, J. Structural Realism. In Stanford Encyclopedia of Philosophy; Zalta, E., Ed. 2020. Available online: https://plato.stanford.edu/archives/win2020/entries/structural-realism/ (accessed on 15 January 2022).
2. Lyre, H. Structural Invariants, structural kinds, structural laws. In *Probabilities, Laws, and Structures*; Dieks, V., Gonzalez, W.J., Hartmann, S., Stöltzner, M., Weber, M., Eds.; Springer: Dordrecht, The Netherlands, 2011; pp. 169–181.
3. Hermens, R. How Real are Quantum States in ψ-Ontic Models? *Found. Phys.* **2021**, *51*, 38. [CrossRef]
4. Jónsson, B.; Tarski, A. Boolean algebras with operators, Part I. *Am. J. Math.* **1952**, *73*, 891–939. [CrossRef]
5. Jónsson, B.; Tarski, A. Boolean algebras with operators, Part II. *Am. J. Math.* **1952**, *74*, 127–152. [CrossRef]
6. Venema, Y. Algebras and co-algebras. In *Handbook of Modal Logic*; Blackburn, P., van Benthem, F.J.F., Wolter, F., Eds.; Elsevier: Amsterdam, The Netherlands, 2007; pp. 331–426.
7. Goldblatt, R.I. *Topoi: The Categorial Analysis of Logic*, 2nd ed.; Elsevier: Amsterdam, The Netherlands, 1984.
8. Anderson, P.W. More is different. *Science* **1972**, *177*, 393–396. [CrossRef] [PubMed]
9. Blasone, M.; Jizba, P.; Vitiello, G. *Quantum Field Theory and Its Macroscopic Manifestations. Boson Condensations, Ordered Patterns and Topological Defects*; Imperial College Press: London, UK, 2011.
10. Basti, G. Foreword: A chronology of the universe evolution. *Aquinas* **2020**, *53*, 33–52.
11. Aczel, P. *Non-Well-Founded Sets*; CLSI Lecture Notes; Stanford UP: Stanford, CA, USA, 1988; Volume 14.
12. Lewis, C.I. *A Survey of Symbolic Logic*; University of California Press: Berkeley, CA, USA, 1918.
13. Lewis, C.I.; Langford, C.H. *Symbolic Logic*, 2nd ed.; Century Company: New York, NY, USA, 1932.
14. Cresswell, M.J.; Huges, G.E. *A New Introduction to Modal Logic*; Routledge: London, UK, 1996.
15. Kripke, S.A. Semantical analysis of modal logic I. Normal modal propositional logic calculi. *Math. Log. Quat.* **1963**, *9*, 67–96. [CrossRef]
16. Kripke, S.A. Semantical analysis of modal logic II. Non-normal modal propositional calculi. In *The Theory of Models*; Addison, J.W., Henkin, L., Tarski, A., Eds.; North Holland: Amsterdam, The Netherlands, 1965; pp. 206–220.
17. Tarski, A. The Concept of Truth in Formalized Languages. In *Logic, Semantics, Metamathematics*, 2nd ed.; Corcoran, J., Ed.; Hackett: Indianapolis, IN, USA, 1983; pp. 152–278.
18. Patton, C.M.; Wheeler, J.A. Is physics legislated by cosmogony. In *Quantum Gravity*; Isham, C.J., Penrose, R., Sciama, D.W., Eds.; Clarendon Press: Oxford, UK, 1975; pp. 538–605.
19. Kripke, S.A. *Naming and Necessity*; Harvard UP: Cambridge, MA, USA, 1980.
20. Galvan, S. *Logiche Intensionali. Sistemi Proposizionali di Logica Modale, Deontica, Epistemica*; Franco Angeli: Milano, Italy, 1991.
21. Von Kutschera, F. *Einfürung in Intensional Semantik*; de Gruyter: Berlin, Germany, 1976.
22. Rescher, N. *Epistemic Logic: Survey of the Logic of Knowledge*; Pittsburgh UP: Pittsburgh, PA, USA, 2005.
23. Von Wright, G.H. *An Essay in Deontic Logic and the General Theory of Action*; North Holland: Amsterdam, The Netherlands, 1968.
24. Horty, J.F. *Agency and Deontic Logic*; Oxford UP: Oxford, UK; New York, NY, USA, 2001.
25. Cocchiarella, N.B.; Freund, M.A. *Modal Logic. An Introduction to Its Syntax and Semantics*; Oxford UP: Oxford, UK; New York, NY, USA, 2008.

26. Husserl, E. *Formale und Transzendentale Logik*; Janssen, P., Ed.; Nijhoff: Den Haag, The Netherlands, 1974.
27. Searle, J.R. *Intentionality. An Essay in the Philosophy of Mind*; Cambridge UP: New York, NY, USA, 1983.
28. Blackburn, P.; De Rijke, M.; Venema, Y. *Modal Logic. Cambridge Tracts in Theoretical Computer Science*; Cambridge UP: Cambridge, UK, 2002.
29. Quine, W.V.O. *Mathematical Logic*; Revised Edition; Harvard UP: Cambridge, MA, USA; London, UK, 1981.
30. Basti, G. From formal logic to formal ontology. The new dual paradigm in natural sciences. In *(Un-)Certainty and (In-)Exactness. Proceedings of the 1st CLE Colloquium for Philosophy and Formal Sciences*; Bertato, F.M., Basti, G., Eds.; Campinas UP-Aracne Edizioni: Campinas, Brazil; Rome, Italy, 2018; pp. 65–110.
31. Basti, G.; Ferrari, F.M. The dual paradigm in quantum physics and its ontology. I Part: Mathematics and phsyics. *Aquinas* **2020**, *53*, 117–162.
32. Basti, G.; Ferrari, F.M. The dual paradigm in quantum physics and its ontology. II Part: Logic and ontology. *Aquinas* **2020**, *53*, 163–194.
33. Weinberg, S. *Foundations of Modern Physics*; Cambridge UP: Cambridge, UK, 2021.
34. Basti, G.; Capolupo, A.; Vitiello, G. Quantum Field Theory and Coalgebraic Logic in Theoretical Computer Science. *Prog. Bioph. Mol. Biol.* **2017**, *123*, 39–52. [CrossRef]
35. Popper, K.R. *The Logic of Scientifc Discovery*; Routledge: London, UK; New York, NY, USA, 2002.
36. Stone, M.H. The theory of representation for Boolean algebras. *Trans. Am. Math. Soc.* **1936**, *40*, 37–111.
37. Blasone, M.; Jizba, P.; Vitiello, G. Dissipation and quantization. *Phys. Lett.* **2001**, *A287*, 205–210. [CrossRef]
38. Ehresmann, A.; Vanbremeersch, J.-P. MES: A Mathematical Model for the Revival of Natural Philosophy. *Philosophies* **2019**, *4*, 9. [CrossRef]
39. Freeman, W.J.; Vitiello, G. Nonlinear brain dynamics as macroscopic manifestation of underlying many-body field dynamics. *Phys. Life Rev.* **2006**, *3*, 93–118. [CrossRef]
40. Freeman, W.J.; Vitiello, G. Dissipation and spontaneous symmetry breaking in brain dynamics. *J. Phys. A Math. Theor.* **2008**, *41*, 304042. [CrossRef]
41. Damasio, A. *The Strange Order of Things. Life, Feeling and the Making of Cultures*; Pantheon Books: New York, NY, USA, 2018.
42. Blackburn, P.; van Benthem, F.J.F. Modal logic: A semantic perspective. In *Handbook of Modal Logic*; Blackburn, P., van Benthem, F.J.F., Wolter, F., Eds.; Elsevier: Amsterdam, The Netherlands, 2007; pp. 2–85.
43. Goranko, V.; Otto, M. Model theory of modal logic. In *Handbook of Modal Logic*; Blackburn, P., van Benthem, F.J.F., Wolter, F., Eds.; Elsevier: Amsterdam, The Netherlands, 2007; pp. 252–331.
44. Landsman, K.N.P. Lecture Notes on Operator Algebras. 14 December 2011. Available online: http://www.math.ru.nl/~{}landsman/OA2011.html (accessed on 11 February 2022).
45. Landsman, K.N.P. *Foundations of Quantum Theory. From Classical Concepts to Operator Algebra*; Springer: Berlin, Germany; New York, NY, USA, 2017.
46. Kant, I. *Prolegomena to Any Future Metaphysics That Will Be Able to Come Forward as Science*; Hatfield, G., Ed.; Cambridge UP: Cambridge, UK, 1997.
47. Smith, B. (Ed.) *Parts and Moments. Studies in Logic and Formal Ontology*; Philosophia: Munich, Germany, 1982.
48. Smith, B. Beyond Concepts, or: Ontology as Reality Representation. In *Formal Ontology and Information Systems, Proceedings of the Third International Conference FOIS 2004, Trento, Italy*; Varzi, A., Vieu, L., Eds.; IOS Press: Amsterdam, The Netherlands, 2004; pp. 1–22.
49. Brentano, F. *Psychologie Vom Empirischen Standpunkt*; Duncker & Humblot: Leipzig, Germany, 1874.
50. Husserl, E. *Logical Investigations*; Moran, D., Ed.; Routledge: London, UK; New York, NY, USA, 1970; Volume 2.
51. Husserl, E. *The Crisis of European Sciences and Transcendental Phenomenology. An Introduction to Phenomenological Philosophy*; Northwestern UP: Evanston, IL, USA, 1970.
52. Husserl, E. Der Folgerungskalkuül und die Inhaltslogik. *Vierteljahr. Für Wissenschafliche Philos.* **1891**, *15*, 168–189.
53. Frege, G. *Begriffsschrift, Eine der Arithmetischen Nachgebildete Formelsprache des Reinen Denkens*; Louis Nebert: Halle, Germany, 1879.
54. Schröder, E. *Vorlesungen über die Algebra der Logik. (Exakte Logik)*; I Band; B. G. Teubner: Leipzig, Germany, 1890.
55. Frege, G. *Grundgesetze der Arithmetik*; Band I (1893), Band II (1903); Verlag Herman Pole: Jena, Germany, 1893; pp. 1893–1903.
56. Husserl, E. Besprechung: Schröder, Ernst, Vorlesungen über die Algebra der Logik. (Exakte Logik) I Band. In *Göttingische Gelehrte Anzeigen*; Leipzig, B.G., Ed.; Teubner: Göttingen, Germany, 1891; pp. 243–278.
57. Peirce, C.S. The logic of relatives. *Monist* **1897**, *7*, 161–217. [CrossRef]
58. Peirce, C.S. *The New Elements of Mathematics*; Arisbe: Brusnswick, NJ, USA, 1902.
59. Peirce, C.S. One, Two, Three: Kantian Categories'. In Peirce's Manuscript Collection; MS [R] 897. 1886. Available online: http://www.commens.org/dictionary/term/firstness (accessed on 15 May 2021).
60. Cocchiarella, N.B. *Formal Ontology and Conceptual Realism*; Springer: Berlin/Heidelberg, Germany, 2007.
61. Cocchiarella, N.B. Logic and Ontology. *Axiomates* **2001**, *12*, 117–150. [CrossRef]
62. Basti, G. The Post-Modern Transcendental of Language in Science and Philosophy. In *Epistemology and Transformation of Knowledge in a Global Age*; Delic, Z., Ed.; InTech Publishing: London, UK, 2017; pp. 35–62.
63. Abramsky, S.; Tzevelekos, N. Introduction to categories and categorical logic. In *New Structures for Physics. Lecture Notes in Physics*; Coecke, B., Ed.; Springer: Berlin/Heidelberg, Germany, 2011; Volume 813, pp. 3–94.

64. Mac Lane, S. *Categories for the Working Mathematician*, 2nd ed.; Springer: Berlin/Heidelberg, Germany, 1998.
65. Awodey, S. *Category Theory*, 2nd ed.; Oxford Logic Guides 52; Oxford UP: Oxford, UK, 2010.
66. Suppes, P. *Axiomatic Set Theory*; Dover Publications: New York, NY, USA, 1972.
67. Abramsky, S. *Logic and Categories as a Tool for Building Theories*, 2nd ed.; Cornell University: Ithaca NY, USA, 2012; Available online: https://arxiv.org/abs/1201.5342v1 (accessed on 31 August 2018).
68. Abramsky, S. Logic and categories as a tool for building theories. *J. Ind. Counc. Phil. Res. Spec. Issue Log. Philos. Today* **2010**, *27*, 277–304.
69. Tarski, A. On the calculus of relations. *J. Symb. Log.* **1941**, *6*, 73–89. [CrossRef]
70. Atiyah, M.F. Duality in Mathematics and Physics, 18 December 2007. Available online: http://fme.upc.edu/ca/arxius/butlleti-digital/riemann/071218_conferencia_atiyah-d_article.pdf (accessed on 20 November 2021).
71. Poinsot, L.; Porst, H.E. Internal coalgebras in cocomplete categories: Generalizing the Eilenberg-Watt theorem. *J. Algebra Its Appl.* **2021**, *20*, 2150165. [CrossRef]
72. Pavlović, D.; Pratt, V. The continuum as a final coalgebra. *Theor. Comput. Sci.* **2002**, *280*, 105–122. [CrossRef]
73. Abramsky, S. Coalgebras, Chu spaces, and representations of physical systems. *J. Phil. Log.* **2013**, *41*, 551–574. [CrossRef]
74. *New Structures of Physics*; Coecke, B. (Ed.) Springer: Berlin/Heidelberg, Germany, 2011; pp. 3–94.
75. Rieger, A.; Paradoxes, Z.F. The axiom of foundation. In *Logic, Mathematics, Philosophy, Vintage Enthusiasms: Essays in Honour of John L. Bell*; Clark, P., Hallett, M., DeVidi, D., Eds.; Springer: Dordrecht, The Netherlands, 2011; pp. 171–187.
76. Hallett, M. *Cantorian Set Theory and Limitation of Size*; Clarendon Press: Oxford, UK, 1996.
77. Boolos, G. The iterative conception of set. *J. Phil.* **1971**, *68*, 215–232. [CrossRef]
78. Mirimanoff, D. Les antinomies de Russell et de Burali-Forti et le problème fondamental de la théorie des ensembles. *L'Enseignement Mathématique* **1917**, *19*, 37–52.
79. Mirimanoff, D. Remarques sur la théorie des ensembles et les antinomies Cantoriennes (I). *L'Enseignement Mathématique* **1917**, *19*, 208–217.
80. Von Neumann, J. Eine Axiomatisierung der Mengenlehre. *J. Für Die Reine Und Angew. Math.* **1925**, *154*, 219–240.
81. Rutten, J.J.M. Universal coalgebra: A theory of systems. *Theor. Comput. Sci.* **2000**, *249*, 3–80. [CrossRef]
82. Aczel, P.; Mendler, N.P. A Final Coalgebra Theorem. In *Category Theory and Computer Science*; Lecture Notes in Computer Science; Pitt, D.H., Rydeheard, D.E., Dybjer, P., Pitts, A.M., Poigné, A., Eds.; Springer: Berlin/Heidelberg, Germany, 1989; Volume 389, pp. 357–365. [CrossRef]
83. Aczel, P. Algebras and coalgebras. In *Algebraic and Coalgebraic Methods in the Mathematics of Program Construction: International Summer School and Workshop, Oxford, UK, April 10–14, 2000*; Revised Lectures (LNCS); Backhouse, R., Crole, R., Jeremy, G., Eds.; Springer: Berlin/Heidelberg, Germany, 2002; Volume 2297, pp. 79–88.
84. Bombelli, L.; Lee, J.; Meyer, D.; Sorkin, R.D. Spacetime as a causal set. *Phys. Rev. Lett.* **1987**, *59*, 521–524. [CrossRef]
85. Sorkin, R.D. A Finitary Substitute for Continuous Topology. *Int. J. Theor. Phys.* **1991**, *30*, 923–947. [CrossRef]
86. Aquinas, T. *Corpus Thomisticum. Subsidia Studii ab Enrique Alarcón Collecta et Edita*; University of Navarra UP: Navarra, Spain, 2000; Available online: http://www.corpusthomisticum.org/ (accessed on 7 January 2022).
87. Hilbert, D. *Grundlagen der Geometrie*; Teubner Verlag: Leipzig, Germany, 1899.
88. Hilbert, D. Mathematical Problems. *Bull. Am. Math. Soc.* **1902**, *8*, 437–479. [CrossRef]
89. Von Neumann, J. *Mathematische Grundlagen der Quantenmechanik*; Springer: Berlin, Germany, 1932.
90. Heisenberg, W. Über quantentheoretische Umdeutung kinematischer und mechanischer Beziehungen. *Zeitschr. Phys.* **1925**, *33*, 879–893. [CrossRef]
91. Born, M.; Jordan, P. Zur Quantenmechanik. *Zeitschr. Phys.* **1925**, *34*, 858–888. [CrossRef]
92. Born, M.; Heisenberg, W.; Jordan, P. Zur Quantenmechanik II. *Zeitschr. Phys.* **1925**, *35*, 557–561. [CrossRef]
93. Schrödinger, E. An ondulatory theory of the mechanics of atoms and molecules. *Phys. Rev.* **1926**, *28*, 1049–1070. [CrossRef]
94. Von Neumann, J. Die Eindeutigkeit der Schrödingerschen Operatoren. *Math. Annal.* **1931**, *104*, 570–578. [CrossRef]
95. Stone, M.H. Linear Transformations in Hilbert Space. III. Operational Methods and Group Theory. *Proc. Nat. Acad. Sci. USA* **1930**, *16*, 172–175. [CrossRef] [PubMed]
96. Von Neumann, J. Ueber Einen Satz Von Herrn, M.H. Stone. *Annal. Math. Second. Ser.* **1932**, *33*, 567–573. [CrossRef]
97. Stone, M.H. On one-parameter unitary groups in Hilbert Space. *Annal. Math. Second. Ser.* **1932**, *33*, 643–648. [CrossRef]
98. Gelfand, I.M.; Naimark, M.A. On the imbedding of normed rings into the ring of operators on a Hilbert space. *Matematiceskij Sbornik.* **1943**, *2*, 197–217.
99. Segal, I.E. Irreducible representations of operator algebras. *Bull. Am. Math. Soc.* **1947**, *53*, 73–88. [CrossRef]
100. Dirac, P.A.M. *Principles of Quantum Mechanics*, 4th ed.; Clarendon Press: Oxford, UK, 1958.
101. Becchi, C.M. Second quantization. *Scholarpedia* **2010**, *5*, 7902. [CrossRef]
102. Haag, R. On quantum field theories. *Mat.-Fys. Medd.* **1955**, *29*, 669.
103. Jordan, P.; Wigner, E. Über Paulisches Äquivalenzverbot. *Zeit. F. Phys.* **1928**, *47*, 631. [CrossRef]
104. Fock, V.A. Konfigurationsraum und zweite quantelung. *Zeit. F. Phys.* **1932**, *75*, 622. [CrossRef]
105. Zuber, J.-B.; Itzykson, C. *Quantum Field Theory*; Dover Pubblication: Berlin, Germany, 1980.
106. Feynman, R. *QED: The Strange Theory of Light and Matter (Princeton Science Library, 90)*; Revised Edition; Princeton UP: Princeton, NJ, USA, 2014.

107. Maldacena, J. The symmetry and simplicity of the laws of physics and the Higgs boson. *Eur. J. Phis.* **2016**, *1*, 015802. [CrossRef]
108. Umezawa, H. *Advanced Field Theory: Micro, Macro and Thermal Concepts*; American Institute of Physics: New York, NY, USA, 1993.
109. Umezawa, H. Development in concepts in quantum field theory in half century. *Math. Jpn.* **1995**, *41*, 109–124.
110. Del Giudice, E.; Pulselli, R.; Tiezzi, E. Thermodynamics of irreversible processes and quantum field theory: An interplay for understanding of ecosystem dynamics. *Ecolog. Model.* **2009**, *220*, 1874–1879. [CrossRef]
111. Bogoliubov, N.N. On a new method in the theory of superconductivity. *Nuovo Cim.* **1958**, *7*, 794–805. [CrossRef]
112. Goldstone, J. Field Theories with Superconductor Solutions. *Nuovo Cim.* **1961**, *19*, 154–164. [CrossRef]
113. Higgs, P. Broken Symmetries and the Masses of Gauge Bosons. *Phys. Rev. Lett.* **1964**, *13*, 508–509. [CrossRef]
114. Goldstone, J.; Salam, A.; Weinberg, S. Broken Symmetries. *Phys. Rev.* **1962**, *127*, 965–970. [CrossRef]
115. Nambu, Y. Quasiparticles and Gauge Invariance in the Theory of Superconductivity. *Phys. Rev.* **1960**, *117*, 648–663. [CrossRef]
116. Nambu, Y.; Jona-Lasinio, G. Dynamical Model of Elementary Particles Based on an Analogy with Superconductivity. I. *Phys. Rev.* **1961**, *122*, 345–358. [CrossRef]
117. Nambu, Y.; Jona-Lasinio, G. Dynamical Model of Elementary Particles Based on an Analogy with Superconductivity. II. *Phys. Rev.* **1961**, *124*, 246–254. [CrossRef]
118. Vitiello, G. Coherent states, fractals and brain waves. *New Math. Nat. Comput.* **2009**, *5*, 245–264. [CrossRef]
119. 'tHooft, G. Quantum gravity as a dissipative deterministic system. *Class. Quant. Grav.* **1999**, *16*, 3263–3279.
120. Brizhik, L.; Del Giudice, E.; Maric-Oehler, W.; Popp, F.-A.; Schlebusch, K.-P. On the dynamics of self-organization in living organisms. *Electromagn. Biol. Med.* **2009**, *28*, 28–40. [CrossRef]
121. Mitchell, M. *Complexity: A Guided Tour*; Oxford UP: Oxford, UK, 2009.
122. Morr, D.K. Lifting the fog of complexity. *Science* **2014**, *343*, 382–383. [CrossRef]
123. Comin, R.; Frano, A.; Yee, M.M.; Yoshida, Y.; Eisaki, H.; Schierle, E.; Weschke, E.; Sutarto, R.; He, F.; Soumyanarayanan, A.; et al. Charge order driven by Fermi-arc instability in $Bi_2Sr_{2-x}La_xCuO_{6+\delta}$. *Science* **2014**, *343*, 390–392. [CrossRef]
124. da Silva Neto, E.H. Ubiquitous Interplay between charge ordering and high-temperature superconductivity in cuprates. *Science* **2014**, *343*, 393–396. [CrossRef]
125. Connes, A.; Rovelli, C. Von Neumann algebra automorphisms and time-thermodynamics relation in general covariant quantum theories. *Class. Quant. Grav.* **1994**, *11*, 2899–2918. [CrossRef]
126. Hollands, S.; Longo, R. Non Equilibrium Thermodynamics in Conformal Field Theory. *Comm. Math. Phys.* **2018**, *357*, 43–60. [CrossRef]
127. Basti, G. The quantum field theory (QFT) dual paradigm in fundamental physics and the semantic information content and measure in cognitive sciences. In *Representation and Reality in Humans, Other Living Organisms, and Intelligent Machine*; Dodig-Crnkovic, G., Giovagnoli, R., Eds.; Springer: Berlin, Germany; New York, NY, USA, 2017; pp. 177–210.
128. Bruno, A.; Capolupo, A.; Kak, S.; Raimondo, G.; Vitiello, G. Geometric phase and gauge theory structure in quantum computing. *J. Phys. Conf. Ser.* **2011**, *306*, 012065. [CrossRef]
129. Nayak, C.; Simon, S.H.; Stern, A.; Freedman, M.; Das Sarma, S. Non-Abelian Anyons and Topological Quantum Computation. *Rev. Mod. Phys.* **2008**, *80*, 1083. [CrossRef]
130. Freeman Walter, J.; Burke, B.C.; Holmes, M.D.; Vanhatalo, S. Spatial spectra of scalp EEG and EMG from awake humans. *Clin. Neurophisiol.* **2003**, *114*, 1053–1060. [CrossRef]
131. Freeman, W.J. Origin, structure, and role of background EEG activity. Part 1. Analytic amplitude. *Clin. Neurophysiol.* **2004**, *115*, 2077–2088. [CrossRef]
132. Freeman, W.J. Origin, structure, and role of background EEG activity. Part 2. Analytic phase. *Clin. Neurophysiol.* **2004**, *115*, 2089–2107. [CrossRef]
133. Freeman, W.J. Origin, structure, and role of background EEG activity. Part 3. Neural frame classification. *Clin. Neurophysiol.* **2005**, *116*, 111–1129. [CrossRef]
134. Freeman, W.J. Origin, structure, and role of background EEG activity. Part 4. Neural frame simulation. *Clin. Neurophysiol.* **2006**, *117*, 572–589. [CrossRef]
135. Wiener, N. *Cybernetics. Or Communication and Control in the Animals and the Machines*, 2nd ed.; MIT Press: Cambridge, MA, USA, 1962.
136. Damasio, A. *Self Comes to Mind: Constructing the Conscious Brain*; Heinemann: London, UK, 2010.
137. Basti, G.; Capolupo, A.; Vitiello, G. The doubling of the Degrees of Freedom in Quantum Dissipative Systems, and the Semantic Information Notion and Measure in Biosemiotics. *Proceedings* **2020**, *47*, 60.
138. Clark, A.; Chalmers, D. The extended mind. *Analysis* **1998**, *58*, 7–19. [CrossRef]
139. Clark, A. *Supersizing the Mind: Embodiment, Action, and Cognitive Extension*; Oxford UP: Oxford, UK, 2008.
140. Adamek, J.; Rosicky, J. *Locally Presentable and Accessible Categories*; Cambridge UP: Cambridge, UK, 1994.
141. Barr, M. The Chu construction. *Theor. Applic. Categ.* **1996**, *2*, 17–35.
142. Rasetti, M.; Marletto, C. Quantum physics, topology, formal languages, computation: A categorial view—an homage to David Hilbert. *Perspect. Sci.* **2014**, *22*, 98–114.
143. Sikorski, R. *Boolean Algebras*; Springer: Berlin/Heidelberg, Germany, 1964.
144. The European Mathematical Society. *Algebra of Sets*; Springer Publishing: Berlin, Germany, 2020; Available online: https://encyclopediaofmath.org/wiki/Algebra_of_sets (accessed on 15 August 2021).

145. Davey, B.A.; Priestley, H.A. *Introduction to Lattices and Order*; Cambridge UP: Cambridge, UK, 1990.
146. Loomis, L.H. On the representation of σ-complete Boolean algebras. *Bull. Amer. Math.* **1947**, *53*, 757–760. [CrossRef]
147. Goldblatt, R. Varieties of complex algebras. *Ann. Pure Appl. Log.* **1989**, *38*, 173–241. [CrossRef]
148. Goldblatt, R. Elementary generation and canonicity for varieties of Boolean algebras with operators. *Algebra Univers.* **1995**, *34*, 551–607. [CrossRef]
149. Bourbaki, N. *Elements of Mathematics: General Topology*; Chapters 1—4; Springer: Berlin, Germany; New York, NY, USA, 1989.
150. Abramsky, S. Abramsky, S. A Cook's Tour of the Finitary Non-Well-Founded Sets (original lecture: 1988). In *Essays in honor of Dov Gabbay*; Artemov, S., Barringer, H., d'Avila, A., Lamb, L.C., Woods, J., Eds.; Imperial College Pubblications: London, UK, 2005; Volume I, pp. 1–18.
151. Paulson, L.C. Final coalgebras as greatest fixed points in ZF set theory. *Math. Struct. Comput. Sci.* **1995**, *9*, 545–567. [CrossRef]
152. Rutten, J.M. *The Method of Coalgebra: Exercises in Coinduction*; CWI: Amsterdam, The Netherlands, 2019.
153. Sangiorgi, D.; Rutten, J.M. (Eds.) *Advanced Topics in Bisimulation and Coinduction*; Cambridge UP: Cmabridge, UK, 2012.
154. Basti, G.; Vitiello, G. A QFT Approach to Data Streaming in Natural and Artificial Neural Networks. *Proceedings* **2022**, *81*, 106.
155. Van Benthem, F.J.F. Modal Correspondence Theory. Ph.D. Thesis, University of Amsterdam, Amsterdam, The Netherlands, 1976.
156. Van Benthem, F.J.F. Correspondence theory. In *Handbook of Philosophical Logic*; Gabbay, D., Guenthner, F., Eds.; Reidel: Dordrecht, The Netherlands, 1984; Volume II, pp. 167–247.
157. Blackburn, P.J.; van Benthem, F.J.F.; Wolter, F. Preface. In *Handbook of Modal Logic*; Blackburn, P., van Benthem, F.J.F., Wolter, F., Eds.; Elsevier: Amsterdam, The Netherlands, 2007; pp. xi–xxiii.
158. Löwenheim, L. On possibilities in the calculus of relatives. In *From Frege to Gödel: A Source Book in Mathematical Logi, 1879–1931*, 3rd ed.; Van Heijenoort, J., Ed.; Harvard UP: Cambridge, MA, USA, 1977; pp. 228–251.
159. Panizzoli, F. *Ontologia Della Partecipazione. Verso Una Formalizzazione Della Metafisica di Tommaso d'Aquino*; Edizioni Aracne: Rome, Italy, 2014.
160. Leech, J. Skew Boolean algebras. *Algebra Univers.* **1990**, *27*, 497–506. [CrossRef]
161. Esfeld, M. Causal realism. In *Probabilities, Laws, and Structures*; Dieks, D., Gonzalez, W.J., Hartmann, S., Stöltzner, M., Weber, M., Eds.; Springer: Dordrecht, The Netherlands, 2011; pp. 157–168.
162. Sen, A.K. *Collective Choice and Social Welfare*; Expanded Edition; Penguin Ltd. Kindle Edition: London, UK, 2017.
163. Peirce, C.S. Nominalism versus Realism. *J. Specul. Phil.* **1868**, *2*, 57–61.
164. Łukasiewicz, J. *Aristotle's Syllogistic from the Standpoint of Modern Formal Logic*; Second Enlarged Edition; Oxford UP: Oxford, UK, 1957.
165. Freeman, W.J. Nonlinear dynamics and the intention of Aquinas. *Mind Matter* **2008**, *6*, 207–234.
166. Berti, E. From physical to metaphysical cosmology: The Aristotelian path. *Aquinas* **2020**, *53*, 281–294.
167. Quine, W.V.O. Two dogmas of empiricism. *Phil. Rev.* **1951**, *6*, 20–43. [CrossRef]
168. Heller, M. Category free category theory and its philosophical implications. *Log. Logic. Phil.* **2016**, *25*, 447–459. [CrossRef]
169. Wang, K. Limits, Colimits and How to Calculate Them in the Category of Modules over a PID, 26 August 2011. Available online: http://www.math.uchicago.edu/~{}may/VIGRE/VIGRE2011/REUPapers/WangK.pdf (accessed on 13 January 2021).
170. Mac Lane, S.; Moerdijk, I. *Sheaves in Geometry and Logic: A First Introduction to Topos Theory*; Springer: Berlin, Germany, 1994.
171. May, J.P. *A Coincise Course in Algebraic Topology*, 2nd ed.; Chicago UP: Chicago, IL, USA, 1999.
172. Armstrong, M.A. *Basic Topology*; Springer: Berlin, Germany; New York, NY, USA, 1983.
173. Johnstone, P.T. *Stone Spaces*; Cambridge Studies in Advances Mathematics, Series Number 3; Cambridge UP: Cambridge, UK, 1982.
174. Quine, W.V.O. *From a Logical Point of View: Nine Logico-Philosophical Essays*; Second Revised Edition; Harvard UP: Cambridge, MA, USA, 1980.
175. Sen, A.K. The Possibility of Social Choice. Nobel Lecture. *Am. Econ. Rev.* **1999**, *89*, 178–215. [CrossRef]
176. *Introduction to Formal Philosophy*; Hansson, S.O.; Hendricks, V.F. (Eds.) Springer: Berlin, Germany; New York, NY, USA, 2018.
177. La Palme Reyes, M.; MacNamara, J.; Reyes, G.E. Functoriality and grammatical role in syllogisms *Notre Dame J. Form. Log.* **1994**, *35*, 41–66.
178. Maddux, R.D. The origin of relation algebras in the development and axiomatization of the calculus of relations. *Studia Log.* **1991**, *50*, 421–455. [CrossRef]
179. Peirce, C.S. On the algebra of logic. A contribution to the philosophy of notation. *Am. J. Math.* **1885**, *7*, 180–196. [CrossRef]
180. Tarski, A.; Givant, S. *A Formalization of Set Theory without Variables*; American Mathematical Society: Providence, RI, USA, 1987.
181. Givant, S. The Calculus of Relations as a Foundation for Mathematics. *J. Autom. Reason.* **2006**, *37*, 277–322. [CrossRef]
182. Nelson, E. *Predicative Arithmetic (Mathematical Notes 32)*; Princeton UP: Princeton, NJ, USA, 1986.
183. Nelson, E. Syntax and Semantics, 7 January 2002. Available online: http://www.irafs.org/irafs_1/cd_irafs02/texts/nelson.pdf (accessed on 20 August 2021).
184. Nelson, E. *Completed versus Incomplete Infinity in Arithmeti. STOQ International Conference on "Infinity in Science Philosophy and Theology"*; Pontifical Lateran University: Vatican City, 2005; Available online: https://web.math.princeton.edu/~{}nelson/papers/e.pd (accessed on 20 August 2021).
185. Ho, M.-W. Illuminating Water and Life. *Entropy* **2014**, *16*, 4874–4891. [CrossRef]
186. Turing, A. The chemical basis of morphogenesis. *Phil. Trans. Royal Soc. Series B Biol. Sci.* **1952**, *237*, 37–72.

187. Rizzolatti, G.; Sinigaglia, C. Mirrors in the Brain. In *How Our Minds Share Actions and Emotions*; Anderson, F., Ed.; Oxford UP: New York, NY, USA, 2008.
188. Cohen, P.J. *Set Theory and the Continuum Hypothesis*; Benjamin Publication: New York, NY, USA, 1966.

philosophies

MDPI

Essay

Scientific Observation Is Socio-Materially Augmented Perception: Toward a Participatory Realism

Tom Froese

Embodied Cognitive Science Unit, Okinawa Institute of Science and Technology Graduate University, 1919-1 Tancha, Onna 904-0495, Okinawa, Japan; tom.froese@oist.jp

Abstract: There is an overlooked similarity between three classic accounts of the conditions of object experience from three distinct disciplines. (1) Sociology: the "inversion" that accompanies discovery in the natural sciences, as local causes of effects are reattributed to an observed object. (2) Psychology: the "externalization" that accompanies mastery of a visual–tactile sensory substitution interface, as tactile sensations of the proximal interface are transformed into vision-like experience of a distal object. (3) Biology: the "projection" that brings forth an animal's *Umwelt*, as impressions on its body's sensory surfaces are reconfigured into perception of an external object. This similarity between the effects of scientific practice and interface-use on the one hand, and of sensorimotor interaction on the other, becomes intelligible once we accept that skillful engagement with instruments and interfaces constitutes a socio-material augmentation of our basic perceptual capacity. This enactive interpretation stands in contrast to anti-realism about science associated with constructivist interpretations of these three phenomena, which are motivated by viewing them as the internal mental construction of the experienced object. Instead, it favors a participatory realism: the sensorimotor basis of perceptual experience loops not only through our body, but also through the external world. This allows us to conceive of object experience in relational terms, i.e., as one or more subjects directly engaging with the world. Consequently, we can appreciate scientific observation in its full complexity: it is a socio-materially augmented process of becoming acquainted with the observed object that—like tool-use and perceiving more generally—is irreducibly self, other-, and world-involving.

Keywords: mind–body problem; sensory substitution; direct perception; enactive cognition; tool-use; fact–value gap; philosophy of mind; cognitive science; philosophy of science; consciousness

Citation: Froese, T. Scientific Observation Is Socio-Materially Augmented Perception: Toward a Participatory Realism. *Philosophies* **2022**, *7*, 37. https://doi.org/10.3390/philosophies7020037

Academic Editors: Gordana Dodig-Crnkovic and Marcin J. Schroeder

Received: 28 February 2022
Accepted: 29 March 2022
Published: 30 March 2022

> *I would just like to leave you with this reminder: when we adopt a paradigm in cognitive science, we enact a shared world—and enact ourselves into the bargain.*
>
> John Stewart (1942–2021)

1. Introduction

In this essay, I pay homage to John Stewart's contributions to the philosophy of science, by tracing the subject–object relationship through his interdisciplinary reflections. Stewart was fond of provocatively claiming that a value-neutral science would be without value (for a recent statement, see [1]). Even the production of mere instrumental value, such as technological gimmicks, would not be enough. Science should aspire to be an intellectual authority that provides guidance as to the layout of reality and affords inspiration for the future of humanity. Striving for these deeper values means that, ultimately, we cannot ignore the most foundational problem of all: how do we, ourselves, as we subjectively live our own existence, fit into this scientific worldview [2]?

In modern science there is little, if any, room for subjectivity in the material universe. In particular, advances in biology and neuroscience are highlighting tensions with our lived experience, including undermining our otherwise recognized status of being persons who act for reasons [3]. However, if a scientific worldview were to eliminate our subjectivity from reality, like a naïve realism that only assigns reality to objects fully describable by physics, it would have the self-defeating consequence that there can be nothing rational about that scientific enterprise! This is because, like the human lifeworld more generally, science depends essentially on our capacity for consciousness—to perceive the world and to collectively reason to decisions accordingly. However, in much of science, there is little recognition of the importance of this foundational problem, i.e., that we still do not understand how subjects and objects can meaningfully coexist, such that our lived mental life *as such* can make a difference in the objective world, and vice versa. Without substantial progress on this core problem, we are faced by a failure of science to understand the conditions of its own possibility. In other words, the stakes are high, both scientifically and existentially.

In response to this situation, this essay aims to motivate an alternative position to naïve realism in the philosophy of science, namely, by assigning a constitutive role to the observing subject, but without thereby falling into the opposite extreme of an agnostic or even anti-realist stance. It aims to overcome the traditional dispute between this kind of anti-realism and realism by developing a middle ground—a kind of participatory realism, which is a realism that is both subject- and world-involving.

2. A Stalemate in the Philosophy of Science

Two major approaches in the philosophy of science, which we will refer to as objectivism and constructivism following Stewart's analysis [4], have been in a long-standing stalemate regarding how to best conceive the subject–object relationship. Worse, when their assumptions are unpacked and pushed to their logical extremes, each approach turns out to have drawbacks that make them less than ideal for scientific worldviews.

Objectivism aims to explain our experience of reality in completely observer-independent terms. However, if the subjectivity of the observer continues to be left out of the natural order—e.g., by being identified with just another objective mechanism, or even eliminated completely—we give up the essentially subjective quality of our own existence too soon. There should also be room in the natural world for human subjects who can make a difference to unfolding events based on their normative evaluations, and who can hence be held accountable for their actions and decisions. This is especially evident if we do not want science to be in tension with practices of responsible decision making, be they in politics, daily life, or even in science itself [1]. Granted, we may never completely understand the basis of our own agency, or fully solve the mind–body problem, but it is also important to note that an acceptance of the possible limits of our scientific understanding is not the source of this tension with lived experience. What we need to avoid is objectivism's overly narrow conception of reality, i.e., a naturalism that excludes subjectivity by definition, and which thereby already rules out in principle the mere possibility of a subject being able to make a difference in their own right [5,6].

Constructivism aims to explain our experience of reality in completely observer-dependent terms [7]. However, if a place for subjectivity in naturalism is bought at the expense of the notion of objectivity, such that it can no longer transcend the domain of our experience, we go too far to the other extreme and undermine science in a different way. It becomes hard to explain why science advances the way it does, why there is a difference between facts and opinions, why there can be broad consensus about scientific knowledge among people with diverse personal and cultural backgrounds, and why scientific knowledge translates into technology that works [8]. Constructivism, especially in its "radical" constructivist formulations, does grant that not everything goes, and that there are observer-independent constraints, yet it remains agnostic about these constraints, nor does it allow that our lived experience can be directly constituted or shaped by the world or by other people [9]. However, if we do not make conceptual space for non-subjective reality to be part of the basis of our experience even in principle, we give up on the hard problem of consciousness prematurely [10]. More disturbingly, we would place an insurmountable gap between the reality our own stream of consciousness—ourselves—and the reality of others and of the world. Again, granted, we may never completely understand the nature and boundaries of consciousness, but this relative ignorance should not impel us to accept an overly narrow conception of subjectivity that ignores the possibility of a world-involving basis of subjective experience; this would be in direct tension with our everyday experience that we can engage with objects (and, more importantly, other people!) that exist beyond our personal experience of them.

In sum, despite the starkly opposing orientations of these two approaches regarding the relationship between subject and world, they share some deeper undesirable premises and implications. Each in their own way fails to do justice to the full complexities of our lived experience of reality—namely, of being conscious subjects embodied in an objective world. In essence, while objectivism rejects the possibility that subjectivity plays a role in the observer-independent world, radical constructivism leaves out the possibility that the observer-independent world plays a role in subjectivity. These approaches thereby promote two distinct one-sided pictures of reality, each with its own theoretical blind spots. Ultimately, their schism prevents them from better grappling with the complex entanglements of natural and human factors that are inherent in reality, and hence makes them unsuitable perspectives for making substantial progress in science.

It is helpful to think of these impoverished visions of reality as two sides of the same old Cartesian dualist coin; they are symptoms arising from the same scientifically unresolved mind–matter problem [11]. *Objectivism* tries to overcome that substance dualism by reducing everything to observer-independent matter, while *constructivism* tries to overcome that substance dualism by reducing everything to the observer-dependent mind; and yet, each side cannot fully reduce the other into itself: objectivism must always access the world from an observer-dependent perspective, while constructivism must always appeal to observer-independent constraints.

Given this shared diagnosis of philosophical ashes of a failed dualism, I propose that we need to step outside of the original dualist binary trap altogether. We need to search for an alternative starting point from which to develop a scientific account that respects the existence of *both* human subjectivity *and* worldly objectivity, whereby they are irreducibly and meaningfully related to one another, jointly participating in shaping reality. In the following I will offer some steps in this direction.

I propose that a fruitful path toward this alternative starting point is to address the unresolved mind–matter dualism indirectly—namely, by way of overcoming its contemporary, metaphysically sanitized heir: cognitivist internalism. Descartes famously thought the linkage between the conscious mind and physical matter was to be found inside the brain, and this localization is still the default assumption for the majority of cognitive scientists today. Both objectivism and constructivism—each in its own way—also assume that the basis of experience is restricted to the inside of the brain/subject. It is this premise of internalism regarding the basis of our experience that motivates objectivism's ideal to recover an external universe that is in itself observer-independent, just as it motivates constructivism's insistence that this is an impossible ideal, given that we could never step outside of the internal to confirm whether it matches the external; but what if the core of the mind–body problem, dating back to the origins of modern science, is the expulsion of the mind from the "external" world into a hidden "internal" realm? What if the basis of mind, and of perceptual experience more specifically, is actually relational and world-involving?

The rest of cognitive science has been developing more inclusive alternatives, commonly known as radical embodied or enactive cognition (e.g., [12–16]). From the perspective of these theoretical advances, the subject's living body is necessary—but not sufficient—for experience, and the environment plays a necessary role as well. This leads to a rejection of brain-centrism in cognitive science, and to the adoption of a more encompassing framework, which treats an agent's behavior as a relational property of the agent's brain–body–environment system as a whole [17]. The rejection of internalism also breaks down the oversimplified separation between observer-dependence and observer-independence, as the mind is instead treated as a regulated form of organism–environment interaction, such that the world itself can make a difference to our experience [18], and our experience can make a difference to the world [5].

3. Three Case Studies of Object Experience

The beginnings of a philosophy of science that seeks to go beyond these classic oppositions can be found in Stewart's contributions to enactive cognitive science (e.g., [6,19]), which centers on the idea that the mind is realized by embodied interaction in the world. Varela had started to develop this new approach, following on from his more constructivist work with Maturana, with a radical guiding vision: "to the extent that we move from an abstract to a fully embodied view of knowledge, facts and values become *inseparable*" ([20], p. 260). In other words, the stated ambition of the enactive approach is to overcome the fact–value gap—which has haunted modernist intellectual life at least since the work of David Hume [21]—while similarly avoiding the collapse of the fact–value distinction altogether, as has occasionally happened in constructivist or postmodern discourse.

However, we can also see in Varela a lingering bias for the internal over the external, as he continues: "To know *is* to evaluate through our living, in a creative circularity" ([20], p. 260). Much will hang on precisely where the boundaries of "our living" are conceived, especially since Maturana and Varela had traditionally identified the boundary of an autopoietic system with its material surface [22]. If so, then brain-centrism would be extended into an organism-centrism, but nonetheless it would stop short of extending into the world and, hence, would remain a form of internalism. Indeed, there has been sufficient ambiguity on this point that some critics have attributed an internalist/subjectivist stance to this line of work [23–25]. More recently, proponents of the enactive approach have started to spell out in what sense this is better understood as a world-involving account, e.g., [3,26–28], and this conceptual clarification has gone hand in hand with a fruitful encounter with the concept of direct realism, as developed by some branches of analytic philosophy and by ecological psychology, e.g., [29,30].

Stewart's philosophy of science defends a stance that "is neither 'internalist,' nor 'externalist,' but rather seeks to go beyond the opposition between them" ([19], p. 18). Indeed, this ambition to go beyond the internalism/externalism dichotomy is one of the key motivations for why the enactive approach adopted phenomenological philosophy as an alternative to radical constructivist philosophy [31,32], and without this dichotomy, nothing stands in the way of the natural world itself participating in the process by which an object shows up in our perceptual experience and, by extension, in scientific observation. In this way, clarifying that perception is world-involving is of broader relevance, as it pertains to how we should understand the scientific process, including our interactions with other people. Indeed, Varela had long emphasized that it is more accurate to explicitly refer to an observer community rather than an observer [33], and as Stewart liked to highlight, cognitive science is inherently reflexive, as we are scientific observers investigating the basis of observation; therefore, a change in paradigm in this field has implications for science as a whole, e.g., [4]. In order to make these reflections more concrete, in the following I will develop this notion of world-involvement as it pertains to our experience of objects in terms of three of Stewart's favorite conceptual "hobby-horses" [34].

3.1. *Latour's Concept of Inversion*

Latour and Woolgar's [35] classic sociological and ethnographic study of modern laboratory life found that the problem faced by scientists is that "at the frontier of science, statements are constantly manifesting a double potential: they are either accounted for in terms of local causes (subjectivity or artefact) or are referred to as a thing 'out there' (objectivity and fact)" (p. 180). Accordingly, they investigated the process of scientific discovery, during which the double potential of a statement becomes stabilized into a fact:

> "Once the statement begins to stabilise, however, an important change takes place. *The statement becomes a split entity.* On the one hand, it is a set of words which represents a statement about an object. On the other hand, it corresponds to an

> object in itself which takes on a life of its own. [...] Consequently, an inversion takes place: the object becomes the reason why the statement was formulated in the first place. [...] Once splitting and inversion have occurred, even the most cynical observers and committed relativists will have difficulty in resisting that the "real" [object] has been found [...]."
>
> ([35], pp. 176–177)

However, Latour and Woolgar's aim is precisely to provide a constructivist position that resists such appeals to reality. A key contribution of their research is to highlight the role of socio-material practices in the scientific creation of facts. They make it evident that facts are the product of a complex process in the lab. However, they emphasize this role of the socio-material practices to the extent that they end up provocatively concluding that factual *statements* match external *objects* because "they are the same thing". Latour and Woolgar realize that this threatens to collapse the objective into the subjective, and they hasten to clarify that they reject a simple relativist position, and do not wish to say "that facts do not exist nor that there is no such thing as reality" (ibid., p. 180). Their main point is, rather, that a scientific object's appearance of "'out-there-ness' is the *consequence* of scientific work rather than its *cause*" (ibid., p. 182). As such, their primary aim is to account for an experiential change during the process of scientific discovery.

It is an important insight that scientific work can change our perception of the world, and that this process is itself amenable to scientific investigation. However, we do not need to fully subscribe to their interpretation. A more reasonable interpretation of the socio-material conditions necessary for the appearance of "out-there-ness", in combination with their concession that facts and reality may exist after all, is that scientific access to objects does not come for free; when science makes contact with aspects of reality outside of our everyday spatiotemporal scales, this is a complex achievement that requires both mastery and the use of the appropriate concepts and equipment. Both the subjective and objective aspects of scientific activity have to come together in just the right ways such that a statement about the world can become recognized as a fact by the observer community.

However, Latour and Woolgar insist on a more extreme interpretation, as they advise that "it is important to eschew arguments about the external reality and outside efficacy of scientific products to account for the stabilization of facts" (p. 183). This kind of bracketing may be a useful ethnographic technique if one's aim is to reveal the socio-material conditions of scientific discovery, but Latour and Woolgar take it too far, and thereby end up with a one-sided account. This prevents them from recognizing that rightfully granting a more prominent role to the "local causes" they identified does not require excluding the world itself from contributing to the establishment of a sense of reality "out there".

In addition, they appeal to the fragility of facts to defend their rejection of such a world-involving account of scientific discovery, but it does not follow from the possibility that statements can be mistaken that facts cannot be world-involving. On the contrary, the fallibility of statements is an essential part of science's capacity for self-correction, which arguably improves its grip on the world and, hence, increases its world-involvement.

Interestingly, Latour and Woolgar's skeptical argument from the fragility of scientific facts has the same form as the classic "argument from illusion" against direct realism in the philosophy of perception. In response to the latter, we can similarly reject its unjustified attempt at a generalization of fallibility: that we sometimes do misperceive does not entail that perception is never world-involving [28]. Conversely, the similarity of these arguments suggests a comparable similarity between observation and perception; scientific observation could then be fruitfully conceived as a socio-material augmentation of the basic perceptual process, rather than only as a socio-material construct.

The development of this conceptual connection could have already occurred in the context of Stewart's work, but unfortunately it remained a latent potential and, hence, is important future work for the enactive approach. Stewart's [36] brief reply to Beaton's [30] world-involving account of the sensorimotor basis of perceptual experience only rehashes

Latour and Woolgar's classic constructivist account of the scientific processes. He thereby misses the opportunity to confront the internalist assumptions of that constructivist account with Beaton's world-involving proposal, which is actually quite similar in spirit to Stewart's [19] own call to go beyond the opposition between internalism and externalism. Stewart and Beaton both reject the postulation of internal constructs as the sole basis of perceived reality, in favor of basing perception directly in distributed organism–environment interaction. Accordingly, we still need to better articulate the consequences of this enactive, world-involving account of perception for our understanding of the scientific process. A step in the right direction is to reconsider the role of technology in perception in a world-involving way.

3.2. *Bach-y-Rita's Concept of Externalization*

It will be useful to further bring out this latent connection between scientific observation and perceptual experience by considering an intermediate case, namely, the way in which tool-use generally tends to shape the user's perceptual experience, as particularly exemplified by sensory substitution devices. Importantly for our comparison with scientific observation, the Compiegne group, of which Stewart was a part, has long argued that such interfaces perform the role of "perceptual supplementation" rather than "sensory substitution" [37], and that this makes them illustrative of a broader class of "mind-enhancing" tools [38], including pen and paper, sketchpads, or calculators [39]. Thus, we have not moved far from the kind of scientific activity analyzed by Latour.

Sensory substitution devices were first studied by Bach-y-Rita et al. [40] using the Tactile Vision Substitution System, which translated an array of black/white pixels obtained from a camera into an array of on/off vibratory actuators placed on their back. They found that a user's mastery of this device gives rise to perceptual experience, whereby the local causes of sensory stimuli are externalized to distal objects:

> "Our subjects spontaneously report the external localization of stimuli, in that sensory information seems to come from in front of the camera, rather than from the vibrotactors on their back. Thus, after sufficient experience, the use of the vision substitution system seems to become an extension of the sensory apparatus."
>
> ([40], p. 964)

This account of the externalization of local stimuli is strikingly similar to Latour and Woolgar's account of scientific observation as involving splitting and inversion. It is suggestive of the possibility that scientific instrumentation may similarly become an extension of our perceptual processes. Indeed, for Stewart [10], the value of sensory substitution devices is precisely that they can be used as a scientific instrument—one that enables us to examine the embodied sensorimotor dynamics involved in perceiving, and to do so under minimalist conditions (see, e.g., Figure 1), thereby allowing us to make an informative comparison with the sensorimotor activity of simpler creatures, such as the tick (more on this below).

For the moment, let us consider in more detail Bach-y-Rita's pioneering research with the Tactile Vision Substitution System, which for the Compiegne group provided two fundamental results that inspired their own research program:

> "(i) If the camera is immobile, placed on a table, the discriminatory capacities of the subjects remain very limited; and the stimuli are perceived on the surface of the skin.
>
> (ii) If the camera is actively manipulated by the subject, the subjects exhibit spectacular capacities to recognize shapes; and the objects are perceived in a distal space, "out there" in front of the subject."
>
> ([31], p. 941)

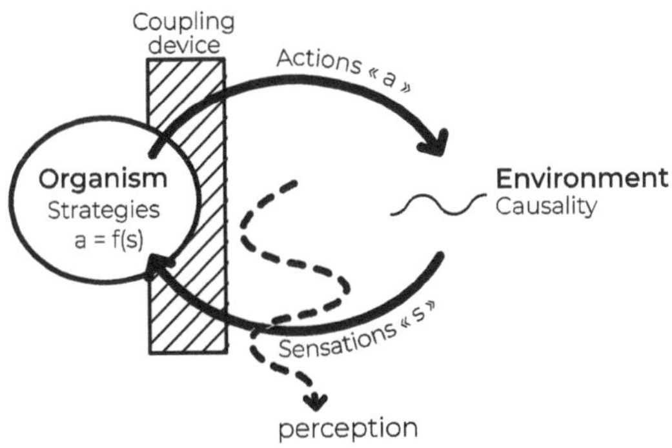

Figure 1. The augmented sensorimotor relation of perception spans the organism, coupling devices, and the environment. The coupling device could be any kind of tool, but the extension of perceptual experience becomes more marked with the skilled use of mind-enhancing tools, including sensory substitution devices and the advanced instrumentation used in scientific observation. Figure redrawn from Lenay and Steiner [31].

This phenomenon of externalization or distal attribution to something "out there" has been systematically studied, and can occur even when observers do not have any prior knowledge of the link between their actions with the device and the resulting variations of stimulations [41]. Accordingly, like Latour and Woolgar's analysis of the discovery of a previously unknown scientific object, we find that in the case of using a sensory substitution interface there are local/subjective aspects that, under suitable conditions, give rise to the experience of an outer/objective entity. Crucially, this distal attribution does not come for free; for a distal object to be perceived as such requires the subject's acquisition and active deployment of user expertise:

> "During the initial phase when the device is first employed, the attention of the user is drawn to the tactile stimuli on the skin. In fact, as long as the stimuli are controlled by the experimenter, the user remains unable to detach his attention from the stimuli. However, if the user himself is able to move the camera, then progressively, after 10–15 h of practice, he comes to perceive objects situated at a distance in front of him. At this point, there is a clear distinction for the subject between the tactile stimuli (which are sometimes a source of irritation) on one hand, and on the other the perception of an object out there in front of him."
>
> ([31], p. 941)

A comparison with Latour's own status as a novice in the lab, and as someone who insisted on remaining a detached observer, is instructive. Latour "was thus in the classic position of the ethnographer sent to a completely foreign environment" ([35], p. 273). As Lynch ([42], p. 503) has argued, Latour's approach will "sever the transivity of technical practices to their real-world objects of study"; hence, we can understand why for Latour it seemed that "a scientist's activity is directed, not toward 'reality,' but toward these operations on statements." ([35], p. 273). In fact, the expert scientists also had this response to his apprenticeship. With respect to his difficulty with understanding reports, "they argued that the observer was baffled because of his obsessive interest in literature had blinded him to the real importance of the papers: only by abandoning his interest in the papers themselves could the observer grasp the 'true meaning' of the 'facts' which the paper contained." (ibid., p. 75). The same consideration of the need for transparency regarding the basis of experience applies to the use of lab equipment: "The material setting both makes possible the phenomenon and is required to be easily forgotten" (ibid., p. 69).

In analogy with the sensory substitution case, if the user is intent on focusing on the local stimulations produced by the interface in a detached manner, and/or puts the control of the interface into the hands of the experimenter, then from the perspective of that naïve user the proximal device will not become transparent, and the distal object will fail to become present in experience. In general, what is required for object perception to occur is that the subject is skillfully engaged in a world-directed interaction. This perceptual process can be supported in various ways, and sensory substitution devices serve as an illustrative example of the role that tools—and our socio-material practices in general—can play in shaping how subject and object relate to one another. Lenay et al. also point to the importance of a user community in constituting a meaningful perceptual experience when using such devices [43], similar to the role of an observer community in science. In this way, we again start to see that the discovery process identified by Latour and Woolgar may be a variation of the externalization process identified by Bach-y-Rita et al. Ultimately, both are forms of socio-materially augmented perception.

3.3. *Von Uexküll's Concept of Projection*

If the preceding analysis is on the right track, then we are close to securing the foundations of an account of science that does justice to both subject and object. Scientific observation is a socio-materially augmented form of object perception, yet for this to be a solid account, more conceptual work still must be done to fully stabilize the foundations, meaning to clarify the world-involving basis of perception.

As we have already discussed, despite the fact that the enactive approach has always emphasized the importance of organism–environment interaction, when it comes to a detailed explanation of the perceptual process, emphasis has typically been placed on the side of the organism [44]. To be fair, the basis of perception is no longer said to be within the brain, but it is often not quite clear how far the basis extends instead. Often it seems as if the basis has only extended to the outer boundary of the body, such that what matters to perception is the internal constitution of the organism—and especially the input–output pattern that plays across its sensorimotor surface. We can see this ambiguity in Stewart's work when he claims that "what the world 'is' *for* the organism amounts to neither more nor less than the consequences of its actions for its sensory inputs; this, in turn, clearly depends on the repertoire of possible actions." ([45], p. 3). The basis of perception certainly includes changes in sensory organs, but if it only includes those changes then it cannot be world-involving in any meaningful sense. It is possible that Stewart inherited this ambiguity regarding the status of the lived world from von Uexküll's [46] classic analysis of the sensorimotor basis of the *Umwelt*. One of his favorite examples was the tick:

> "Here is the story. The female tick climbs to the end of a branch, and ... waits. If she gets a whiff of butyric acid, she lets herself fall; if she does not fall onto a hairy surface, she climbs up again onto a branch and starts over. If she does fall onto a hairy surface, she crawls until she finds a smooth surface. When she does find a smooth surface, she sticks her proboscis into the surface. If she finds underneath a liquid at roughly 37 °C, she sucks up the liquid to satiety."
>
> ([47], p. 57)

Interestingly, for von Uexküll each of these different functional circles involves a process of externalization (in German: *Hinausverlegung*, occasionally also translated as transposition, reassignment, or projection). Here is an example:

> "The skin glands of the mammal are the bearers of perceptual meaning in the first cycle, since the stimulus of butyric acid releases specific receptor signs in the tick's receptor organ, and these receptor signs are projected outside as an olfactory cue."
>
> ([48], p. 324)

Exactly how von Uexküll conceived of this process of projection is ambiguous, and the status of his concept of the *Umwelt* therefore remains an active area of debate [49]. Certainly, interpretations of the *Umwelt* that are compatible with the participatory realism promoted by the enactive approach and ecological psychology are possible [29]. On the other hand, it could also be interpreted in purely internalist, subjectivist terms. As Fultot and Turvey [50] argue, it is likely that von Uexküll considered externalization to consist of a two-step process, whereby local activation of a receptor sign leads to the internal creation of a perceptual cue, which is then attributed to the environment in such a way that the organism can use it as standing for that external stimulus.

The form of this two-step process is similar to Bach-y-Rita et al.'s account of the change in experience associated with mastering a sensory substitution device, and it matches even more closely Latour and Woolgar's account of the change in status of a statement into a fact during the process of scientific discovery, which also involves a two-step process of the "doubling" of a scientific statement into a scientific object, followed by an "inversion" of the causal order such that the object "out there" ends up being given precedence over the local causes. Thus, we arrive again at the conclusion that these accounts derive their argumentative force from their logical consistency with an internalist starting point, rather than from doing justice to the empirical phenomena that they aim to describe.

However, once we reject internalism, an alternative account suggests itself: during these successful perceptual processes, the quality of the agent–environment interaction is transformed; the agent's grip on a specific aspect of the world is improved, and this allows that aspect to be grasped as a distal object. From this point of view, the local activity is still a necessary part of the process, but its role is fundamentally different: it no longer serves as the input for the internal construction of a putative object, but rather becomes part of the coupling through which the object is disclosed to experience. Of course, precisely how the object will make its appearance in experience will depend on how it is approached; experience is always perspectival, and this much of constructivism is retained. However, as Stewart also recognizes, this is not the only relevant dependence: "A point worth making here is that an *Umwelt* is not created by the organism alone, nor by the environment, but through the characteristic *relation* between the organism and its environment." ([47], p. 58). It is this insistence on an organism–environment *relation*, without an internal doubling or other intermediaries, that enables us to conceive of an enactive, world-involving, and world-directed account of perception without double-talk [51].

Noë [18] has long been developing such an enactive account; he argues that a perceptual relation must satisfy the conditions of both movement-dependence and object-dependence—that is, "we are perceptually in touch with an object when our relation to the object is highly sensitive to how things are with the object and to the way what we do changes our relation to the object" (pp. 22–23). In this way, the subjective and objective aspects of perception can be integrated into a unified account of how we find ourselves in the world: perceptual experience not only discloses how things are (objectivity), but also reflects how the perceiver relates to how things are (subjectivity): "When we encounter the world, we do so by encountering how it perceptually appears *from here*. We experience how things are, and we experience how they merely seem to be." (p. 68). For example, we perceive a whole coffee cup even if we always only ever immediately see one of its profiles from our perspective. It is our skillful engagement with the world that enables this contribution of subject- and world-involvement to become integrated into object perception, and it is our awareness of others' possible complementary perspectives that plays a constitutive role in our perception of the object being "out there" in the world [52].

4. From Perception to Scientific Practice

This enactive theory of perception is a productive middle ground for the philosophy of science—a sensorimotor basis of perception that loops through the body and the world makes it reasonable to accept the constructivist insight that object perception is always perspectival, while also accommodating the objectivist's appeal to the world's reality such that the object transcends this perspective. In addition, the basic perceptual relation through which we get a grip on the world can be transformed and empowered in different ways, which is a consideration especially relevant for a theory of scientific practice. It is beyond the scope of this essay to systematically develop this consideration in detail, but there is room for a few pointers for future work.

For the case of human perception, technics plays a particularly important constitutive role, but one that is already prefigured in other forms of life in the inventive creativity at play in organismic adaptation [53]. For our purposes, we are specifically interested in how the scope of the perceptual relation can be augmented with the skilled use of certain kinds of coupling devices, as illustrated in Figure 1.

Despite their many differences, sensory substitution devices and scientific instruments belong to the same broad category of coupling devices that augment our perception of the world. They do so by enabling skilled users to become sensitive to changes in their relation to the world that would otherwise remain inaccessible to them, and when these changes are responded to appropriately, they permit the user to get a grip on a previously inaccessible aspect of the world. This contact is experienced qualitatively as more perceptual when the contact is more direct, but even highly spatiotemporally, socially, and technologically mediated forms of sensorimotor interaction are never completely severed from the world. On the contrary, the kind of detached object perception that is most relevant for scientific observation is arguably a social achievement, which essentially depends on participatory sense-making [54]. In addition, it seems advisable to consider the scientific instruments supporting observation as participating in the coupled agent–environment system, and to treat the resulting observation as a relational property of that system [55]. The mediation afforded by scientific tools has long been a topic of post-phenomenology [56], but it remains to be seen how the enactive approach can address these specific forms of mediation on its own terms. More generally, it is crucial to explicitly include this irreducibly relational perspective in the interpretation of observations; the enactive approach to the philosophy of science is a situated approach [57].

Another important outstanding topic is how to include linguistic thought processes into such a world-involving account of the scientific process. One attractive possibility is to scale up basic perception–action loops by appealing to the right kind of socio-material affordances, such that they become linguistically enabled [58]. Noë similarly argues that some forms of thinking, at their core, retain a perceptual relation to the world: "Thought can be extended perception when one deploys sensorimotor skills (presumably in conjunction with other sorts of skills and knowledge) to achieve access to something or someone very remote." (2012, p. 27). An intriguing implication of conceptualizing thought in this manner is that, for Noë, such world-involvement places strict constraints on the form that the thought process can take. For example, he observes that "in the way I am now thinking of [something absent], it would not be possible to think of something nonactual." (p. 27). If so, then we could expect that scientists are forced to change their way of thinking once a statement changes its status from positing a possibly nonactual object to connecting with an actual object, and this is indeed what Latour and Woolgar observed to occur on such occasions, when "even the most cynical observers and committed relativists will have difficulty in resisting that the 'real' [object] has been found" (p. 177). This possibility of world-involving thought as another form of augmented perception is an exciting topic for future research.

5. Conclusions and Outlook

In the end, human experience of the world is a subject-, other-, and world-involving achievement, and this co-dependent basis of experience means that its limits are not within our brain–body and its supposed internal constructs—we genuinely relate to the worldly objects of our perception. We are at home in the world, and science is a sophisticated socio-material elaboration of this more basic situatedness.

Accordingly, no one-dimensional account of scientific facts will do; the scientific process cannot be reduced to the subjective perspectives and social conditions that enable it, nor can we take ourselves completely out of this process in the hope of elevating the targets of scientific investigation to observer-independent objects, as if they were uncovered by a "view from nowhere". This enactive approach to the philosophy of science is more humane because it does not undermine the lifeworld that makes science possible in the first place, and it does not stop there, given the reflexivity of the sciences of the mind. If this world-involving theory of observation is on the right track, then it also promises to lead to better science in practice. We need to explicitly incorporate the fact that our experience of the world—including during scientific observation—is always perspectival:

> "When we try to understand reality by focusing only on physical things outside of us, we lose sight of the experiences they point back to. The deepest puzzles can't be solved in purely physical terms, because they all involve the unavoidable presence of experience in the equation." [59]

Importantly, an anti-realist interpretation of this irreducible presence of experience is blocked by the complementary insight that experience is also world-involving. The next step is to put this enactive approach to observation to the test and to turn it into a workable research program, including stepping outside of the traditional disciplines of cognitive science altogether—it is time for an enactive approach to physics.

For example, following phenomenological philosophy, Noë highlights that a defining characteristic of the reality of the world is that it is inexhaustible in principle from any and all perspectives, no matter the scale: "In the large, in the small, no experienced quality is so simple that it can be taken in all at once. The world is structured and complex and it always outstrips what can be taken in a glance" ([18], p. 95). The fact that the objects of experience transcend our perspective in this way is most parsimoniously explained by appealing to their dual status as being a part of the world while participating in the basis of experience. This transcendence of the real is characteristic of our everyday perceptual experience, but it similarly applies to scientific observation.

Consequently, it becomes intelligible why physics runs up against irreducible unobservables and indeterminacy when it socio-materially augments our perceptual relation with the world to its smallest and largest scales. At these scales, the limitations on experience imposed by the transcendence of the real can no longer be ignored, such as at the scale of cosmology when we try to grasp the entire universe as a whole [60], or at the quantum scale when we try to grasp the properties of an elementary particle all at once [61]. The discovery that the world in principle escapes our grasp in these extreme situations of scientific observation ceases to be mysterious. Instead, such observational limitations are an expected consequence of the world-involving basis of perception; hence, they come to be seen as reassuring evidence in support of a participatory realism [62]: we are real, so is the world, and so is our interaction.

Funding: This research received no external funding.

Institutional Review Board Statement: Not applicable.

Informed Consent Statement: Not applicable.

Data Availability Statement: Not applicable.

Acknowledgments: Maria Gohlke drew the illustration of Figure 1. Pierre Steiner provided detailed constructive feedback on an earlier draft of this manuscript. Members of the Embodied Cognitive Science Unit helped to fine-tune the final version.

Conflicts of Interest: The authors declare no conflict of interest.

References

1. Stewart, J. Science is not value-free. *Constr. Found.* **2014**, *10*, 28–29.
2. Froese, T. Epilogue to "Questioning Life and Cognition" by John Stewart. *Adapt. Behav.* **2021**, *29*, 523–527. [CrossRef]
3. Fuchs, T. *Ecology of the Brain: The Phenomenology and Biology of the Embodied Mind*; Oxford University Press: Oxford, UK, 2018.
4. Stewart, J. Radical constructivism in biology and cognitive science. *Found. Sci.* **2001**, *6*, 99–124. [CrossRef]
5. Froese, T.; Taguchi, S. The problem of meaning in AI and robotics: Still with us after all these years. *Philosophies* **2019**, *4*, 14. [CrossRef]
6. Stewart, J. Neurophenomenology, enaction, and autopoïesis. In *Behavioral Neuroscience*; Palermo, S., Morese, R., Eds.; IntechOpen: London, UK, 2019; pp. 1–8.
7. Von Glasersfeld, E. The radical constructivist view of science. *Found. Sci.* **2001**, *6*, 31–43. [CrossRef]

8. Stewart, J. Life as a process of bringing forth a world. *Constr. Found.* **2011**, *7*, 21–22.
9. Di Paolo, E.A. A mind of many. *Constr. Found.* **2008**, *3*, 89–91.
10. Stewart, J. What is it like to be conscious? Towards solving the hard problem. *Constr. Found.* **2017**, *12*, 155–156.
11. Dreyfus, H.; Taylor, C. *Retrieving Realism*; Harvard University Press: Cambridge, MA, USA, 2015.
12. Chemero, A. *Radical Embodied Cognitive Science*; The MIT Press: Cambridge, MA, USA, 2009.
13. Hutto, D.D.; Myin, E. *Radicalizing Enactivism: Basic Minds without Content*; The MIT Press: Cambridge, MA, USA, 2013.
14. Di Paolo, E.A.; Buhrmann, T.; Barandiaran, X. *Sensorimotor Life: An Enactive Proposal*; Oxford University Press: Oxford, UK, 2017.
15. Noë, A. *Out of Our Heads: Why You Are Not Your Brain, and Other Lessons from the Biology of Consciousness*; Hill and Wang: New York, NY, USA, 2009.
16. Kirchhoff, M.; Kiverstein, J. *Extended Consciousness and Predictive Processing: A Third Wave View*; Routledge: Abingdon, UK, 2019.
17. Beer, R.D. Dynamical approaches to cognitive science. *Trends Cogn. Sci.* **2000**, *4*, 91–99. [CrossRef]
18. Noë, A. *Varieties of Presence*; Harvard University Press: Cambridge, MA, USA, 2012.
19. Stewart, J. The mind is not in the brain. *Constr. Found.* **2008**, *4*, 17–18.
20. Varela, F.J. Whence perceptual meaning? A cartography of current ideas. In *Understanding Origins*; Varela, F.J., Dupuy, J.-P., Eds.; Springer: Dordrecht, The Netherlands, 1992; pp. 235–263.
21. Froese, T. Hume and the enactive approach to mind. *Phenomenol. Cogn. Sci.* **2009**, *8*, 95–133. [CrossRef]
22. Villalobos, M.; Razeto-Barry, P. Are living beings extended autopoietic systems? An embodied reply. *Adapt. Behav.* **2019**, *28*, 3–13. [CrossRef]
23. Wheeler, M. Mind in life or life in mind? Making sense of deep continuity. *J. Conscious. Stud.* **2011**, *18*, 148–168.
24. Pascal, F.; O'Regan, K. Commentary on Mossio and Taraborelli: Is the enactive approach really sensorimotor? *Conscious. Cogn.* **2008**, *17*, 1341–1342. [CrossRef] [PubMed]
25. Seth, A.K. A predictive processing theory of sensorimotor contingencies: Explaining the puzzle of perceptual presence and its absence in synaesthesia. *Cogn. Neurosci.* **2014**, *5*, 97–118. [CrossRef] [PubMed]
26. Di Paolo, E.A. The worldly constituents of perceptual presence. *Front. Psychol.* **2014**, *5*, 450. [CrossRef]
27. Ramirez-Vizcaya, S. Behavioral autonomy and a meshwork of habits: Review of *Sensorimotor Life: An Enactive Proposal* by Ezequiel Di Paolo, Thomas Buhrmann, and Xabier Barandiaran. *Adapt. Behav.* **2022**, *30*, 205–208.
28. Beaton, M. Phenomenology and embodied action. *Constr. Found.* **2013**, *8*, 298–313.
29. Baggs, E.; Chemero, A. Radical embodiment in two directions. *Synthese* **2021**, *198*, 2175–2190. [CrossRef]
30. Beaton, M. Sensorimotor direct realism: How we enact our world. *Constr. Found.* **2016**, *11*, 265–297.
31. Lenay, C.; Steiner, P. Beyond the internalism/externalism debate: The constitution of the space of perception. *Conscious. Cogn.* **2010**, *19*, 938–952. [CrossRef] [PubMed]
32. Froese, T.; Stewart, J. Enactive cognitive science and biology of cognition: A response to Humberto Maturana. *Cybern. Hum. Knowing* **2012**, *19*, 61–74.
33. Froese, T. From second-order cybernetics to enactive cognitive science: Varela's turn from epistemology to phenomenology. *Syst. Res. Behav. Sci.* **2011**, *28*, 631–645. [CrossRef]
34. Di Paolo, E.A. Bridges and hobby-horses: John Stewart's adventure of ideas. *Adapt. Behav.* **2021**, *29*, 437–440. [CrossRef]
35. Latour, B.; Woolgar, S. *Laboratory Life: The Construction of Scientific Facts*; Princeton University Press: Princeton, NJ, USA, 1986.
36. Stewart, J. Realities in the plural. *Constr. Found.* **2016**, *11*, 277–278.
37. Lenay, C.; Tixier, M. From sensory substitution to perceptual supplementation. In *Living Machines: A Handbook of Research in Biomimetics and Biohybrid Systems*; Prescott, T.J., Lepora, N., Verschure, P.F.M.J., Eds.; Oxford University Press: Oxford, UK, 2018; pp. 552–559.
38. Clark, A. *Natural-Born Cyborgs: Minds, Technologies, and the Future of Human Intelligence*; Oxford University Press: New York, NY, USA, 2003.
39. Auvray, M.; Myin, E. Perception with compensatory devices: From sensory substitution to sensorimotor extension. *Cogn. Sci.* **2009**, *33*, 1036–1058. [CrossRef]
40. Bach-y-Rita, P.; Collins, C.C.; Saunders, F.A.; White, B.; Scadden, L. Vision substitution by tactile image projection. *Nature* **1969**, *221*, 963–964. [CrossRef]
41. Auvray, M.; Hanneton, S.; Lenay, C.; O'Regan, J.K. There is something out there: Distal attribution in sensory substitution, twenty years later. *J. Integr. Neurosci.* **2005**, *4*, 505–521. [CrossRef]
42. Lynch, M.E. Technical work and critical inquiry: Investigations in a scientific laboratory. *Soc. Stud. Sci.* **1982**, *12*, 499–533. [CrossRef]
43. Lenay, C.; Gapenne, O.; Hanneton, S.; Marque, C.; Genouëlle, C. Sensory substitution: Limits and perspectives. In *Touching for Knowing: Cognitive Psychology of Haptic Manual Perception*; Hatwell, Y., Streri, A., Gentaz, E., Eds.; John Benjamins: Amsterdam, The Netherlands, 2003; pp. 275–292.
44. Heft, H. Ecological psychology and enaction theory: Divergent groundings. *Front. Psychol.* **2020**, *11*, 991. [CrossRef]
45. Stewart, J. Foundational issues in enaction as a paradigm for cognitive science: From the origin of life to consciousness and writing. In *Enaction: Toward a New Paradigm for Cognitive Science*; Stewart, J., Gapenne, O., Di Paolo, E.A., Eds.; The MIT Press: Cambridge, MA, USA, 2010; pp. 1–31.
46. Von Uexküll, J. *Umwelt und Innenwelt der Tiere*; Julius Springer: Berlin, Germany, 1909.

47. Stewart, J. *Breathing Life into Biology;* Cambridge Scholars Publishing: Newcastle upon Tyne, UK, 2019.
48. Von Uexküll, J. A stroll through the worlds of animals and men: A picture book of invisible worlds. *Semiotica* **1992**, *89*, 319–391. [CrossRef]
49. Feiten, T.E. Mind after Uexküll: A foray into the worlds of ecological psychologists and enactivists. *Front. Psychol.* **2020**, *11*, 480. [CrossRef] [PubMed]
50. Fultot, M.; Turvey, M.T. von Uexküll's theory of meaning and Gibson's organism-environment reciprocity. *Ecol. Psychol.* **2019**, *31*, 289–315. [CrossRef]
51. Steiner, P. Steering a middle course between intentionality and representation: Some remarks about John Stewart's enactive stance. *Adapt. Behav.* **2021**, *29*, 471–483. [CrossRef]
52. Gallagher, S. Intersubjectivity in perception. *Cont. Philos. Rev.* **2008**, *41*, 163–178. [CrossRef]
53. Dereclenne, E. Simondon and enaction: The articulation of life, subjectivity, and technics. *Adapt. Behav.* **2021**, *29*, 449–458. [CrossRef]
54. Di Paolo, E.A. Participatory object perception. *J. Conscious. Stud.* **2016**, *23*, 228–258.
55. Barad, K. *Meeting the Universe Halfway: Quantum Physics and the Entanglement of Matter and Meaning;* Duke University Press: Durham, UK, 2007.
56. Ihde, D. Husserl's Galileo Needed a Telescope! *Philos. Technol.* **2011**, *24*, 69–82. [CrossRef]
57. Bitbol, M. Science as if situation mattered. *Phenomenol. Cogn. Sci.* **2002**, *1*, 181–224. [CrossRef]
58. Kiverstein, J.; Rietveld, E. Scaling-up skilled intentionality to linguistic thought. *Synthese* **2021**, *198*, 175–194. [CrossRef]
59. Frank, A.; Gleiser, M.; Thompson, E. The Blind Spot. In *Aeon;* 2019. Available online: https://aeon.co/essays/the-blind-spot-of-science-is-the-neglect-of-lived-experience (accessed on 27 March 2022).
60. Unger, R.M.; Smolin, L. *The Singular Universe and the Reality of Time: A Proposal in Natural Philosophy;* Cambridge University Press: Cambridge, UK, 2015.
61. Smolin, L. *Einstein's Unfinished Revolution: The Search for What Lies Beyond the Quantum;* Penguin Press: New York, NY, USA, 2019.
62. Fuchs, C.A. On participatory realism. In *Information and Interaction;* Durham, I.T., Rickles, D., Eds.; Springer: Cham, Switzerland, 2017; pp. 113–134.

philosophies

MDPI

Article

Multidisciplinarity, Interdisciplinarity, and Transdisciplinarity: The Tower of Babel in the Age of Two Cultures

Marcin J. Schroeder

Global Learning Center, Institute for Excellence in Higher Education, Tohoku University, 41 Kawauchi, Sendai 980-8576, Japan; mjs@gl.aiu.ac.jp

Abstract: Despite the continuous emphasis on globalization, we witness increasing divisions and divisiveness in all domains of human activities. One of the reasons, if not the main one, is the intellectual fragmentation of humanity, compared in the title to the failed attempt at building the Biblical Tower of Babel. The attempts to reintegrate worldview, fragmented by the specialization of education (C.P. Snow's *The Two Cultures*) and expected to be achieved through reforms in curricula at all levels of education, were based on the assumption that the design of a curriculum should focus on the wide distribution of subjects of study, as if the distribution was the goal. The key point is not the distribution of themes, but the development of skills in the integration of knowledge. The quantitative assessment of the width of knowledge by the number of disciplines is of secondary importance. We cannot expect the miracle that students without any intellectual tools developed for this purpose would perform the job of integration, which their teachers do not promote or demonstrate, and which they cannot achieve for themselves. There are many other reasons for the increasing interest in making inquiries interdisciplinary, but there is little progress in the methodology of the integration of knowledge. This paper is a study of the transition from multidisciplinarity to interdisciplinarity, and further, to transdisciplinarity, with some suggestions regarding the use of methodological tools of structuralism and the choice of a conceptual framework.

Keywords: integration of knowledge; multidisciplinarity; interdisciplinarity; transdisciplinarity; education; structuralism

check for updates

Citation: Schroeder, M.J. Multidisciplinarity, Interdisciplinarity, and Transdisciplinarity: The Tower of Babel in the Age of Two Cultures. *Philosophies* **2022**, *7*, 26. https://doi.org/10.3390/philosophies7020026

Academic Editor: Christopher Winch

Received: 13 February 2022
Accepted: 28 February 2022
Published: 7 March 2022

1. Introduction

1.1. Motivations for the Paper

The subtitle of this paper refers to two memes: The Biblical Tower of Babel, and *The Two Cultures* by C.P. Snow. Both are expressions of concern about damage brought by the intellectual fragmentation of humanity. This concern is not new. The first of the memes belongs to ancient Biblical tradition, and the second was born of reflection on the Second World War and the inhumane aspects of modern technology. The revival of the interest in problems they represent comes with the transformation of the world, generated by information technology—in particular, the rise of Artificial Intelligence (AI) and automation. AI and automation do not have any direct impact on the intellectual fragmentation of humanity, but they amplify the problems created by this fragmentation.

There is one obvious reason why we should worry about the divisions of human collectives large and small. We all face an global ecological crisis of multiple dimensions. The most dramatic is the change in climate, but this is only one facet of the complex problem of the interaction between humanity and its ecosphere. Others, such as the depletion of resources, overpopulation, pollution, destruction of natural habitats, suppression of biodiversity, etc., are less obvious; however, they are equally dangerous threats to the future of humanity. It is not clear whether we can prevent the most serious consequences of these challenges; however, it is very clear that without global cooperation, there is no chance of success. This global cooperation requires some forms of political and economic unity;

however, it also requires a united vision of reality, and common values and norms, forming an inclusive global culture of humanity overarching local cultures associated with diverse ethnic, religious, regional identities.

The task of achieving even a rudimentary level of the global unity necessary for the coordinated effort to solve the problems of the global ecological crisis is very difficult. There are too many local, short-term interests involved and too many centrifugal forces opposing unification. However, there is no alternative to the continuing effort of propagating the smaller-scale forms of unification. One of them is education. The role of education is not as obvious and simple as might be expected. The traditional role of educational systems is primarily the perpetuation of existing culture and the promotion of the political interests of those who control these systems. Fortunately, education has become interdependent with a more general endeavor of scholarship, engaging people who want to learn and to share what they have learned. The young audience of education has always been more than eager to challenge the inherited products of the culture of their predecessors and to pursue their own goals. This has created an unusual role for higher education, not only as a mode of transmission from the past to the future, but also from the future to the present.

1.2. Objectives of the Paper

One of the goals of this paper is a broad reflection on the conditions necessary for the future of education, education for the future, and above all, the direction of its evolution in the context of the need to build a common stage for a coordinated effort to resolve the problems faced by humanity. This common stage is a unified view of reality for which we can use The Tower of Babel as a symbol, and which is the main goal of this paper. No discussion of the future can be fruitful without reference to the present. To think about the ways of unifying human intellect, we have to consider the reasons for the centrifugal forces that push in the direction of fragmentation, symbolized here by *The Two Cultures*. Thus, instead of speculation about the future, this paper is an attempt to assess the past and present status of intellectual centrifugal and centripetal forces, which can inform us in attempts to design education that supports the goal of achieving unity.

The centripetal forces, not as new as usually reported, can be aligned with the three stages of inquiries transcending the borders of disciplinary divisions; these are identified in the literature as multidisciplinary, interdisciplinary, and transdisciplinary studies. An attempt will be made to clarify their meanings and each of their roles. Finally, some suggestions will be presented regarding the methodology of the highest of the three forms of inquiry, in particular regarding the intellectual experience of structuralism. The purpose of invoking structuralism is to demonstrate the feasibility of a transdisciplinary methodology bridging the Two (or more than Two) Cultures.

2. The Two Cultures or Many More?

2.1. The Two Cultures and Other Divisions

The story of the Tower of Babel, explaining the diversity of human languages as a means to disrupt by confusion and mutual misunderstanding of the blasphemous attempt to reach Heaven, can be applied to every domain of current human endeavors. Differences in religious beliefs stimulate violence and hatred, contradicting the precepts of religious conduct. Claims of unfair international treatises stimulate withdrawals from cooperation and reversal of the rule of military or economic power, completely ignoring the interests of weaker societies. Democratic forms of governance are blamed without any rational explanation for the allegedly dysfunctional economic or social organizations of society; this is in order to justify the exclusion of minorities and other vulnerable communities from access to resources such as education, work, health care, and income, to solidify political power gained through the support of those who are privileged or who are promised privileges. Accusations of lying, directed at opponents, are used as justification for the inconsistency of one's own statements. It seems that everything that made humanity closer to Heaven, or that was intended as a path in its direction, is now disrupted by the fragmentation of interests, mistrust, and complete lack of mutual understanding.

Of course, many people and organizations oppose these frightening tendencies, and the tendencies themselves are not new. The principle of divide et impera (divide and conquer) goes back at least to Julius Caesar, if not to Philip II of Macedon; since those times, it has been the main political tool for European (and not only European) sovereigns to maintain their monopoly of power. Others have not needed to learn this doctrine to use it effectively. History has a long record of the calamities caused by allowing enemies to destroy alliances, but this does not make the principle less effective in its destructive mission.

In the story of the Tower of Babel from Genesis, the division was intentional to prevent the insubordination of the human race. Politicians use this division with the clear intention to gain or to maintain power. This is why the second meme of *The Two Cultures* is more appropriate when we consider another form of fragmentation among contemporary intellectual elites. In this case, the process is not driven by someone's intention, although it is sometimes used for political purposes. This meme was born from the title of the article under this title, written by Charles Percy Snow in 1956 for *The New Statesman*. It was later developed into his 1959 Rede lecture at the University of Cambridge, and finally, into his famous book. These were all on the same subject of the division and growing distance between The Two Cultures: that of the humanities and that of science-technology [1,2]. The complicated background of educational and cultural politics in post-war Britain that motivated Snow, and that made his publications so prominent, is irrelevant here [3]. After all, Snow's critique of the increasing separation of The Two Cultures was not very innovative when we look at the developments in the world.

2.2. Elusive Goals of General Education

The famous 1945 *Red Book of Harvard* (*General Education in a Free Society: Report of the Harvard Committee)* blamed the atrocities of the Second World War on the deficiencies of education; in the pursuit of the depth of knowledge, it pushed for a higher level of specialization implemented by the cost of neglecting alternative domains of study [4]. The faculty of Harvard University proposed General Education as a required component of the lower level curriculum, in which students study topics complementary to their future major concentration to broaden their intellectual perspective. By now, General Education (GE) is a fixed element of American college education in virtually every undergraduate level institution of higher education. The ideals of education promoted a century earlier by Wilhelm von Humboldt were already deeply ingrained in American higher education, even before the initiative of the Harvard Committee. However, this common but purely formal GE requirement of completing a couple of credits in the introductory science courses or the humanities is only a faint shadow of the original idea. This is especially true of the innovative content of *The Red Book*, which planned specially designed courses, introducing students to alternative ways of inquiry and inviting them to reflect on the similarities and differences in ways of thinking. An example of this type of education can be found in the 1957 book *The Copernican Revolution: Planetary Astronomy in the Development of Western Thought*, containing lecture notes from the course taught at Harvard in the late 1940s by the young Thomas Kuhn [5].

On both sides of the Atlantic Ocean, a cure was seen in the distribution of required subjects at some stage of education, especially in the education of future elites. Within the next fifty years, General Education in several countries in the world went through periods of enthusiastic support and periods of negligence and suppression. The latter were associated with calls to educate highly qualified specialists necessary for the high-tech industry, management of production, or health care. After all, the time spent on studying literature or art could be used to lift the level of professional education of the constantly increasing level of specialization, which required additional time in the curriculum.

The pressure on students in the humanities or social sciences was also increasing, but the bigger problem here was the ineffectiveness of the attempts to increase the level of scientific literacy. The main reason for this failure was, and still is, the necessity to involve a higher level of abstraction and mathematical skills; the development of this at the college level in students who do not major in science requires too much time to be acceptable for typical curriculum development.

2.3. *Error of Waiting for the Miricle*

The actual source of the problems in achieving the ideal of comprehensive and unified education was always overlooked. All attempts were based on the assumption that the design of curricula should focus on the wide distribution of subjects of study, as if the distribution was the goal. The key point is the development of skills in the integration of knowledge. The quantitative assessment of the width of knowledge measured by the number of disciplines covered in the curriculum is of secondary importance. We cannot expect the miracle that students without any intellectual tools would perform the job of integration, which their teachers do not promote and which they most likely cannot achieve for themselves.

This is the reason we have to start from the identification and development of intellectual tools for the unification of knowledge. Only after these tools are available can we think about a design for education in which students acquire the ability to integrate the pieces of the curriculum knowledge, competencies, and abilities into their individual intellectual armor, making them autonomous.

The unresolved issue of the increasing intellectual fragmentation of humanity received an additional dimension with the entirely new sources of social, cultural, economic, and technological transformation, brought by computer technology and based on the communication technology of the Internet. These new, revolutionary technologies changed the meaning of the idea of cultural fragmentation, making it much more complex. The Internet accelerated the processes of globalization, i.e., to a high degree, it eliminated and abolished the established manners of spatial divisions, such as political borders or geographic regionalization. However, the idea that the world became more uniform or more connected is a gross oversimplification. The divisions did not disappear, but they changed.

2.4. *Borgesian Library of the Internet*

At first sight, the Internet revolutionized access to information, and therefore, created unprecedented learning opportunities. Today, the majority of people in industrialized societies have knowledge at hand accumulated through centuries, including the most recent results of research. Browsing machines make it easy to access immense intellectual resources in the search for the answers to all possible questions. Does it bring closer the dream of informed and educated citizenship?

Liberal arts education, which shaped the ideal of learning in Western civilization, had, as its core, the development of intellectual autonomy; this was understood as the ability to make informed, rational judgments and decisions, independently of the opinions or authority of others. When someone has direct access to such extensive resources, this ideal seems easy to achieve. However, this is only an illusion of accessibility. The external barrier to the record of knowledge was eliminated. Everyone can enter the Internet, which is a slightly simplified version of Jorge Luis Borges' repository of entire possible knowledge, described in his short novel *The Library of Babel* [6]. The tower of the library, with its innumerable stories, had the collection of all possible 410-page books. The only task in the search for the answer to any question was to find the book that provides it, which of course, was impossible. This does not mean that the task is much easier on the Internet, with its overflow of information and promotion of sources willing to pay for the display on the top of the list provided by browsers.

The Internet created a highly democratized "market" of free expression. Everyone can post their works and make them accessible to a very wide audience. This cyberspace Hyde Park Corner has many advantages. For instance, the perspective to become a contributor to collective knowledge stimulates some people to conduct research in their leisure time and to share its results. This comes with the price of making it very difficult for someone without extensive experience to distinguish between rigorous, well-organized, and well-documented research, and the products of ignorant self-confidence. To be sure, this is not a new phenomenon that has come with Internet blogs and influencers. Alexander Pope in his 1709 *Essay on Criticism* warned us that "A little learning is a dangerous thing" and "Fools rush in where angels fear to tread" [7]. It is questionable that the academic affiliation of the author gives more credibility to the content of blogs. Very often, extensive blogging is a substitute activity for the members of the academic world who are frustrated by their inability to pass through (not necessarily fair) peer reviews in research journals. Even those who pass through this obstacle sometimes easily trespass into domains where they are not qualified to claim authority.

There is no simple recipe for the selection of research resources. This is one of the tools acquired through academic apprenticeship in the work of developing knowledge. Authentic knowledge requires some form of structural unity which cannot be found in or on the Internet. This does not mean that Internet resources are useless, but that they become useful only for someone with sufficiently structured intellectual capacities. Information technology changed the range of skills necessary for being an autonomous thinker. Contemporary education can eliminate the exercises that develop skills in memorization, and it can focus on skills for the integration of information. Now, the question is whether the only problem in designing educational programs is in overcoming the separation of Snow's Two Cultures. Do we have the two cohesive domains of intellectual activities with organized systems of values, norms, and procedures, as was suggested by Snow? Or is the division into the Two Cultures a tip-of-the-iceberg symptom of deeper and more fundamental fragmentation. When we use the term "science", does it have a delineated meaning of commonly agreed identity? Or, rather, maybe it is just a convenient short description of a variety of human activities with some similarities and some differences of equal importance.

3. One, Two, or More Sciences?

3.1. The Lost Unity of Natural Philosophy

The idea of the unification of knowledge accompanied European intellectual tradition from its beginnings in the Mediterranean Antiquity, in the form of attempts to compile all knowledge and organize it in some way. An example could be the work of Aristotle, Simplicius of Cilicia, and Albert the Great who was called, in his time, *Doctor Universalis* for his extensive knowledge. The individual efforts of distinguished philosophers—whose knowledge spanned the entire scientific knowledge of their generations—had to be replaced by the collective work initiated and coordinated by the few enthusiasts (for example, Denis Diderot and Jean le Rond d'Alembert who edited *Encyclopédie*, which had many contributors).

The task itself of the collective work on scientific programs, originally promoted by Francis Bacon, was more oriented towards the effective sharing of the effort of creating knowledge within separate domains of individual interests than an organization of diverse disciplines with separate conceptual frameworks into structured units. The task of unification, even in the time of Immanuel Kant, belonged to the individual person equipped with the exceptional ability to span various directions of inquiry. Only in the 19th century has the specialization of scientific work become reflected not only in the choice of the direction of inquiries, but also in their consolidation into the separate independent scientific disciplines. rather than the general idea of Natural Philosophy. The separation was stimulated by the organizational changes in European universities, initiated by Wilhelm von Humboldt, who propagated the idea of education through engagement in research. Universities gradually transformed from institutions where students learned about science to those where they participated in its creation or at least observed its creation.

The division of Natural Philosophy into scientific disciplines, although not invented by Auguste Comte, was propagated by him in his writings from the 1830s and collected into *The Course in Positive Philosophy*; this occurred together with its organization into a linearly ordered ladder hierarchy, with the first rung of mathematics followed by astronomy, physics, chemistry, biology, up to *physique sociale* (sociology). They were arranged according to the perceived levels of exactness (positivity) reflecting the reversed order of their complexity.

Comte's classification of the disciplines of science was epistemic rather than ontological (these present standard philosophical terms, are used here in an anachronic way, as they were introduced later in 1854 by James Fredrick Ferrier in his *Institutes of Metaphysic: The theory of knowing and being*) [8]. The order of the ladder of disciplines was based on the qualification of knowledge corresponding to the complexity of phenomena, but without the claim that phenomena at one level can be reduced to the other level. He did not claim that astronomy or physics can be reduced to mathematics, but that mathematical methods are tools for these disciplines. The reductionist view of reality that emerged in the 19th century required an additional ontological step, which came with the attempt to justify synthesis. Its unity was based on the assumption that phenomena of a higher level can be reduced to phenomena of the lower levels.

The emergence of separate disciplines of science and specialization within them, from the very beginning, raised concerns about the loss of the meaning of knowledge fragmented by specialization. These concerns were already explicitly expressed by William Whewell in the March 1834 issue of the *Quarterly Review* within an enthusiastic commentary on *On the Connextion of the Physical Sciences* published by Margaret Somerville. It was this commentary where he introduced, in written form, the term "scientist"; he sometimes used this term in discussions as a rather satirical name for those engaged in the pursuit of truncated specialized knowledge, long before it started to be used with a more respectful meaning [9].

The sense of the lost unity of Natural Philosophy generated several different attitudes to the issue. This is too extensive a subject to elaborate on here. Comte's hierarchic structure of disciplines initiated diverse forms of reductionism, including its ontological forms. Its extreme form, which acquired its name only in the 20th century, was physicalism. It can be identified by the use of the terminology of physics to address philosophical ideas outside of the discourse on phenomena directly studied in the theories of physics (e.g., energy, force, matter), and even more clearly in the use of the adjective "physical" as an ontological qualification of existence (e.g., "physical space"). This ontological reductionism and its extreme version of physicalism can be understood as one of the possible ways in which to re-establish the unity of the scientific vision of reality.

It is one of the ironies in intellectual history that after a century of attempts to eliminate physicalism from the philosophical vision of reality, according to a recent survey, it is the dominating position among philosophers [10]. Surprisingly, these are physicists who, with full confidence, object physicalistic reduction, or at least its traditional forms. This critique of physicalism is not always completely free from reduction, but it is free from reductionism in its classical understanding, as presented by Nagel [11]. The best example would be the view presented by Phillip W. Anderson in his excellent explanation of the non-reductive forms of complexity, in the short but very influential 1972 article *More is Different* [12]. By now, the emergent character of the higher forms of complexity, not only in the domain of phenomena studied by physics, is commonly acknowledged by physicists. However, this does not mean that overcoming physicalism resolves the issue of the fragmentation of the scientific vision of reality.

Yet another irony of the intellectual history is in the role of the theory of biological evolution of species, published by Charles Darwin in his 1859 *On the Origin of Species*. It stimulated interpretations aligned with reductionism for those who believed that life can be explained in terms of physicochemical processes, by removing the barrier separating humanity, identified with the mental aspects of reality, from other forms of life identified with the material aspects. On the other hand, the theory of evolution stimulated interest in the unity of biological systems with its symbolic concept of an organism that was absent in the subject of studies in physical sciences. Positivistic thought moved from Comte's "physics of society" to "society as an organism".

The extraordinary popularity of the views of Herbert Spencer, in particular, his Synthetic Philosophy, is an example of confusion in the intellectual circles of the 19th century. At first glance, the Synthetic approach, as presented in his *First Principles* (1862), seems an attempt to restore the unity of the scientific vision of the entirety of reality governed by a few universal laws.

Spencer proposed three principles: the Law of the Persistence of Force, the Law of the Instability of the Homogeneous, and the Law of the Multiplicity of Effects [13]. The universe is a subject of permanent action of non-uniform forces driven by energy. An originally homogeneous universe is in a constant transition to an increasingly heterogeneous state, because force acts in an unstable and variable way in different parts of the universe. The third law, the Law of the Multiplicity of Effects, states that the heterogeneity of the universe grows exponentially, accelerating the transition from homogeneity up to some state of equilibrium, after which it will return to homogeneity.

We know, in hindsight, that Spencer's Principles were doomed by the arrival, around the same time, of the Second Law of Thermodynamics; this established the equally universal but opposite direction of the transformation of all closed systems, including the universe. Moreover, the Principles involved many inconsistencies. For instance, we can identify one in force, which acts un-evenly and non-uniformly but directs the universe in some specific direction. If we want to avoid inconsistency, we have to question the possibility of any science, as the action of force makes it impossible to discover regularities. Some followers of Spencer (progressionists and supporters of orthogenesis) maintained that force has some teleological meaning. Whatever the interpretation of the Principles, they have no more explanatory power than his (in)famous "survival of the fittest".

The great popularity of Spencer's ideas can be interpreted as the result of the need for systematic organization of the view of reality, fragmented by diverging scientific disciplines and; at the same time, it can be interpreted as a defense against reductionist tendencies that seemed to oversimplify relations between different domains of study. In particular, we can see in it the need to find tools for the understanding of complexity.

3.2. From Unity of Science to Holism

Probably the most radical opposition to the reductionist vision of reality, accompanied by proposals of radically new methodology, came from biology. After the initial success in merging some methods of research on life with chemistry into biochemistry, more and more problems appeared in the reduction of the study of life to the concepts of disciplines from the lower rungs of the traditional ladder of science. It is not an accident that the calls for an entirely new methodology focused on the organic unity of complex systems came from biologists.

The concept of holism was introduced not by a biologist, but by a lawyer and politician with interests in botany. Jan Smuts in his 1926 book *Holism and Evolution* was influenced by biology [14]; however, his holism was intended as a general methodology for the study of complex systems that form wholes irreducible to their components. Smuts did not attempt to provide philosophical foundations or a clearly defined methodology for holism. He wrote about it in the preface to the second 1927 edition of his book. He wrote: "I recognize that there is a Metaphysic or Logic of Holism which has still to be written; but it is not for me to write it" [14]. The task of building philosophical foundations for holism was undertaken more than two decades later by Ludwig von Bertalanffy in his General Systems Theory (GST). His work generated great hope for a new chapter in the philosophy and methodology of science [15]. It is not a surprise that von Bertalanffy was a biologist working on mathematical models of the growth of organisms. His work in biology retained its relevance and value over the decades, but his General Systems Theory did not acquire a clearly defined methodology that could unify studies of complex systems. Its importance manifested more in the stimulation of searches for new methodologies than in being, itself, a clearly defined and codified methodology that guides and organizes research.

The General Systems Theory had a very wide range of followers; some of them, such as Anatol Rapoport, Kenneth E. Boulding, William Ross Ashby, Margaret Mead, Gregory Bateson, and more recently, Erwin Laszlo, made important contributions to their fields of studie inspired by the idea of the holistic approach. Unfortunately, GST has been frequently reduced to the trivial apocryphal quotation, supposedly from the *Metaphysics* of Aristotle: "The whole is more than the sum of its parts." This easy-to-remember and easy-to-repeat slogan attracted the attention of people who searched for shortcuts to philosophy and wisdom, giving GST its prominence; however, it also gained bad publicity when the authors of wild speculations tried to support their ideas by referencing the work of Bertalanffy. Of course, we cannot blame him for giving inspiration to questionable intellectual enterprises. After all, he promoted the idea that opened a new chapter in scientific inquiry.

The idea of the reorientation of scientific methodology to include holistic aspects of complex systems, initiated by Bertalanffy in the 1950s, resurfaced later in many different ways. It was not always authors of similar scientific or philosophical initiatives who referred to GST or gave credit to Bertalanffy. They actually might have not been familiar with his written works. However, it is hard to believe that they could not be influenced by the intellectual trend of a holistic way of thinking.

The philosophical reflection on the part–whole relation was the central theme of the 1967 book by Arthur Koestler, *The Ghost in The Machine* [16], which introduced the concept of a holon, something which is simultaneously a part and a whole. Neither a part nor a whole exists anywhere. We just have a holarchy, understood as a hierarchy of self-regulating holons which have some level of stability and autonomy. As an example of holon, Koestler gives the example of a human being, stating: "No man is an island—he is a holon. A Janus—faced entity who looking inward, sees himself as a self-contained unique whole, looking outward as a dependent part" [16].

The systemic way of thinking was not restricted to philosophy. We can find its reflection, for instance, in the work of Robert Rosen, who followed his teacher, Nicolas Rashevsky, in the development of relational biology. Rashevsky engaged in the development of a new mathematical approach to biology, starting with the change in the language of biological discourse [17]. His Generalized Postulate of Relation Forces was as follows: "The development of organismic set proceeds in such a manner as to maximize the total number of relations and the number of different kinds of relations during the total course of development" [18]. The novelty of this approach was in the direct involvement of mathematics in the study of living organisms and in the structural analysis describing, in a clearly defined way, the idea of the unity of biological systems.

Rosen's approach was similar, although he involved a different mathematical formalism of the category theory [19]. The choice of the category theory was dictated by Rosen's interest in self-reference as the main characteristic of life, distinguishing it from the subjects of study in physics [20,21]. He believed that category theory could help him to avoid the destructive logical consequences of self-reference.

Philosophy and science of life received an intellectual tool of exceptional importance in the concept of autopoiesis, introduced by Humberto Maturana and Francisco Varela [22]. The idea of the fundamental and distinctive characteristic of life expressed as a self-creation goes much further than the cybernetic concept of self-regulation, and its influence on the understanding of life was, and is, immense. However, we can ask whether the entire direction of holistic, systemic, and autopoietic studies takes us closer to a cohesive and uniform scientific worldview. The fact that we are studying complex phenomena, which require consideration of wholes irreducible to their parts, does not mean that our view of reality is more cohesive, or that our methodology is more uniform and consistent.

There is no simple, straightforward answer to the question of whether the holistic, systemic approach belongs to centrifugal or centripetal tendencies in human knowledge. On one hand, we get closer to resolving one of the most fundamental and challenging problems of the mind–body duality, which opened the precipice between the humanities on the mental side and the natural sciences on the side of the body. The embodied cognition described in terms of life and its scientific characteristics, which escape the epistemic reduction to physics, brings us closer to the uniform vision of reality studied by uniform science. However, this uniformity of science can be, and actually is, questioned. What types of contribution have this systemic direction of studies made to physics? Can we bridge the mind–body precipice when we still claim that to cross another border between physics and life sciences, we have to change our intellectual tools? What does constitute the authentic unity of knowledge?

The last question is sometimes answered without direct dependence on the systemic or holistic way of thinking. An example can be found in the two books published by Edward O. Wilson, the first in 1998, *Consilience: The Unity of Knowledge* [23], and the second in 2011, *The Meaning of Human Existence* [24]. Both are more attempts at the reconciliation of sciences than methodological proposals of the integration (as the title of the first states). The motive of the unification of knowledge is too wide to discuss or review in this paper more elaborately. However, even this fragmentary account of centripetal tendencies for the study of transdisciplinarity would have been incomplete without mentioning Gregory Bateson and his 1979 *Mind and Nature: A Necessary Unity* [25].

4. Multidisciplinary, Interdisciplinary, and Transdisciplinary Studies

4.1. Interdisciplinary vs. Multidisciplinary Studies

The typical answer to the question about achieving unity of knowledge is that the solution can be found in the so-called interdisciplinary studies. There were examples of successful (to some degree) merges of disciplines, such as those mentioned before, establishing the discipline of biochemistry. There were diverse research organizations that included the expression "interdisciplinary" in their names. More recently, the term "interdisciplinary" has become fashionable and is applied in a wide range of contexts, always in the sense of the broadening of the perspective of knowledge; however, there are rarely any attempts to define it or to make the claim of their interdisciplinarity accountable.

The reflection on the experience from these attempts at interdisciplinarity produced the distinction of three forms of crossing the borders of disciplines: multidisciplinarity, interdisciplinarity, and transdisciplinarity. The distinction of the first two became a standard and commonly accepted recognition of the failed programs of interdisciplinarity that promised too much. For instance, Peter van den Besselaar and Gaston Heimerics write that in multidisciplinary research, "the subject under study is approached from different angles, using different disciplinary perspectives. However, neither the theoretical perspectives nor the findings of the various disciplines are integrated in the end"; meanwhile, interdisciplinary research "creates its own theoretical, conceptual and methodological identity. Consequently, the results of an interdisciplinary study [...] are more coherent and integrated" [26].

Bernard C.K. Choi and Anita W.P. Pak express the distinction that "[Multidisciplinarity] draws on knowledge from different disciplines but stays within their boundaries. [...] In contrast, interdisciplinarity "analyzes, synthesizes and harmonizes links between disciplines into a coordinated and coherent whole" [27].

Similar but different distinctions, based more on the level of the mutual interdependence of the subjects of the study, were already present for a long time in the discussions of the differences between the cross-disciplinary studies and the interdisciplinary studies. These were understood as a distinction between studying diverse instances of phenomena and the studies of these phenomena in mutual interactions, for instance, the distinction between cross-cultural analysis and intercultural analysis, which require different methodologies.

4.2. Transdisciplinary Inquiries

Transdisciplinarity is also not a new concept and was, from the beginning, more of a methodological program for the future than the existing methodology. It appeared first in the early 1970s in a vision of such a program propagated by Jean Piaget: "Finally, we hope to see succeeding to the stage of interdisciplinary relations a superior stage, which should be 'transdisciplinary,' i.e., which will not be limited to recognize the interactions and/or reciprocities between the specialized researches, but which will locate these links inside a total system without stable boundaries between the disciplines" [28].

The triad of the approaches to crossing disciplinary borders was described by Basarab Nicolescu as follows: "Multidisciplinarity concerns itself with studying a research topic in not just one discipline but in several simultaneously. From this perspective, any topic will ultimately be enriched by incorporating the perspectives of several disciplines. Multidisciplinarity brings a plus to the discipline in question, but this "plus" is always in the exclusive service of the home discipline. In other words, the multidisciplinary approach overflows disciplinary boundaries while its goal remains limited to the framework of disciplinary research. Interdisciplinarity has a different goal than multidisciplinarity. It concerns the transfer of methods from one discipline to another. Like multidisciplinarity, interdisciplinarity overflows the disciplines, but its goal remains within the framework of disciplinary research. Interdisciplinarity even has the capacity of generating new disciplines, such as quantum cosmology and chaos theory.

Transdisciplinarity concerns that which is at once between the disciplines, across the different disciplines, and beyond all disciplines. Its goal is the understanding of the present world, of which one of the imperatives is the unity of knowledge" [29,30]. We can see that Nicolescu's view of transdisciplinarity is strikingly close to what we are looking for in the present paper. However, even if the goals are similar, there are substantial differences. As in a popular saying "the devil is in the detail."

Nicolescu presents his program of transdisciplinarity in what he claims to be an axiomatic approach: "After many years of research, we have arrived at the following three axioms of the methodology of transdisciplinarity:

1. The ontological axiom: There are, in Nature and society and in our knowledge of Nature and society, different levels of Reality of the Object and, correspondingly, different levels of Reality of the Subject.
2. The logical axiom: The passage from one level of Reality to another is ensured by the logic of the included middle.
3. The complexity axiom: The structure of the totality of levels of Reality or perception is a complex structure: every level is what it is because all the levels exist at the same time" [30].

The program proposed by Nicolescu requires very strong and radical commitments, and at the same time, does not offer much beyond very general principles that refer to undefined concepts of the levels of reality, multivalued logic, and complexity. Without these missing specifics, it cannot be considered a methodology of transdisciplinarity. Axiomatic systems require a much higher level of specific identification of their primitive concepts, or alternatively, very precise definitions of the conceptual framework. On the other hand, we cannot expect that a very disruptive approach excluding the accumulated knowledge of the past acquired within disciplines can serve this purpose.

Another example of a program intended as a support for transdisciplinarity can be found in the works of Søren Brier on cybersemiotics. Brier refers directly to transdisciplinarity in his articles, and provides a comprehensive description of his vision of the role of cybersemiotics in overcoming the barriers between the biological and social realms [31–34]. In *Can Cybersemiotics Solve the Paradox of Transdisciplinary Knowing?* Brier writes: "My major claim is that combining Luhmann's system theory with biosemiotics provides a new transdisciplinary framework, which is an alternative to 'the unity science' of positivism on one hand, and post-modernism on the other. I advocate Cybersemiotics as a multidimensional semiotic constructive realism, the point of which is that signs as concepts and classifications arise in our embodied biological and social 'life forms'. For our understanding of meaning production, a concept has to have a phenomenological and emotional constitution; there is therefore no good reason why the inner world of cognition, emotions, and volition should not be accepted as just as real as the physical world as well as our cultural world of signs and meaning" [34]. In a slightly different context of the informational transdisciplinarity in the article *Can Cybersemiotics Solve the Problem of Informational Transdisciplinarity?*, he writes: "Cybersemiotics attempts to combine a systemic and a semiotic view trying to amend the shortcoming of the above described transdisciplinary models into a model that is not totalitarian mechanistic, algorithmic or physicalistic reductionism and on the other hands is not a constructivist relativism giving up any scientific truth claims. Cybernetics and systems science attempts to overcome these problems through its dynamics theory of emergence, where like in dialectical materialism now qualities arise in systems development or when to types of systems are integrated" [34].

We can see that Brier's work is focused on bridging the social and biological inquiries to be achieved by finding an alternative method, avoiding the extremes of reductionism or constructivist relativism. However, Brier does not provide a distinction between interdisciplinarity and transdisciplinarity, and does not present methodological tools that can be used outside of the context of socio–biological research. He considers the two forms of synthesis just a part of the same tendency, stating that "There is a long history of striving towards inter- and trans-disciplinarity in the sciences, from Newton and Laplace through Comte and the logical positivists" [34]. It is not clear whether the cybersemiotic approach can benefit other disciplines of science, or what methodologies it can offer for a unified view of reality.

4.3. From Multidisciplinarity and Interdisciplinarity to Transdisciplinarity

Someone could question the importance of the distinctions between different forms or levels of synthesis. However, the distinction has practical meaning. It gives both a direction for the search of methodological transformations and a measure of the achievement. Catherine Scott and Anne Hofmeyer give their very practical perspective on this issue: "It is timely to develop improved understandings about strengthening interdisciplinary contexts to guide effective and quality healthcare research; contexts in which health and social issues occur do not recognize disciplinary boundaries. Similar to the notion of "partnership", the terms multidisciplinary, interdisciplinary and transdisciplinary are in danger of becoming conceptually indistinct and thus of limited usefulness for researchers, practitioners and teams" [35].

Without a consensus about the meaning of transdisciplinarity, despite the frequent calls for its establishment, we can conclude that this is an expected highest stage of research activity that requires intellectual tools common for all disciplines, serving the purpose of unification. The most important feature of these tools should be that they facilitate the transfer of results and methods between different disciplines, and that they have an integrative power to unify fragmentary models of reality into a larger whole.

We could see that the third, highest form of crossing the borders of disciplines is desired and expected, but it is far from being implemented. This does not mean that the idea of interdisciplinarity can be easily achieved and that we can consider science unified, at least at this level. Here, we can use as an example the case of the cognitive science analyzed in an excellent study by Rafael Núñez, Michael Allen, Richard Gao, Carson Miller Rigoli, Josephine Relaford-Doyle, and Arturs Semenuks, with the title *What Happened to Cognitive Science?* [36] .

When you look up a dictionary definition, for instance, in the Merriam-Webster Dictionary, cognitive science is defined as "an interdisciplinary science that draws on many fields (such as psychology, artificial intelligence, linguistics, and philosophy) in developing theories about human perception, thinking, and learning." There are institutes and departments of cognitive science, and academic organizations with this science in their names. However, the identity of this science is not as obvious as it may seem. In 2003, George A. Miller, commonly considered one of the founders of this discipline, recalled the early times when cognitive science emerged in the 1950s, and when, in the 1970's, it received formal recognition in the program sponsored by the Sloan Foundation: "I argued that at least six disciplines were involved: psychology, linguistics, neuroscience, computer science, anthropology and philosophy. I saw psychology, linguistics and computer science as central, the other three as peripheral. These fields represented, and still represent, an institutionally convenient but intellectually awkward division. Each, by historical accident, had inherited a particular way of looking at cognition and each had progressed far enough to recognize that the solution to some of its problems depended crucially on the solution of problems traditionally allocated to other disciplines. The Sloan Foundation accepted my argument [...]" [37].

In the conclusion of his reminiscences, Miller wrote about his view of cognitive science in 2003: "Some veterans of those days question whether the program was successful and whether there really is something now that we can call 'cognitive science'. For myself, I prefer to speak of the cognitive sciences, in the plural. But the original dream of a unified science that would discover the representational and computational capacities of the human mind and their structural and functional realization in the human brain still has an appeal that I cannot resist" [37]. Thus, it is not cognitive science, but cognitive sciences. The extensive study carried out by Núñez and his colleagues gives very clear confirmation of this diagnosis.

> "More than a half-century ago, the 'cognitive revolution', with the influential tenet 'cognition is computation', launched the investigation of the mind through a multidisciplinary endeavour called cognitive science. Despite significant diversity of views regarding its definition and intended scope, this new science, explicitly named in the singular, was meant to have a cohesive subject matter, complementary methods and integrated theories. Multiple signs, however, suggest that over time the prospect of an integrated cohesive science has not materialized. Here we investigate the status of the field in a data-informed manner, focusing on four indicators, two bibliometric and two socio-institutional. These indicators consistently show that the devised multi-disciplinary program failed to transition to a mature inter-disciplinary coherent field. Bibliometrically, the field has been largely subsumed by (cognitive) psychology, and educationally, it exhibits a striking lack of curricular consensus, raising questions about the future of the cognitive science enterprise" [34].

To obtain a more comprehensive view of the present time, science, and its cohesion through multidisciplinary, interdisciplinary, and transdisciplinary studies, we need many more studies of this type about attempts to form interdisciplinary fields. However, even this single example shows that overcoming the fragmentation of scientific inquiries is a formidable task that requires very broad and deep foundations, especially in the form of clearly defined methodology.

In the absence of an example of a mature, fully developed transdisciplinary study, we can only try to conceive an outline of the necessary methodological tools. The starting point could be the question about the conceptual and methodological framework of Natural Philosophy. What was the glue that held it united and gave it an identity? There are many possible answers. For instance, the role of empirical methods could be considered, combined with increasing use of the quantitative description of phenomena. However, empirical methods do not have much sense without some form of realism. Quantitative description of phenomena, perceived by Comte and many others as evidence for exactness, can impress only novices. In my view, the key is in the balance and correspondence between ontology and epistemology, more specifically, the correspondence between structures that define and describe entities and epistemological structures (logical, mathematical, and other). Empirical methods are just ways to control this correspondence.

This is the reason why in the following, we will explore the intellectual experience brought by structuralism in its specific version associated with symmetry. After all, in the attempt to use structuralism as the common methodology of *The Two Cultures*, the Tower of Babel of human knowledge was closest to Heaven.

5. Revisiting the Lost Paradise: Symmetry and Structure

5.1. Symmetry Study as Methodology for Structural Analysis in Science

About half a century ago, for the first and last time, there was hope for building a bridge across all disciplines including the humanities, science, and mathematics in the form of structuralism. The story of structuralism is probably the closest to the Biblical story of The Tower of Babel.

It is difficult to find a term of more universal use in all contemporary domains of human inquiry than "structure". This does not mean that people who use this term can define it or explain its meaning. If asked about the meaning, people would refer to the explanation from a dictionary that structure is an organization, arrangement of parts, elements, or constituents of a complex system, considered from the point of view of the whole rather than of any particular part, etc. The terms "organization" or "arrangement", in turn, are explained in dictionaries with the use of the word "structure". Once we restrict the term structure to a particular discipline, we can find an increasingly precise definition of its meaning. The concept of a structure is convoluted with another concept of symmetry. Both were born in the mid 19th century to replace the traditional general ideas of form and harmony, respectively. The marriage of these concepts had, as its offspring, the formalization of the concept of structure in terms of symmetry, understood as invariance with respect to transformations.

The question about the common meaning of geometric structures considered in the different forms of geometry led to the modern general methodology for structural analysis in mathematics and mathematical sciences. This methodology involved the concept of symmetry. Up to the late 18th century, symmetry was usually conceptualized as a harmony of proportions [38]. It can be a surprise that although its modern understanding can be traced to earlier, the precise definition of symmetry in its simplest mirror type was provided by Ernst Mach in 1872 [38], in the same year that the entire methodology of the symmetry study was born in the Erlangen Program of Felix Klein [39]. Klein proposed a new paradigm of mathematical study focusing not on its objects, but their transformations. His mathematical theory of geometric symmetry was understood as an investigation of invariance, with respect to transformations of the geometric space (two-dimensional plane or higher dimensional space).

Klein used this very general concept of geometric symmetry for the unification of different types of geometries (Euclidean and non-Euclidean), and the classification of different geometric structures within these geometries. The fundamental conceptual framework of Klein's Program (which was intended by Klein as a paradigm of study, and became such paradigm on a scale not expected by him) was based on the scheme of (1) space as a collection of points $\rightarrow$ (2) the algebraic structure (group) of its transformations $\rightarrow$ and (3) invariants of the transformations, i.e., configurations of points that do not change as a whole, while their points can be permuted by transformations. Selections of algebraic substructures (subgroups) of transformations correspond to different types and levels of invariant configurations, allowing the differentiation and comparison of structural properties associated with symmetry. The classical example of mirror symmetry (symmetry with respect to the surface of the mirror) can be identified with invariance with respect to the mirror reflection, understood as a transformation of the entire space.

Klein's work applied a new theory of groups which in the works of Arthur Cayley [40] and Camille Jordan [41] became a part of algebra. Klein's Erlangen Program to classify geometries has been extended to many other disciplines of mathematics, becoming one of the most common methods of in all mathematical research. The new analytical tool was very soon adapted to studies in theoretical physics.

Group theory was originally "preconceived" in the early 19th-century work of young Evarist Galois in his study of the impossibility of solving polynomial equations of degrees higher than four (the actual definition of a group was introduced several decades later). Galois considered invariance of the solutions for equations under transformations of the set of numbers that can be substituted for the variable. This line of thinking was investigated later in the mid 19th century by James Joseph Sylvester and Arthur Cayley in the context of solving equations. This led to the natural question about the invariance of physical equations under transformations of physical coordinates.

It was quickly recognized, under the influence of the Klein Program, that in classical mechanics, the equations have to be invariant with respect to the group of transformations of coordinates, which was called the Galilean group. This group was, and is, called Galilean because it consisted of the transformations which Galileo considered several centuries earlier when he observed that the description of reality has to be independent of the choice of the position, orientation, linear motion with a constant velocity of the observer, and of the choice of the time for observation. It turns out that the equations of Newtonian mechanics are invariant with respect to the Galilean group of transformations, but curiously, Maxwell's equations for electrodynamics are not. The latter equations turned out to be invariant with respect to another group of transformations recognized in 1905 by Henry Poincare, which he called the Lorentz group. This was the group with respect to which relativistic mechanical equations are invariant (the full group of transformations, including the Lorentz group, for which they are invariant was, in turn, called the Poincare group by Herman Minkowski in 1908).

The transition from classical to relativistic mechanics started to make sense as a change in the type of symmetry with different geometry (change from Euclidean geometry to the geometry of Minkowski space) and with the change in invariant physical magnitudes. In this new description of reality, the separate Galilean group invariant magnitudes of mass and energy are combined into the one Lorentz group invariant magnitude of mass-energy.

However, a bigger, Copernican-type conceptual revolution came a little bit later. The transition from classical to relativistic mechanics, in terms of the change in symmetry groups from the Galilean to the Lorentz or Poincare group, was a realization of the Klein Program in physics regarding the change in geometry. The change in the invariant physical magnitudes from separate mass and energy to united mass–energy was a welcome gift of great theoretical and practical importance (its consequence is the most famous equation of physics: $E = mc^2$).

5.2. Copernican Revolution of Symmetry in Physics

One of the most celebrated achievements of 19th-century physics, expressed as the Law of Conservation of Energy (i.e., the First Law of Thermodynamics), was interpreted as an expression of the ontological status for an alternative form of physical entities. Energy was an alternative to matter, associated with the Law of the Conservation of Mass. Until the 19th century with the dominating Newtonian corpuscular theory of light, the concept of matter was associated with the atomistic tradition founded on the distinction matter-vacuum. It was meant, to belong in the realm of the Cartesian res extensa. The wave theory of light and the development of electrodynamics, which introduced electromagnetic waves transmitted through a vacuum, made the earlier atomistic distinction matter-vacuum meaningless.

At first, there was hope that this distinction could be retained if electromagnetic waves were the waves of the aether, an exotic form of matter filling out the entire space. The end of the 19th century was the end of this hope. Special Relativity theory eliminated aether, but also united mass and energy into mass–energy, restoring the uniform ontological status. This may explain why even today, anachronisms such as "matter and energy" finds their place in philosophical discussions when in physics, the actual ontological distinctions are along the dualisms of wave-particle or field-particle.

A Copernican revolution in the understanding of conservation laws was a consequence of one of the most important contributions to mathematical physics of all time, published in 1918 by Emmy Noether [42] and stating that every differentiable symmetry of the action of a physical system has a corresponding conservation law. Noether showed that the conservation laws for physical magnitudes such as energy, momentum, and angular momentum are associated with transformations describing changes of reference frames, i.e., observers. Thus, it is not true that we have given, in advance, distinct physical entities with corresponding magnitudes that obey conservation laws, and for which we can find fundamental equations that are invariant with respect to some groups of symmetries.

The roles are reversed. If we have a general description of a physical system and we want to render this description objective in the sense that every theoretical observer of the system, free from the action of external forces, describes the reality the same way (i.e., the description is the same in every inertial reference-frame equivalent with respect to symmetry group transformation), then with each type of symmetry, there is a corresponding magnitude conserved in time. Conservation of energy is a consequence of invariance with respect to the time shift. Conservation of momentum is a consequence of invariance with respect to space shift. Conservation of angular momentum is a result of invariance with respect to the rotation of the reference frame.

This, in fact, answers a naïve but legitimate question: "What does make $1/2mv^2$ better than $1/2mv^3$ as the description of kinetic energy?" The answer is that the former is invariant with respect to theoretical changes of observers (i.e., reference frames) in the case of an isolated physical system in the absence of forces, while the latter is not invariant at all.

Therefore the distinguished physical magnitudes satisfying the corresponding conservation rules are determined by the choice of symmetry transformations, and those are determined by the condition of equivalence of all possible observers. Naturally, this makes the study of symmetry a central tool for scientific methodology. Physics (and science in general) looks for an objective description of reality, i.e., a description that is invariant or covariant with changes of observers. Noether's theorem tells us that such a description can be carried out with the conserved magnitudes. This is why the expression "matter and energy" does not make sense in physics. What is the group of symmetry for which matter is invariant? What does it mean that matter is invariant?

5.3. *Symmetry Climbs Comte's Ladder*

The year 1872, when the Erlangen Program was published, can be considered a starting point for the study of symmetries; this was clearly defined in terms of group theory, but not in terms of the scientific exploration of symmetries studied before the concept of symmetry was formalized. The intuitive recognition of the similarities and differences between some configurations of points and their mirror reflections generated the interest of the greatest minds for quite a long time. For instance, Immanuel Kant tried to rationalize the distinction between human left and right hands (although they are different, they are mutual mirror images), but his hypothetical argumentation that in a world in which God would have created humans with the one hand only, this hand is neither left nor right, is not very convincing. Sooner or later, humans could invent a mirror and could realize that there is an alternative form for their single hand. More important is that Kant's reflection did not contribute much to the understanding of mirror symmetry.

The structural characteristic that gives the distinction of left- and right-handedness was given the name of chirality by Lord Kelvin much later, but their distinction could be recognized easily thanks to our everyday experience with our hands and the gloves that have to match them. Thus, either of our hands is chiral, and they are enantiomorphs of each other, while the majority of simple organisms are symmetric with respect to all rotations and reflections, and therefore achiral.

In 1848, Louis Pasteur published one of his most important papers, explaining the isomerism of tartrates, more specifically of tartaric acid by molecular chirality ("left- or right-handedness" of molecules) [43]. He showed that the differences in the optical properties of the solutions of this organic compound between samples synthesized in living organisms and samples synthesized artificially result from the fact that artificially synthesized molecules—although constructed from the same atoms as those in natural synthesis—have two geometric configurations; he concluded that they are symmetric with respect to the mirror reflection, but not exchangeable by spatial translations or rotations (in the same way as left and right palms of human hands); conversely, he showed that in the nature, only left-handed configurations occur. Later, it turned out that almost exclusively naturally synthesized amino acids (and therefore proteins) are "left-handed", and sugars are "right-handed". Artificial synthesis, if not constrained by special procedures, leads to the equal production of left and right-handedness. There is no commonly accepted explanation of this mysterious phenomenon, even today.

The chirality of organic molecules became one of the most important subjects of 19th-century biochemistry, leading to the discovery of the role of atoms of carbon in the formation of chiral molecules formulated into the Le Bel–van 't Hoff Rule, published by these two researchers independently in 1874.

The study of symmetry in biology, particularly of chirality in complex organisms, could not have been explained in the 19th century; however, researchers published some phenomenological laws of evolution and phenotypic development of organisms, such as Bateson's Rule. Much later Bateson's son Gregory explained this rule in terms of information science [44,45].

A similar interpretation can be given to Curie's Dissymmetry Principle. Pierre Curie made so many important contributions to physics and chemistry that this fundamental principle of great philosophical importance is rarely invoked. The outdated original formulation, using the term "dissymmetry" instead of the now commonly used "asymmetry", was: A physical effect cannot have a dissymmetry absent from its efficient cause [46]. This rather unintuitive principle has very important consequences in biology and chemistry. The real importance of these early developments could be fully appreciated half a century later when it became fully clear, thanks to advances in physics (elementary particle theory), that the study of the conditions for maintaining symmetry is no more important than the study of breaking symmetry.

By the mid 20th century, the study of symmetry became a fundamental tool for mathematics, physics, chemistry, and several branches of biology. This can explain the explosion of the interest in symmetry among philosophers. The swing of the pendulum of the dominating philosophical interests between the tendency to seek an objective methodology for philosophical inquiry, inspired by scientific methodology, and the calls for freedom of the use of introspective, and therefore subjective, phenomenal experience, reached the apex of the former. The most influential expression of the alignment of the humanities with science was in structuralism.

5.4. Structuralism or Many Structuralisms?

Although the generic term "structuralism" was already in use in the late 19th century in the context of mental structures in the psychology of Wilhelm Wundt (for instance, in the description of Wundt's position by his student Edward B. Titchener), the beginnings of structuralism, understood as a broad methodological perspective, can be traced most directly to the works of Ferdinand de Saussure on linguistics (more specifically his lectures from 1907–1911, posthumously published by his disciples in 1916 as *Course in General Linguistics* [47]). The emphasis on the structural characteristics of language and their synchronous analysis prompted increased interest in the meaning of the concept of structure, although de Saussure himself used the term "system" rather than "structure". It is only speculation, but it seems that the words "structure" and "system" gained their popularity in the 19th century in parallel to the decline in the use of the term "form", due to the latter's luggage of associations to its use in diverse meanings in philosophy through the centuries. Through the association with de Saussure, the concept of a structure acquired an implicit characteristic of synchrony. In disciplines in which diachrony was fundamental, such as biology, the preferred dynamical concept of morphogenesis, popularized by Goethe, appended the idea of structure expressed as morphology. We should not be deceived by terminological preferences. In all cases, the central concept was of a structure, viewed either statically or dynamically; by the end of the 19th century, this concept had replaced in science the concept of form. Klein's Erlangen Program, originally formulated for geometry, provided the pattern of a methodological tool for mathematics, and soon later, for physics.

It was a natural consequence that the tools used in science for structural analysis in terms of symmetry found their way to psychology, anthropology, and philosophy.

The clearest programmatic work *Structuralism* by Jean Piaget, published originally in 1968, refers explicitly to the concept of the group of transformations, although very little of the formal apparatus was presented there [48,49]. On the other hand, Piaget, in his work on developmental psychology, used this methodology explicitly. For instance, he based his theory of child development on the so-called Klein's "Four-group". The works of others, for instance, Claude Levi-Strauss, also directly employed the methods developed as a consequence of the Erlangen Program, and included the use of Klein's group too [50].

This was the time in which structuralism was triumphant, but also a time of great confusion. The popularity of structuralism made its name a buzzword, and everything written at that time with the word "structure" was (and unfortunately still frequently is) associated with structuralism. Of course, nobody owns the name of structuralism and there is nothing wrong with using it for different purposes. However, mixing these different uses and the resulting misattributions of views are errors.

Piaget should be prized for the popularization of the idea of structuralism as a broad philosophical direction of thought, but his postulate to use the concept of a structure as a bridge between the scientific and humanistic forms of inquiry was preceded by the short but very influential 1952 book *Symmetry*, written by Hermann Weyl [51]. Weyl did not use the name structuralism, but demonstrated the use of the method of Klein's Erlangen Program in studying structures from mathematics, physics, crystallography, and biology, to art, design, etc. In his view, the study of structures was the study of the invariants of transformations (more exactly, groups of transformations).

It was a time when both terms "structure" and "symmetry" had already established fundamental roles in mathematics, physics, chemistry, and biology. A more elaborate exposition of these roles in the present context of their unifying power is presented by the author elsewhere, in the study of the question "What is a structure?", carried out with the use of, or reference to, rather advanced mathematical formalisms [52]. However, to prevent confusion, it is necessary to disentangle some ideas before we proceed to further discussion of the methodological tools that structuralism, based on the symmetry concept, can offer for transdisciplinary studies.

Piaget's book *Structuralism* generated enormous interest among readers belonging to a very wide audience, although in its Conclusion, he already complains that "[...] one can only be disturbed by the current modishness of structuralism, which weakens and distorts it" [49] (p. 137). Thus, the book was intended as a means to clarify the confusion that often comes with popularity. Unfortunately, it also generated a lot of misunderstandings, possibly partially because of the differences between its original French edition and its English translation. In the following, I will refer to the English version. This could have been prevented if Piaget gave a reference to the much earlier book *Symmetry* by Herman Weyl (whose different book on a different subject he quoted).

Piaget prizes Klein's Erlangen Program as "a prime example of the scientific fruitfulness of structuralism" [49] (p. 22) and provides its explanation, but makes it so over-simplified that it does not make much sense. Statements such as "Groups are systems of transformations; but more important, groups are so defined that transformation can, so to say, be administered in small doses, for any group can be divided into subgroups and the avenues of approach from any one to any other can be marked out" [49] (p. 21), or "Group structure and transformation go together. However, when we speak of transformation, we mean an intelligible change, which does not transform things beyond recognition at one stroke, and which always preserves invariance in certain respects" [49] (p. 20) are nonsensical. What is meant by "preserves invariance in certain respects"?

Piaget made it clear that structuralism, or to use his words, "structuralism in general", is a descendant of Klein's Erlangen Program: "In this little book we shall, therefore, confine ourselves to the kinds of structuralism that are to be met in mathematics and the several empirical sciences, already a sufficiently venturesome undertaking. [...] But first we must elaborate somewhat on the definition of the structuralism in general that is here proposed, else it will be hard to understand why a notion as abstract as that of a 'system closed under transformation' should raise such high hopes in all domains of inquiry" [49] (p. 6).

This may explain the bizarre formulation of his presentation of Klein's ideas, which he might have considered easier to understand for those who abhor abstraction. Whatever his intention, some of his statements are confusing, and some are confused. For instance, in the passage "Indeed, all known structures – from mathematical groups to kinship systems — are, without exception, systems of transformation. But transformation need not be a temporal process {...}" [49] (p. 11), he is obviously right in that symmetry is an invariance with respect to transformations that do not have to be temporal, and that what is invariant is structure; however, it is the invariant of a group of transformations, not a "system of transformations."

Thus far, we can talk about the confusing formulation of some claims. The real problem starts when Piaget makes claims that are meaningless or explicitly inconsistent with the mathematical description of structures as invariants of transformations: "In short, the notion of structure is comprised of three key ideas: the idea of wholeness, the idea of transformation, and the idea of self-regulation" (p. 6). While the idea of transformation (in the mathematical description realized by transformations understood as functions of a specific type) is fundamental for structures, and we can interpret wholeness as the result of invariance, the idea of self-regulation does not have any meaning in the structures of mathematics or physics. By introducing the idea of self-regulation, Piaget resurrects the ghost of the systemic, organismic conceptual framework without giving self-regulation any formal meaning.

A similar problem of the error of commission can be identified in the Conclusion, where Piaget writes, in the context of all possible structures: "*There is no structure apart from construction,* either abstract or genetic" [49] (original emphasis, p. 140) and "The problem of genesis is not just a question of psychology; its framing and its solution determine the very meaning of the idea of structure. The basic epistemological alternatives are predestination or some sort of constructivism" [49] (p. 141). Here, Piaget makes the mistake of mixing two levels of the discourse.

There is nothing in the general inquiry of the structures defined as invariants of transformations that commits us to a particular epistemological or ontological position. These commitments in works of contributors are always posterior to the study of structures. Naturally, from the position of his Genetic Epistemology, transformations can be interpreted as types of constructions, and with this interpretation comes the interpretation of structures as constructs [49,53]; however, this is the result of his commitment, not an inherent feature of structures. In any case, Piaget's interpretation seems artificial when we consider the symmetries of physics. For instance, transformations from the Galileo group or Lorentz group are transitions between potential observers (reference frames). Not only can these transitions hardly be considered constructions, but the transition from Newtonian to Relativistic Mechanics—which, in hindsight, we can associate with the transition from the Galilean to the Lorentz group—was a discovery made against the expectations and the will of physicists involved. Later, we had a string of major discoveries in physics that consisted of surprising cases of breaking symmetry.

Thus far we had examples of ramifications in the understanding of structures and related forms of structuralism. There is no reason to claim that one form is better than the other. We just have to choose one and, in this paper, it is the one that is based on the concept of symmetry, initiated by the Erlangen Program of Klein, and is free from any additional assumptions or interpretations. The reason for the choice is that the task for this paper is to build a bridge between the Two Cultures, and to search for patterns in developing a transdisciplinary methodology.

Thus, it was clarified that we have more than one structuralism, and the one that serves our purpose the best is derived from the Erlangen Program. However, this does not eliminate the confusion. Another source of misunderstanding is a false belief that our preference excludes directions of inquiry of special importance. An example of the apparent contestant to the role of a methodological tool for transdisciplinarity is category theory, introduced in 1945 by Samuel Eilenberg and Saunders MacLane in a long, epoch-making paper, *General Theory of Natural Equivalences* [54]. The misunderstanding is in the relationship between these two directions of thought. First, it should be made clear that the more important idea of this famous paper is that of a functor, understood as a transition between categories that are, themselves, auxiliary concepts.

For those who are familiar with the research genealogy of the subject in which Emmy Noether was a mentor for MacLane, and Felix Klein for Emmy Noether, it should not be a surprise that the work of Eilenberg and MacLane was intended as a continuation of the Erlangen, as this quotation from the Introduction tells us: "The invariant character of a mathematical discipline can be formulated in these terms. Thus, in group theory all the basic constructions can be regarded as the definitions of co- or contravariant functors, so we may formulate the dictum: The subject of group theory is essentially the study of those constructions of groups which behave in a covariant or contravariant manner under induced homomorphisms. More precisely, group theory studies functors defined on well specified categories of groups, with values in another such category. This may be regarded as a continuation of the Klein Erlanger Programm, in the sense that a geometrical space with its group of transformations is generalized to a category with its algebra of mappings" [54].

Thus, the two directions of thought are not only not in competition, but one is an extension of the other. This may generate a question about whether the the Category and Functor Theory is a better choice for a methodological tool for transdisciplinarity. The answer is that it is a matter of preference. The theory of structures based on symmetry is just less abstract. Most of the research in physical sciences and, of course, in other disciplines is carried out at this lower level of abstraction, so there is no compelling reason to go further. There is a close analogy of the relationship between these two programs of inquiry and the pair of algebraic structures of groups and monoids (the former are special cases of the latter type). In mathematics, if you can carry out something using exclusively groups, you do not introduce the concept of a monoid. If it turns out that there is a need for generalization, you can always achieve it; however, you then have to pay the price of some lost tools that require the eliminated assumptions (in this case, the reversibility of morphisms and the methods of set theory).

There is one aspect of symmetry and structure studies that is worth mentioning here. The Theory of Categories and Functors has its role in mending the historical division of research on symmetries between Felix Klein and Sophus Lie. Klein, in his Erlangen Program, explicitly renounced the exploration of symmetries described by continuous groups studied by Lie, and left it to his friend [39]. This led to the study of continuous symmetries in terms of the so-called pseudogroups. Emmy Noether's two famous theorems regarding the relationship between symmetries and conservation laws of physics were in the conceptual framework of continuous symmetries [42]. Moreover, there is a parallel direction of research in physics initiated by the work of Lie within mathematics—the study of dynamic systems, culminating in the works of René Thom—especially his 1972 *Structural Stability and Morphogenesis* [55]—and in the complexity studies of the members of Santa Fe Institute focusing on complex adaptive systems. Although the two directions of research initiated by Klein and Lie have many differences in their methodologies, they do not compete but complement each other. Unforunately, the perception of the general audience of their supposed opposition is biased by the differences in terminology (e.g., structure vs. complex system).

5.5. Is the Paradise of Structuralism Lost?

The swing of the pendulum reversed its direction and in the late 20th century, structuralism lost its dominating position to competitors; however, its importance can be seen in the name of this reversed swing as "Post-structuralism". Some of this criticism is naïve.

For instance, structuralism was criticized as "ahistorical," "static," "too much formalized," and "too much restrictive for the freedom of expression".

The view of the ahistorical characteristic is most likely a result of mistaken association with the views of Ferdinand de Saussure. In the context of linguistics, de Saussure distinguished the two modes of inquiry: diachronic and synchronic. The synchronic perspective focuses on the structure at some particular moment. However, in a more general context, there is nothing precluding the evolution of structures, as is commonly done in physics. We can see here why the inquiry of structures using the methodology of symmetry is so effective. Structures are invariants of transformations and they are distinguished from their environment by being invariant. Their existence is a resolution of the opposition between change (diachrony) and identity (synchrony). This is the key distinction between structuralism based on symmetry invariance and its other types or versions.

More justified is the objection to the lack of interest in the explanation of the origin of structures considered in the studies of Levi-Strauss and others. The missing evolutionary or dynamic theory of structures can be blamed on these authors, but it is more a matter of the misunderstanding of the mathematical tools than of their absence. Physics and chemistry possess powerful, exact dynamic theories of their structures in terms of group theory and symmetry, so there is no good reason to believe that such a dynamic approach is impossible in other disciplines of philosophy. The most convincing explanation of the shortcomings identified in the applications to the study of culture and society is probably that the mathematical tools of symmetry theory found little use in works of the most prominent propagators of structuralism. Symmetry was more a metaphor for the literary treatment of the subject than an actual study of the invariants of groups of transformations [56].

Symmetry can easily be identified in the studies of visual arts and music. The structural study of music initiated by Pythagoreans found its way to medieval philosophy via Neoplatonic authors, and then to the works of the founders of modern science such as Johannes Kepler. The music of heavens, understood literally as music produced by the motion of the planets, was a mathematical model of the universe. An example of the highest-quality contemporary study of symmetries in art in a cross-cultural perspective can be found in the book *Symmetries of Culture: Theory and Practice of Plane Pattern Analysis* by Dorothy K. Washburn and Donald W. Crowe [57].

At this time, group theory in the context of symmetries had already become an everyday tool for all physicists and had assumed a permanent place in university curricula for studies in physics, chemistry, and biology [58]. A statement from an article published in Science in 1972 by a future Nobel Prize laureate in Physics Philip Warren Anderson stating that "It is only slightly overstating the case to say that physics is the study of symmetry" was an expression of a commonly accepted truth [12].

The study of symmetry became a fundamental methodological tool. Anderson's article not only closed the century of its development, but also included another very important message. Anderson emphasized the role of "breaking symmetry" and of the hierarchical structure of reality. He demonstrated that, at least in the perspective of physics, reality has a hierarchic structure of increasing complexity and that the transition from one level of complexity to the next is associated with breaking symmetry, understood as a transition from one group of symmetry to another of a lower level. Thus, not only is the study of symmetry important, but so are the ways in which it changes.

5.6. Symmetry as a Unification Tool

The role of symmetry and its breaking is equally fundamental in physics and other scientific disciplines today as it was half a century before, while structuralism in the humanities has gone through a period of strong denial (seen in the proudly declared dissents of those who, like Umberto Eco, were considered structuralists, or in the frequent denunciations of its ineffectiveness). However, recently, there has been increased interest in structuralism, not only in philosophy but also in social sciences and economics [59].

One of the reasons for the revival of the interest in structuralism and symmetry outside of scientific disciplines—where they both remained at the center of attention without any decline in interest—was the growing recognition that practically every complex system retains its identity only as a structure, not as an aggregation of elements. For instance, every living organism replaces its chemical substrates in a time incomparably shorter than its life span, which applies, of course, to the human organism. Every social organization goes through a similar process of exchange. This naturally led to the claim that, because of this universal feature of complex systems and because of deficiency in the description of complexes in terms of their simple components, the actual, real status should be given to structures, not their substrates.

We can already find the same way of thinking in Weyl's book which, in 1952, initiated an interest in symmetry: "We found that objectivity means invariance with respect to the group of automorphisms. Reality may not always give a clear answer to the question of what the actual group of automorphisms is, and for the purpose of some investigations, it may be quite useful to replace it by a wider group" [51].

There is possibly a legitimate concern regarding backlash against structuralism, which generated such strong polarization of views within The Two Cultures. If we want to use the methodology that was denounced in the past as faulty, this may lead to yet another story of The Tower of Babel. The answer is that the use of symmetry in humanistic or cultural contexts was misguided by the lack of appropriate methodological tools.

The most typical misunderstanding in attempts to extend the methodology of symmetry studies, in geometry to other contexts, is a consequence of misinterpretation of Klein's Program. Klein did not consider arbitrary transformations of the plane (or set of points on which geometry is defined), but only those that preserve the underlying geometric structure. This very important but very frequently ignored aspect of the Program was clearly described in Weyl's book popularizing symmetry in the general audience: "What has all this to do with symmetry? It provides the adequate mathematical language to define it. Given a spatial configuration $\mathfrak{F}$, those automorphisms of space which leave $\mathfrak{F}$ unchanged form a group Γ, and *this group describes exactly the symmetry possessed by* $\mathfrak{F}$. Space itself has the full symmetry corresponding to the group of all automorphisms, of all similarities. The symmetry of any figure in space is described by a subgroup of that group." [51]

Even in recent books popularizing symmetry studies within the restricted domain of geometry, we can find statements exhibiting a lack of understanding of this aspect of Klein's Program. Therefore, everywhere in textbooks we find statements such as "Symmetry of a geometric object consisting of some set of points A is every transformation of a space S, i.e., bijective function from S to itself, that leaves A unchanged." In these cases, the authors are talking about "groups of symmetries" as groups of all arbitrary transformations leaving object A unchanged. Of course, these "symmetries" and "groups of symmetries" would only be useful in very limited situations. If symmetry is just one particular collection of transformations, then every two squares of the different centers would have different symmetries. In addition, we have to consider separate "symmetries" transformations that leave all points of a square identical, but arbitrarily permute all other points.

The beauty and power of Klein's Program are in the recognition of what is important for the study of symmetry. We have a more general group of transformations of a particular type, i.e., determined by a specific type of the structure (geometric, topological, algebraic, etc.); then, we look for the subgroup of transformations that leave our object unchanged, even if particular points within the object have different images through the transformations. The difference would emphasize the importance of the pre-defined total group of transformations which typically is a proper subgroup of the group of all transformations. Weyl calls it the group of symmetries for the entire space: "Space itself has the full symmetry" [51]. Only then we can make a selection of the subgroup describing a specific symmetry.

This is the point where we can find the sources of the doubts about the applications of symmetry in cultural studies. In the geometric context, everyone, even those who did not understand the method in its generality, automatically considered only subgroups of the group of all isometries, i.e., transformations preserving the metric (distance) characterizing a particular type of geometry. In application to the humanities, the choice of the symmetry group was arbitrary, guided only by the desired result. The presence of arbitrary choices of transformations generated resentment, expressed in the form of claims that Levi-Strauss and others using this methodology could not get anything new beyond that which they entered into consideration. However, this is more of an aberration of the structuralistic way of thinking than the norm [60].

6. Expulsion from Paradise?

The title of the preceding section included the expression "Lost Paradise" with an explicit intention. In this paper, there is not much about the intellectual movement of postmodernism, which is openly hostile not only towards structuralism but also towards any tendencies to look for general syntheses of knowledge. The initiator of this movement, Jean-François Lyotard, in his programmatic work *The Postmodern Condition: A Report on Knowledge*, declares the war against totalitarian syntheses: "Let us wage a war on totality; let us be witnesses to the unpresentable; let us activate the differences and save the honor of the name" [61] The war was, and still is, in the name of the postmodern: "Simplifying to the extreme, I define postmodern as incredulity toward metanarratives. This incredulity is undoubtedly a product of progress in the sciences: but that progress in turn presupposes it. To the obsolescence of the metanarrative apparatus of legitimation corresponds, most notably, the crisis of metaphysical philosophy and of the university institution which in the past relied on it" [61]

The book exemplifies the dangers of the external view of science acquired from the usually unsuccessful attempts at popularization. Lyotard gives the evidence for his external view, writing: "A crude proof of this: what do scientists do when they appear on television or are interviewed in the newspapers after making a 'discovery'? They recount an epic of knowledge that is in fact wholly unepic. They play by the rules of the narrative game; [...]" [61]. What scientists do in front of TV cameras has nothing to do with science. Lyotard's reflection on science is not only external but also very naive: "I have already made the point that the question of proof is problematical since proof needs to be proven. One can begin by publishing a description of how the proof was obtained, so other scientists can check the result by repeating the same process. But the fact still has to be observed in order to stand proven. What constitutes a scientific observation? A fact that has been registered by an eye, an ear, a sense organ? Senses are deceptive, and their range and powers of discrimination are limited. This is where technology comes in" [61] Is this supposed to be a competent critique of science? It is not necessary to publish spoof articles in postmodernist journals to ridicule the movement [62]. Postmodernists perform this job better by publishing their own articles.

There are many other statements in Lyotard's book that show his disarmingly naive view of science, its methodology, and the actual problems that can be rightly interpreted as the crisis. These problems cannot, and should not, be mixed with the crisis in modern societies that do not understand science, the needs of people contributing to it, or their organizations. Among the other misconceptions, there are some clearly mistaken views on what science is. Science is not a narrative, not because it is descriptive (as Lyotard claims), but because it is a dialog that never ends. The crucial point of every scientific activity is the formulation of the research question, which initiates the dialog and which often is more difficult than finding the answer. The popularization of science (admittedly not of the highest quality) frequently has the form of a narrative. The most important feature of science that makes it different from the narratives of culture and religion is that while the latter make all effort to suspend disbelief, the former starts from generating disbelief and keeping disbelief as a self-control tool.

How is the story of postmodernism related to John Milton's Paradise Lost? postmodernists fell to the temptation of Satan to eat the fruit from the tree of the knowledge of good and evil. postmodernists believe that they tasted the fruit and that this gives them the power to judge what is good or evil in knowledge or philosophy. This makes them so interested in the issue of legitimization because they reject the criteria of objective reality. There is no objective reality, so there is no objective truth. Reality consists of social constructs, so the task is to achieve legitimization of the constructions.

An example of the postmodernist way of thinking in the subject of this paper is the book *The Unity of Nature: Wholeness and Disintegration in Ecology and Science* by Alan Marshall [63]. The author explicitly identifies his postmodernist position: "[T]here is an ongoing debate between naturalism and realism on the one hand and social constructivism on the other; and I should state at the onset that I am, myself, more allied to the latter than to the former" [63]. The following quotation effectively presents the way of thinking of the author, who opposes the idea of the unity of nature: "Natural unity, we shell see, does not just exists on its own as an independent idea. It has a whole attendant army of supporting concepts, narratives and metaphors from which it gains its strength. For example, unity has attachments to the ideas of 'balance', 'order', 'hierarchy', 'stability', and the concept of the 'system'. In this book, these companion ideas are filtered out from the unity of nature idea and then, one by one, distilled so as to expose their unfortunate philosophical side-effects." We ought to be thankful to the author that he did not dispose of the entire idea of 'philosophy' because of its unfortunate side effects. This is a good example of a postmodernist meta-narrative which, with arrogance typical of this movement, claims possession of the fruits from the tree of the knowledge of good and evil to judge that which is fortunate or unfortunate. I am afraid that the followers of postmodernism are not aware of their affinity to those who, like Kellyane Conway, Senior Counselor to the President of the US, promote such social constructs as "alternative facts" [64].

7. Conclusions

To go beyond the declarations of the importance of transdisciplinarity for the reunification of knowledge dispersed through diverse disciplines of inquiry, it is necessary to develop a clearly defined transdisciplinary methodology based not on the reduction in the conceptual frameworks of component domains to a specific framework of a dominant discipline, but on the concepts that are central for all contexts.

In this article, we revisited the concept of structure, which was of great importance in modern intellectual history. This concept raised hopes for reconciliation between diverse studies, but was rejected on the humanistic side because of misunderstandings and attempts to directly transmit a highly abstract version of its methodology of the study of symmetry. The revolt against structuralism and symmetry may be puzzling when we realize that both structure and symmetry are descendants of the ideas of form and harmony, without which no humanistic discourse is possible.

Even if we witnessed how Paradise was Lost, there is still a chance for Paradise Regained. In this paper, diverse tendencies in searching for a unified view of reality were reviewed, and the most promising direction seems to be in the attempt to revive, extend, and develop the methodology of structuralism. I am sure, at least, that structuralism is more universal and no more totalitarian than postmodernism. I believe that there is no other methodological framework that could compete with it. However, more important than my belief is the demonstration that there are available methodological tools for transdisciplinary studies.

Funding: This research received no external funding.

Acknowledgments: The author appreciates comments from the audience of the PHAEDE 2019 Conference, where an early version of this paper was presented.

Conflicts of Interest: The author declares no conflict of interest.

References

1. Snow, C.P. The Two Cultures. In *The New Statesman*; 1956; pp. 413–414. Available online: newstatesman.com (accessed on 12 February 2022).
2. Snow, C.P. *The Two Cultures and the Scientific Revolution*; Cambridge University Press: Cambridge, UK, 1959.
3. Ortolano, G. *The Two Cultures Controversy: Science, Literature and Cultural Politics in Postwar Britain*; Cambridge University Press: Cambridge, UK, 2009.
4. Conant, J.B. *General Education in a Free Society: Report of the Harvard Committee*; Harvard University Press: Cambridge, UK, 1946.
5. Kuhn, T.S. *The Copernican Revolution: Planetary Astronomy in the Development of Western Thought*; Harvard University Press: Cambridge, UK, 1957.
6. Borges, J.L. The Library of Babel. In *Labyrinths: Selected Stories and Other Writings*; New Directions: New York, NY, USA, 1962.
7. Pope, A. Essay on Criticism. Poetry Foundation. Available online: https://www.poetryfoundation.org/articles/69379/an-essay-on-criticism (accessed on 15 August 2020).
8. Ferrier, J.F. *Institutes of Metaphysic: The Theory of Knowing and Being*; Blackwood: Edinburgh, UK, 1854.
9. Elder, R.B. Some Comments on Renato Barilli's 'The Medium in the Medium'. *New Explor. Stud. Cult. Commun.* **2020**, *1*, 85–99. Available online: file:///C:/Users/MDPI/AppData/Local/Temp/cignatov,+08_Elder_OnBarilli.pdf (accessed on 12 February 2022).
10. Bourget, D.; Chalmers, D.J. What Do Philosophers Believe? *Philos. Stud.* **2013**, *170*, 465–500. [CrossRef]
11. Nagel, E. *The Structure of Science*; Harcourt, Brace, and World: New York, NY, USA, 1961.
12. Anderson, P.W. More is Different. *Science* **1972**, *177*, 393–396. [CrossRef] [PubMed]
13. Weinstein, D. Herbert Spencer. The Stanford Encyclopedia of Philosophy. Available online: https://plato.stanford.edu/archives/fall2019/entries/spencer/ (accessed on 12 February 2022).
14. Smuts, J.C. *Holism and Evolution*; Macmillan: London, UK, 1926.
15. Von Bertalanffy, L. An Outline of General System Theory. *Br. J. Philos. Sci.* **1950**, *1*, 134–165. [CrossRef]
16. Koestler, A. *The Ghost in The Machine*; Hutchinson: London, UK, 1967.
17. Rashevsky, N. The Representation of Organisms in Terms of (logical) Predicates. *Bull. Math. Biophys.* **1965**, *27*, 477–491. [CrossRef] [PubMed]
18. Rashevsky, N. *Organismic Sets: Some Reflections on the Nature of Life and Society*; Mathematical Biology, Inc.: Holland, MI, USA, 1972.
19. Rosen, R. The Representation of Biological Systems from the Standpoint of the Theory of Categories. *Bull. Math. Biophys.* **1958**, *20*, 317–341. [CrossRef]
20. Rosen, R. Some epistemological issues in physics and biology. In *Quantum Implications: Essays in Honour of David Bohm*; Hiley, B.J., Peat, F.D., Eds.; Routledge & Kegan Paul: London, UK, 1987.
21. Rosen, R. *Life Itself: A Comprehensive Inquiry into the Nature, Origin, and Fabrication of Life*; Columbia University Press: New York, NY, USA, 1991.
22. Maturana, H.R.; Varela, F.J. *Autopoiesis and Cognition: The Realization of the Living*; Boston Studies in the Philosophy of Science: Dordrecht, The Netherlands, 1980; Volume 42.
23. Wilson, E.O. *Consilience: The Unity of Knowledge*; Vintage Books: New York, NY, USA, 1999.
24. Wilson, E.O. *The Meaning of Human Existence*; W.W. Norton: New York, NY, USA, 2014.
25. Bateson, G. *Mind and Nature: A Necessary Unity*; Dutton: New York, NY, USA, 1979.
26. Van den Besselaar, P.; Heimeriks, G. Disciplinary, Multidisciplinary, Interdisciplinary: Concepts and Indicators. In Proceedings of the 8th International Conference on Scientometrics & Informetrics (ISSI), Sydney, Australia, 16–20 July 2001; Davis, M., Wilson, C.S., Eds.; University of New South Wales: Kensington, Australia, 2001; pp. 705–716.
27. Choi, B.C.K.; Pak, A.W.P. Multidisciplinarity, Interdisciplinarity and Transdisciplinarity in Health Research, Services, Education, and Policy: 1. Definitions, Objectives, and Evidence of Effectiveness. *Clin. Investig. Med.* **2006**, *29*, 351–364. [PubMed]
28. Piaget, J. L'épistémologie des relations interdisciplinaires. In *L'interdisciplinarité–Problèmes D'enseignement et de Recherche, Centre pour la Recherche et L'innovation dans l'Enseignement*; Apostel, L., Berger, G., Briggs, A., Michaud, G., Eds.; Organisation de Coopération et de développement économique: Paris, France, 1972.
29. Nicolescu, B. Methodology of Transdisciplinarity—Levels of Reality, Logic of the Included Middle and Complexity. *Transdiscipl. J. Eng. Sci.* **2010**, *1*, 19–38. [CrossRef]
30. Nicolescu, B. *Manifesto of Transdisciplinarity*; Transl, K.-C., Ed.; State University of New York (SUNY) Press: New York, NY, USA, 2002.
31. Brier, S. Cybersemiotics: A New Foundation for Transdisciplinary Theory of Consciousness, Cognition, Meaning and Communication. In *Origins of Mind*; Swan, L., Ed.; Biosemiotics, Springer: Berlin/Heidelberg, Germany, 2013.
32. Brier, S. Transdisciplinary View of Information Theory Seen from a Cybersemiotics Point of View. In *Fundamental Notions of Information, Communication, and Knowledge: Its Effects on Scientific Research and Inter–Disciplinarity*; Ibekwe–San, F.J., Dousa, T., Eds.; Springer: Berlin/Heidelberg, Germany, 2013; pp. 23–49.
33. Brier, S. Can Cybersemiotics Solve the Paradox of Transdisciplinary Knowing? In *Semiotics 2014: The Semiotics of Paradox*; Owens, G.R., Katic, E.K., Eds.; Legas Publishing: Nepean, ON, Canada, 2014; pp. 611–641.
34. Brier, S. Can Cybersemiotics Solve the Problem of Informational Transdisciplinarity? In *Philosophy and Methodology of Information*; World Scientific Series in Information Studies; World Scientific: Singapore, 2017; Volume 1, p. 196.
35. Scott, C.; Hofmeyer, A. Acknowledging complexity: Critically analyzing context to understand interdisciplinary research. *J. Interpr. Care* **2007**, *21*, 491–501. [CrossRef] [PubMed]

36. Núñez, R.; Allen, M.; Gao, R.; Miller Rigoli, C.; Relaford-Doyle, J.; Semenuks, A. What happened to cognitive science? *Nat. Hum. Behav.* **2019**, *3*, 782–791. [CrossRef] [PubMed]
37. Miller, G.A. The Cognitive Revolution: A Historical Perspective. *Trends Cogn. Sci.* **2003**, *7*, 141–144. [CrossRef]
38. Hon, G.; Goldstein, B.R. *From Summetria to Symmetry: The Making of a Revolutionary Scientific Concept*; New Studies in the History of Science and Technology (Archimedes 20); Springer: Berlin/Heidelberg, Germany, 2008.
39. Klein, F.C. A Comparative Review of Recent Researches in Geometry. *arXiv* **2008**, arXiv:0807.3161. [CrossRef]
40. Cayley, A. On the theory of groups as depending on the symbolic equation $\theta^n = 1$. *Philos. Mag.* **1854**, *7*, 40–47. [CrossRef]
41. Jordan, C. *Traite de Substititions et des Equations Algebraiques*; Gauther-Villars: Paris, France, 1870.
42. Noether., E. Invariante Variationsprobleme. *Nachr. D. König. Gesellsch. D. Wiss. Zu Gött. Math-Phys. Klasse* **1918**, *1918*, 235–257.
43. Pasteur, L. Sur les relations qui peuvent exister entre la forme cristalline, la composition chimique et le sens de la polarisation rotatoire (On the relations that can exist between crystalline form, and chemical composition, and the sense of rotary polarization). *Ann. Chim. Phys.* **1848**, *24*, 442–459.
44. Bateson, W. *Materials for the Study of Variation*; Macmillan: London, UK, 1894.
45. Bateson, G. A Re-examination of 'Bateson's Rule'. In *Steps to an Ecology of Mind: Selected Essays*; Bateson, G., Ed.; Ballentine Books: New York, NY, USA, 1990; pp. 379–396.
46. Curie, P. Sur la symetrie dans les phenomenes physiques: Symetrie d'un champ electrique et d'un champ magnetique. *J. Phys.* **1894**, *3*, 393–415. [CrossRef]
47. de Saussure, F. *Course in General Linguistics*; Baskin, W., Ed.; Columbia University Press: New York, NY, USA, 1916.
48. Piaget, J. *Le Structuralism*; Presses Universitaires de France: Paris, France, 1968.
49. Piaget, J. *Structuralism*; Maschler, C., Ed.; Harper & Row: New York, NY, USA, 1972.
50. Lévi-Strauss, C. *Structural Anthropology*; Jacobson, C., Grundfest Schoepf, B., Eds.; Doubleday Anchor Books: New York, NY, USA, 1967.
51. Weyl, H. *Symmetry*; Princeton University Press: Princeton, NJ, USA, 1952.
52. Schroeder, M.J. Search for an Algebraic Structure of Not-Necessarily Algebraic Structures. In *Algebras, Logics, Languages and Related Areas, RIMS Kokyuroku, Kyoto: Research Institute for Mathematical Sciences*; Kyoto University: Kyoto, Japan, 2018; No. 2096; pp. 59–68. Available online: www.kurims.kyoto-u.ac.jp/~{}kyodo/kokyuroku/contents/2096.html (accessed on 4 March 2022).
53. Piaget, J. *Genetic Epistemology*; Columbia University Press: New York, NY, USA, 1970.
54. Eilenberg, S.; MacLane, S. General theory of natural equivalences. *Trans. Amer. Math. Soc.* **1945**, *58*, 231–294. [CrossRef]
55. Thom, R. *Structural Stability and Morphogenesis*; Fowler, D.H., Ed.; Benjamin: Reading, MA, USA, 1975.
56. Barbosa de Almeida, M.W. Symmetry and Entropy: Mathematical Metaphors in the Work of Levi-Strauss. *Curr. Anthropol.* **1990**, *31*, 367–385.
57. Washburn, D.K.; Crowe, D.W. *Symmetries of Culture: Theory and Practice of Plane Pattern Analysis*; University of Washington Press: Seattle, WA, USA, 1988.
58. Hamermesh, M. *Group Theory and Its Application to Physical Problems*; Addison-Wesley: Reading, MA, USA, 1962.
59. Gibson, B. An Essay on Late Structuralism. Available online: http://www.uvm.edu/~{}wgibson/Research/tay.pdf (accessed on 12 February 2022).
60. Schroeder, M.J. Invariance as a Tool for Ontology of Information. *Information* **2016**, *7*, 11. [CrossRef]
61. Lyotard, J.-F. *The Postmodern Condition: A Report on Knowledge*; Theory and History of Literature; Bennington, G., Massumi, B., Eds.; University of Minnesota Press: Minneapolis, MI, USA, 1984; Volume 10.
62. Lindsay, J.A.; Boghossian, P. The conceptual penis as a social construct. *Cogent Social Studies* **2017**, *3*, Retracted. [CrossRef]
63. Marshall, A. *The Unity of Nature: Wholeness and Disintegration in Ecology And Science*; Imperial College Press: London, UK, 2002.
64. Conway, K. Meet with the Press Interview on 22 January 2017. Available online: https://www.huffpost.com/entry/1984-sales-spike-alternative-facts_n_5887c8cce4b0441a8f71871d (accessed on 12 February 2022).

philosophies

MDPI

Article

Naturalizing Morality to Unveil the Status of Violence: Coalition Enforcement, Cognitive Moral Niches, and Moral Bubbles in an Evolutionary Perspective

Lorenzo Magnani

Philosophy Section and Computational Philosophy Laboratory, Department of Humanities, University of Pavia, 27100 Pavia, Italy; lmagnani@unipv.it

Abstract: I propose that the relationship between moral and violent behavior is overlooked in current philosophical, epistemological, and cognitive studies. To the aim of clarifying the complex dynamics of this interplay, I will describe, adopting an evolutionary perspective, the concepts of *coalition enforcement*, *cognitive moral niche*, and of what I call *moral bubbles*. Showing the interesting relationships between these three basic concepts, I will explain the role of morality in causing and justifying violence. The main theoretical merit of the concept of coalition enforcement is that it permits the naturalization of morality that is the only conceptual means to unveil, in a naturalized way, the status of violence beyond the constraints generated by the so-called moral bubbles that prevent agents from seeing the potential violence generated by their own moral acts.

Keywords: morality; violence; coalition enforcement; cognitive niches; moral bubbles; Moral Niches; free-riders

Citation: Magnani, L. Naturalizing Morality to Unveil the Status of Violence: Coalition Enforcement, Cognitive Moral Niches, and Moral Bubbles in an Evolutionary Perspective. *Philosophies* **2022**, *7*, 39. https://doi.org/10.3390/philosophies7020039

Academic Editors: Gordana Dodig-Crnkovic and Marcin J. Schroeder

Received: 7 February 2022
Accepted: 30 March 2022
Published: 2 April 2022

1. Can Violence Be Turned into an Autonomous Object of Philosophical Reflection?

In a previous book of mine, *Understanding Violence. The Intertwining of Morality, Religion, and Violence: A Philosophical Stance*, published by Springer, Heidelberg/Berlin, 2011 [1], I started research aiming at showing that violence can become an important object of philosophical reflection. A typical tendency of modernity leads to avoiding the analysis of violence (especially visible, strong, bloody violence), liquidating it through a sort of easy "psychiatrization": violent people are people who "are not well"—that is, crazy people. In most cases, however, psychiatry has nothing to do with it. I remember an episode that occurred in 2013: the attacker Luigi Preiti, who had shot two policemen, was immediately classified by the media and public opinion as a lunatic, a madman[1].

The hidden causes of that terrible violence were thus obscured, taking advantage of a kind of medicalization/psychiatrization. This common way of reacting ends up attenuating and putting the violence perpetrated in the background, classifying it as the result of something "sick". It thus evades the task of saying more about the roots of much of the violence. However, the attacker had immediately declared—showing a standard moral perspective—that he wanted to "hit" (and therefore "punish") politicians, identified as being guilty of causing his desperate situation of being unemployed and separated, which was experienced as unfair. One can therefore interpret his behavior in moral terms: every moral rule that is adopted provides for the potential punishment of violators.

Transforming violence into an autonomous object of reflection means avoiding talking about violence using shortcuts (such as the one just mentioned of psychiatrization) and adopting, in my case, a naturalistic approach, as I will explain in this article, seeing its various facets, keeping in mind the results of various disciplines, and enucleating the indissoluble interweaving of morality and violence between evolutionism, cognitive science, mathematical catastrophe theory, cognitive niche theory, logic and informal logic, psychology, psychoanalysis, and semiophysics.

Every human cooperation, since prehistory, is based on the sharing of moral (or proto-moral) rules (even those incorporated in laws since a certain moment of the civilization process); however, the breaking of moral rules involves possible punishments that are more or less violent (or rather that are experienced as such by the affected subjects). The contact with people and communities that share different moral horizons also generates conflicts that can result in very violent outcomes (think of wars or terrorism of religion, where the conflict caused by different moral rules embedded in religious frameworks is powerful).

Everyone witnesses violence of all kinds every day and it is talked about it in the media and at home; however, social ignorance about the status of violence is great and, even among scholars and philosophers, things are not better. As I already observed, we often hear the emphasis on physical violence as the only violence worth mentioning, while it is passed over in silence and not reported as violent, for example, certain behaviors of the members of the parliament, when there is an exchange of votes and favors. We are in a society that aspired to become a "knowledge society" and instead has become an "ignorant society", where everyone is entitled to speak their mind, which too often is the ignorant opinion of individualistic narcissists who "know" nothing or know very little, whether they are ministers, journalists, housewives, or scientists.

Even scholars and intellectuals do not shine with "open-mindedness" because they usually know only specialized areas; if they possess a broader culture and express themselves, then they are attacked with the verbal violence that for decades now has classified them as abstract and useless in various media and "socially" ignorant and boorish. This contributes to an ignorant society (too proud to be such) that, in my opinion, is the bitter fruit of thirty years of violent and obtuse stubborn neoliberalism, which is undermining the foundations of the civilization that, in many ways, we inherited from ancient Greece (and with the point of view of having seen the Greek economic catastrophe as an ominous symbol).

Fortunately, philosophy has been helping us to interpret the world for more than twenty centuries, and today it also wants to understand violence. The naturalization of morality I will describe in this article, taking advantage of a few examples, aims at producing that intelligibility of violence that only philosophy can give. For example, I will describe how human beings can ignore their own violence, thanks to what I called "moral embubblement" (so to speak, violence is never mine but always that of others because I conceal the violence I commit by never considering it as such).

Morality, and therefore also religion (which, first among all cultural creations of humanity, has played the role of "moral carrier") and violence are strongly intertwined. It would seem a paradox, given that human beings are endowed with morality precisely to defend themselves from evil and violence and to foster cooperation. However, it is not a paradox: (1) every morality potentially conflicts in a violent way with other moralities; and (2) every morality implies, more or less, the violent punishment of transgressors. Two aspects that the analysis of the so-called coalition enforcement that I will describe in this article will explain this in a clear way[2].

I believe that it is necessary, first, to have "respect" for violence and to attribute to it once and for all the "moral dignity" of becoming a philosophical/knowledgeable topic, extracting it from the restricted circuit of futile daily chatter, the statistics provided by the human sciences, and easy psychiatry. Philosophers have always dealt with important topics, such as rationality, science, knowledge, and ethics, which are generally thought of by everyone as having an intellectual dignity in themselves. They have commonly thought that violence, precisely because it is such, shows itself as something trivial, bad, intolerable, confusing, inescapable, and marginal and thus not sufficiently interesting to them. Violence has therefore been considered more suitable to be studied as a fact: history, sociology, psychology, criminology, anthropology, to name a few disciplines, have always seemed more appropriate to analyze it and to provide data, explanations, and causes.

My naturalistic approach to morality moves from the conviction that, at least in our time, philosophy possesses the style of intelligence and intelligibility suitable for a new, impertinent and profound understanding of a theme that is so intellectually neglected and disrespected. When it comes to violence, philosophy, while still remaining an abstract discipline as we know it, paradoxically acquires the mark of a sort of indispensable and irreplaceable "applied science". I plan to attribute more philosophical dignity to violence because it is extremely important in the lives of human beings, whether we want to accept it or not.

We will see that violence is therefore usually generated by moral reasons: moral conflict is at the basis of violent outcomes. People who, at first glance, appear to be perfectly decent (as Hannah Arendt previously observed with regard to the good fathers of families who were at the same time Nazi criminals), are capable of violent and even bloody acts in the name of their morals, which they consider as safe and certain.

If one acts violently for moral reasons, to defend oneself and preserve one's moral point of view, or to punish those who have broken the moral rules of the group one adheres to, how is it possible to think that one has been violent? One says to oneself, for example, "I am a follower of the law of honor, which I consider just and moral, and thus revenge is 'justice' even when it results in the extreme violence of murder". I call this kind of inability to recognize possible violence arising from our moral beliefs a "moral bubble", which will be illustrated in the second part of this article.

Morality and Violence Entangled—Epistemology and Ethics Entangled

The fact that morality and violence are *entangled* means that not only can morality and violence be studied together but also that doing so has advantages. Even if those two human different qualities have their own theoretical dignity, many of the actions they deal with are deeply interwoven; thus, ignoring one component or the other may produce a philosophical misconception of the topic at stake[3].

Unfortunately, philosophers have a history of studying detailed problems that are firmly related to a specific profession and avoiding wider integrative commitments. This is, of course, related to the right need to deepen a specific problem, thus, mimicking science; however, we also need to move away from certain excessively negative aspects of contemporary philosophical practices, dubbed "analytic metaphysics" by Daniel Dennett, who sees it as a "naïve auto-anthropology" in which research participants appear to be convinced that their program actually gets at something true, and not only believed to be true by a "particular subclass of human beings (philosophers of the analytic metaphysics persuasion)" [5] (p. 98).

Dealing with the active critical role of reason, as suggested by the cognitive style of modern and contemporary sciences, highlights the current relevance of adopting a refurbished "openness" of philosophical research: "science instructs reason", Gaston Bachelard once remarked. A philosophical openness also favored, following Kant's teaching, by the attention attributed to the "constitutive" role played by intellectual creativity. An openness that is critical in the case of addressing the neglected problem of the status of violence and that can only be obtained through a process of *naturalization of morality*, in this case outside of the analytic tradition.

The entanglement between morality and violence that I introduced above takes advantage of a more fundamental one—between epistemology and ethics—that has arisen invisibly in recent years, overcoming the philosophical impasses of the past and the annoying *is/ought* debate. Clarifying the relationship and entanglement between epistemology and ethics aids in illuminating, in a naturalistic perspective, the intertwined relationship I mentioned earlier, namely the one between morality and violence. Indeed, because the four poles are related in a twofold system that I shall explore in this essay, each theoretical entanglement (epistemology–ethics and morality–violence) depends on the comprehension of the other.

To reach the envisaged these results, first, a naturalization of morality is necessary: in this article, this task will be achieved due to the adoption (and conceptual exploitation) of an evolutionary perspective that will take advantage of the concepts of *coalition enforcement*, *cognitive niche*, and of what I call a *moral bubble*. After all, an evolutionary perspective possesses a kind of priority in the case of an author that aims at offering a naturalistic account of the behavior of actual human beings. No one should question that biological concepts provide a kind of privileged access to "practical" and "natural" subjects.

2. Naturalizing Morality in an Evolutionary Perspective

2.1. Coalition Enforcement: Morality and Violence

The coalition enforcement hypothesis, proposed by Bingham [6,7], aims at explaining the "human uniqueness" that is at the root of human communication and language, in a strict relationship with *Homo Sapiens'* spectacular ecological dominance and the role of cultural heritage. In this perspective, human beings are animals that *domesticated themselves* exactly for two million years of *Homo* self-enforcement history. It is thanks to this hypothesis that *cooperation* is presented as a basic feature reached due to the effect of *moral* rules accompanied—as I will soon explain—by inescapable violent outcomes.

Following Boyd [8], individual learning in the framework of the transmission of cultural contents is responsible for the reaching of adaptive rules capable of counteracting more instinctive inclinations. In sum, morality refers to all those rules that grant cooperation and the possibility of the ownership of their destinies for human beings. It is clear that when you can expect that other human beings follow the rule of the shared morality, it is easy to predict their behavior and thus cooperation to various kinds of plans and projects is granted.

We will soon see that, in human collectives, *moral* and *violent* behaviors are interrelated, and are of course linked to the continuous construction and modification of *cognitive niches*: the key term that explains the nature of this entanglement is—as I will soon explain—"punishment". Indeed, we can say that coalition enforcement is executed in an evolutionary sense by the establishment of social cognitive niches as a new manner of diversified human adaptability[4].

The concept of coalition enforcement basically states that cooperation between related and unrelated animals produces considerable reciprocal advantages that outbalance the costs and are possibly adaptive for the involved individuals. It seems mandatory to take advantage of a reference to the cognitive activities of individuals as well as of groups. Indeed, by referring to the "group mind" hypothesis, whose role would be fundamental in social cognition and group adaptation, the evolutionary scientists Wilson, Timmel, and Miller [12] contended that groups play a crucial role in cooperative behaviors because they are capable of strongly improving the capabilities of individual performances as a direct fruit of Darwinian mechanisms.

In the perspective of the theory of cognitive niches, we can say that groups "socially" build cognitive niches that incorporate several kinds of rules, including, of course, the moral ones. This framework permits avoiding seeing the adaptation in a direct Darwinian way, such as in the case of Wilson's theory. In this last case, cognitive niches substantiate a change of the environment that "can" modify selective pressure in a strict Darwinian sense, producing both adaptations and maladaptations[5].

In hominids, group cooperation (which, unlike in non-human animals, is largely independent of kinship) arose from the need to detect, control, and punish social parasites who, for example, did not share the meat they hunted or partook of the food without joining the hunting party[6] (those parasites are also known as "free-riders"). These social parasites were dealt with in many ways, including by killing or wounding them (as well as cooperators who refused to punish them) from afar with projectile and clubbing weapons.

In this example, harming and killing are both cooperative and remote (while of course also being "cognitive" tasks). Individual risks are reduced by avoiding proximal conflict, according to the coalition enforcement hypothesis (thus, the importance of emphasizing remote killing). Of course, cooperative morality that generates "violence" against unusually "violent" and aggressive free-riders and parasites can be carried out in weaker ways, such as denying future access to the resource, injuring a juvenile relative, gossiping to persecute dishonest communication and manipulative in-group behaviors, or waging war against less cooperative groups, among other things[7].

At least in the case of modern humans, it is exactly the multiplicity of the various forms of punishment that are responsible for the potential generation of violent acts. Human beings subjected to punishment can (1) share another moral framework or a slightly different one so that they see punishment as violent and unjust, and (2) they can simply feel that punishment is appropriate but excessively violent and intolerable.

Indeed, in modern human collectives, various moralities act together in an efficient way, often at the level of the same individual: imagine, to make an example among the many, the contradictory coexistence in some collectives of at least three moral frameworks, the one related to a religious system, the one embedded in the laws, and finally, the one informed by honor culture. This situation generates conflicts, for example between religious morality against abortion and the moral rule incorporated in a law that permits it.

2.2. *Cooperation, Docility, and Punishment*

Group cooperation has been able to evolve adaptively in this fashion, rendering parasitic behaviors no longer systematically adaptive (for instance, for effective group hunting and meat sharing thanks to the "supervision" of free-riders). The individual costs of punishment, as well as individual aggressiveness and violence, are greatly reduced through cooperation and remote killing, possibly because violence is morally "disseminated" in a more durable way: "Consistent with this view, contemporary humans are unique among top predators in being relatively placid in dealing with unrelated conspecific nonmates under a wide variety of circumstances" (cf. Bingham [6] (p. 140)). [I must add, "contrary to common sense conviction", given to the massive quantity of violence that humans face on a daily basis!].

As a result, we can say that, in contrast to other animals, humans share a significant quantity of relatively trustworthy information with not germane conspecifics[8]. In other words, people rely on external input gathered through their senses from their social context to support their restricted decision-making abilities.

That is, docility possesses an adaptive character and can be seen as the consequence or direct outcome (that is selective pressure) of the increased quantity of available cognitive information caused by the incessant construction of the cognitive niches. To put it another way, docility allows a great amount of beneficial knowledge to be passed along while lowering the costs of (individual) learning. Docility is linked to the concepts of *socializability*, in Simon's work, as well as altruism in the sense that when an individual is an altruist, he is also docile: docility, not altruism, is the most significant term in this perspective because docility allows—from a cognitive point of view—for the genesis of altruism.

In light of the coalition enforcement theory, I believe moral altruism may be legitimately seen as a byproduct of—or at least tied to—the violent acts necessary to "morally" sustain and enforce coalitions. The altruist is frequently generous with those who are charged with punishing free-riders, and as a result, they may inadvertently be involved in possible acts of violence. I previously stated that groups must identify, control, and punish social parasites by murdering or damaging them (and also cooperators who do not agree to perform the punishment) and that they must enlist the help of other possible punishers in order to accomplish this goal.

Punishment itself (and thus the potential violence perceived by the people subjected to it) can be classified as altruistic because it is performed to help and favor the other members of the collective (and also to the aim of correcting the behavior of the targeted person); according to research on chimp behavior, the process is also frequently combined with the activity of preserving the position of the top-ranking males, so that the entire group is inordinately subordinated to the interest of the minority that is on top (cf. Rohwer [18] (p. 805)).

As I previously stated, organizations must identify and punish parasites by murdering or hurting them (as well as the cooperators who renounce to perform punishment), and to do so, they must enlist the help of other possible punishers. As I illustrated, this process describes altruistic behavior and those cognitive aspects that are the condition of the possibility of behavior itself: certain emotions, affectivity, empathy, that are fundamental to trigger cooperation.

Human coalitions, as the most gregarious animal groups, must take care of the individuals who cooperate in order to manage free-riders that infest a certain collective and protect them when foreign groups are aggressive and threatening. Again, it is from this perspective that we can see why modern human beings are, at the same time, certainly violent and also extremely docile and calm if compared to the top predators, as I indicated above, quoting Bingham. It is also necessary to add that in the case of modern humans the violence that originates from moral punishment is not necessarily and always perceived as such. Various degrees can occur in this case, depending on individuals' moral attitudes, emotions, preferences, biological endowments: not every child perceives as violent and offensive a slap given to punish him.

Furthermore, Lahti and Weinstein [19] described a kind of "group stability insurance" that is related to their concept of moral "viscosity", that is to the fact that moral rules can remain stable and efficacious even if more or less frequently disregarded. Morality is accompanied by an aura of absolutism; however, in reality, its flexibility is granted not only by viscosity but also by the "embubblement"—I will illustrate below in Section 4—that is related to the "cancellation" of the "actual" violent results of the agent's moral actions at the level of the perpetrator's awareness[9].

More words can be added concerning the issue of docility, which plays a fundamental role in the formation of moralities. First of all, docility explains why humans externalize a great deal of cognitive information outside, in the environment, thanks to the building of cognitive niches and in other human beings, as "biological" repositories, so to speak. Second, it also reminds us that humans tend to trust others. As I better explained in a previous book [21] (chapter three), it is only thanks to an already developed docility that human beings were able to build their *minds* as "universal machines", to adopt Turing's term.

Thus, this process was favored together by the presence of a large cortex and a rudimentary speech capacity, of small collectives minimally organized from a social point of view, and by the birth of the so-called "material culture", considered a kind of Big Bang that started the cultural evolution of *Homo*. A large cortex was not sufficient in itself as an evolutionary gain: delegating cognitions to external supports and building artifacts and the presence of "society" were fundamental. It is in this sense that docile engagement is at the heart of the development of both societies and large brains, which grew due to a clear process of co-evolution.

Docility is clearly linked to cultural growth, morality, and the status of cross-cultural intertwining. In human groups, there are various chances of taking advantage of docility: they are related to the levels of cultural transfers and their related modifications and improvements, aimed at augmenting or diminishing non-Darwinian fitness.

As I previously mentioned, the formation and vital function of cultural heritage (including morality and a sense of guilt) is a direct result of what has been called coalition enforcement: in other words, I emphasized the significance of cultural *cognitive niches* as novel ways of arriving at diverse human adaptations (not necessarily in a direct Darwinian sense, see below). The long-lived and yet abstract human sensation of guilt, in this view, is a psychological adaptation *to render it almost impossible to become the objective of violent coalitional enforcement*, thanks to *abductively* hypothesizing the assessment of a moral circumstance and thus acting consequently[10].

Again, we must keep in mind that Darwinian mechanisms are working not only at the genetic level but also (albeit with less accurateness and indirectly) at the cultural level, as a result of selection pressure caused by environmental changes. The collective human coalition as a crucial cognitive niche built by human beings is destined to realize a substantial aspect of the Darwinian selection, imposing additional limitations on its members (created by extragenetic information)[11].

2.3. The Role of Genetic and Extragenetic Information

The coalition enforcement concept appears to be supported by some empirical evidence (given by Bingham [7]). Selection created the human capacity to manage projectiles and clubbing weapons (thanks to motor actions favored by bipedalism and the development of the *gluteus maximus muscle* and its role in rotational acceleration, etc.) based on the examination of skeletal adaptations in *Homo* (but not in Australopithecines). When extragenetic information is sufficiently stored, used, and communicated, social cooperation grows together with brain size, as already stated above. It is important to note that biological, evolutionary, and obstetric constraints on brain dimensions suggest that humans can only absorb a limited amount of extragenetic knowledge, which must be massively stored and made available in the external environment.

Typically, Darwinian processes operating on genetic information are said to develop human minds whose qualities include the generation of innovative, complex adaptive design represented in human material products that are *sui generis*. These explanations, however, are insufficient. These explanations fall short of explaining human uniqueness. If Darwinian selection of genetic information could be used to create such minds, this adaptation would likely be commonplace. Humans appear to be the only ones who have it.

Before moving on to a possible answer to this conundrum, two further characteristics of human technical creativity should be remembered. First, humans capable of complex behaviors appeared about 40,000 years ago and rapidly exploded. Second, the velocity of the increase of the creative "abductive" capacity of modern humans is extraordinary, and it sometimes seems to overturn speeds that Darwinian classical selection at the genetic level could achieve [6].

As a result, the evolution of "non-genetic" information is extremely important. Coalition enforcement is an important factor in promoting the production of new extragenetic information—for example, due to further information transmission and exchange and thanks to both language and model-based communication between people unrelated from the genetic point of view[12]. Of course, extragenetic information is also stored in artifacts of various types, which boost the communication of concepts and ideas (cultural, moral, etc.), thus, creating a real situation of availability of ecological inheritance.

This information can be, in part and occasionally, stored in brain memory—for example as a more or less permanent configuration of neural networks (that, in turn, can govern actions) and at the level of external props, devices, and artifacts (to build cognitive niches, as I said), which can be transferred endlessly and so potentially eternally and outside of groups of individuals who are genetically related.

3. The Roots of Moral/Social Norms and the Related Violence

It is critical to demonstrate how moral norms, cooperation, and social dominance hierarchies may be explained in evolutionary terms. In this field, a large amount of relevant research has lately been conducted. The evolutionary framework of cognitive niches is extremely appropriate to solve the problem of the distinction between genetic and extragenetic information. The evolutionary account of social norms in terms of cognitive niches allows us to attempt to solve the still-unsolved question of the genetic or extragenetic source of social norms: even if some norms (such as the prohibition of incest) are surely not learned, a considerable quantity of norms are clearly learned, as contended by Bandura [26].

O'Gorman, Wilson and Miller [27] see social norms as derived from a very old phylogenetic history: many animal species show various conforming behaviors—that is the creation of a multitude of individuals—to better defend themselves from predators; other researchers think that social norms come from the evolved efforts to overcome the high costs related to the necessity of updating knowledge to the aim of affording environments that are always subject to mutations; finally, other consider the role of social norms as related to the urgent necessity of certain collectives of achieving a representative status: in this situation, being trained in a certain rule-based behavior is fundamental to being recognized as a member of the collective itself.

Recent research has also emphasized the relevance of dominance hierarchies in the establishment of social cooperation both in human and animal mammals; they are capable of affecting the evolution of both our minds and social organizations. Basic concepts and cognitive devices (which are not interpreted as innate modules but rather as favored by a kind of "biological preparedness", a propensity to develop them in a combination of genetic and environmental circumstances) were shaped by pressures derived from behaving in hierarchical collectives.

These ideas and cognitive tools are linked to the many stages of cognitive niche development, which are critical to surviving in those hierarchical environments: (1) being able to detect and evaluate dominance relationships; (2) being quick in incorporating norms, such as permissions and prohibitions; (3) being able to recognize violations of norms, codes, and rules; and (4) reading other minds to envisage intentions and predicting related behaviors also to the aim of identifying transgressions (as well as to perform commiseration, when needed). In monkey communities, perceived infractions have already been examined as the most common source of aggression[13]. High-ranking persons cannot monopolize resources, build profitable partnerships, or keep the peace without the ability to notice violations. Violation detection is the most valuable instrument for ensuring that social norms (implicit and explicit) are respected, allowing social control, and promoting the emergence of altruism as a stable strategy, especially in human collectives.

Transgressions and Violence

I contend that transgressions are clearly commonly perceived as *moral* violations by the people that identify them but not necessarily by the transgressors themselves. Modern humans regard violence generated by detectors to limit transgression as morally legitimate, whereas transgressors frequently regard it as plain violence, as I already indicated above. Is this not what is occurring in today's collectives when a killer believes he "did the right thing", for instance in retaliating, in contrast to "other" individuals who believe it is moral to put them to death following the law of capital punishment? The murderer believes he has committed a "moral" killing; in turn, precise retribution (the death penalty in that case) is a part of a variety of human moral acts when seen in the light of the law.

In the field of evolutionary and cognitive studies, other elements of the relevance of cooperation (and thus of violent punishment) are also examined. A cognitive paleoanthropological study recently stressed the cognitive role of the internalization of phonemes for collaboration[14]. The authors show that the "enhanced working memory" (and its organizational functions) can be dated 30,000 years ago in hominids: it appeared to coevolve with the birth of a *phonological storage capacity*, as well as language and other modern reasoning abilities, such as planning, problem-solving/algorithm manipulation, analogy, modeling, holding inner representations, tool-use, and tool-making.

An improvement of phonological storage, in particular, could have favored cross-modal methods of cognition, such as abductive hypothetical cognition, as well as the social actions rendered fundamental by coalition enforcement. Increased phonological storage may have rendered language free from the more rudimentary forms, such as the mere use of present tense and rough imperatives to favor the exploitation of future tense and of the subjunctive.

Although real opponents' activities may be predicted, phantom enemies and other intangible terrors can be brought to life. With novel views (e.g., the purpose of life, thoughts of death, life after death, etc.) great anxiety may arise. As a result, there is greater room for morality and punishment, as well as more complicated opportunities for performing violent acts.

Castro and Toro [30] suggested that the emergence of moral judgments is linked to the whole cultural evolution as an extragenetic inheritance process. They also consider the improvement of the imitation capacities a central aspect that is able to explain the transmutation of primate collective learning in a cumulative extragenetic inheritance cultural process as it occurs during *hominization*, based on the model of dual inheritance theory and gene-culture coevolution. The authors argue that better imitation is essential but not sufficient for this shift to occur and that the key component enabling it is that some hominids gained the deontic capacity to accept or disapprove of their offspring's learned behavior[15]. This ability to approve or disapprove of one's offspring's conduct reduces the cost and improves the accuracy of learning, and thus changed hominid civilization into a system of cumulative extragenetic inheritance cultural process comparable to that of human beings; however, the system would still be proto-linguistic. We have to emphasize that, in this view, axiological and moral components are totally linked to the evolution of culture as a whole.

4. Moral Bubbles Protect Moral Frameworks

In 1999, Justin Kruger and David Dunning published in the *Journal of Personality and Social Psychology* the important paper "Unskilled and Unaware of It: How Difficulties in Recognizing One's Own Incompetence Lead to Inflated Self-Assessments" [32] which illustrated the so-called Dunning–Kruger effect. In the perspective of the naturalization of morality[16], I stress the attention to something analogous to that effect that I call the *moral bubble effect*, related to the deficient (or very precarious) awareness of humans regarding their own violent behaviors—often human beings perform violent acts but do not detect their effects so that they ignore the imposed harm.

It is important to observe that this cognitive peculiarity of humans is central to securing moralities. The act of turning violence invisible and accepting it is based on a common psychological phenomenon known as "embubblement". As I have already anticipated in the previous sections, human behavior is enslaved by what I called *moral bubbles*, which regularly conceal violence: this is also related to the common knowledge that, in our society, many violent behaviors are generally treated as if they were something different.

Such widespread concealed violence, which is often also excused or justified, leads to the heart of my conviction that a naturalization of morality is the only way to increase our knowledge on violence beyond the common repetitive stereotypes, such as seeing violence as something exceptional: the main example is the psychiatrization of all kinds of violent behaviors.

Indeed, in the second section of this paper, I discussed the *coalition enforcement hypothesis*, which combines a philosophical view emphasizing the inherent *moral* (and simultaneously *violent*) constitution of behaviors in a paleoanthropological and evolutionary perspective. It was through coalition enforcement, as I have illustrated, that the forefathers built groups characterized by cooperative modalities granted by the existence of proto-moral elementary rules, of course establishing as right possible violence against free-riders and groups carrying different (proto-moral/proto-religious) rules.

Let us begin by focusing on the concept of a *epistemic bubble*. Woods claims that a "cognitive agent is in an epistemic bubble with respect to proposition *a* if he is in a *k*-state with respect to *a* and the distinction between his knowing that *a* and his experiencing himself as knowing it is phenomenally inapparent to him in the there and now" [33] (p. 162). In sum, we know less than we believe we know. Of course, a related consequence is that it is impossible to discriminate between a real correction and an apparent one, from the first-person perspective. It is necessary to have a third-person perspective to detect an error.

Woods adds: "The first-person/third-person asymmetry bites hard here. For the person who brings it off, error detection is a kind of coming to their senses. He comes to their senses in recognizing the incompatibility of what he now sees to be true with what he used to think was true. However, the asymmetry is such that what is experienced in these ways may not be as those ways suggest" [33] (p. 163). As clearly indicated by Woods, when in an epistemic bubble, cognitive agents, being in the ambiguous situation in which it is difficult to discriminate between the apparently true and the actually true, privilege the genuinely true.

Hence, truth is "fugitive", and the process of embubblement emphasizes the need for corrigibility of our ideas, particularly those that have undesirable consequences during problem-solving and decision-making processes. Woods is correct when he asserts that the cognitive mechanism of embubblement is not reversible unless the agent abandons a bubble in order to embrace a new one. The issue of corrigibility is certainly related to the one of de-biasing that, analogously to epistemic bubble, is impermeable to correction.

Autoimmunity and Embubblement

The embubblement process is likewise self-sustaining because it supports the agent's knowledge expansion without making the agent conscious of his self-delusion. As the agent uses the same cognitive mechanisms to acquire knowledge and control its validity, it produces a cognitive autoimmune system (cf. Arfini [34]). When Woods [33] explained the peculiar status of the so-called epistemic bubble, which I previously discussed, he coined the term autoimmunity. I am suggesting an expanded form of the expression to describe the omnipresent and self-sustaining character of the "embubblement" process. Indeed, the initial motif for presenting "autoimmunity", a concept with apparent negative connotations, is the examination of human skills in dealing with mistakes.

The concept of "autoimmunity" is used in biomedical sciences to refer to an agent's diminished well-being caused by a faulty immune system reaction. It is used to represent an aberrant event harming an individual's normal health because it denotes a group of disorders. In the context of this article, however, the term does not refer to a biological disorder because human cognition is not primarily defined as "healthy", as envisioned by the intents of the so-called "naturalization of logic" and from the eco-cognitive approach that I introduced in my research on abductive cognition [23].

The expression still refers to a troublesome and inadvertent reaction of the agent against themself; however, it no longer denotes an abnormal state. The autoimmune processes allow the human cognizer to enter a more relaxed state, such as when it comes to adopting decisions and to emotionally responding to troubling situations. The idea of cognitive autoimmunity is based on the intertwining of the agent's epistemological standing and their related cognitive and emotional condition.

5. The Moral Bubble Effect, Fallacies, and Moral Viscosity

Studies in logic, informal logic, and rhetoric always stress that fallacies, which are typical of human language at work especially in everyday situations, and that are prone to errors of various types, even concealed, possess what René Thom called "military intelligence" [35], in the framework of the catastrophe theory. The *softness* and *gentleness* that often accompany fallacies render them particularly efficient in intelligent strategies to protect groups, to affirm moral frameworks, and thus to generate possible more or less invisible violent effects. Moral bubbles constitute an important part of these processes of dissimulation because being unaware of our mistakes and/or violence in a fundamental and spontaneous way is often entwined with our own "certitude" that the arguments we are pushing and the related actions are absolutely not carriers of possible violent outcomes[17].

We must keep in mind that people use language and the so-called fallacies buried in it to achieve positive and crucial outcomes even though they can have violent consequences at the same time; fallacious expressions frequently have a violent effect on the eventual target agent. Very often errors are eco-cognitively fruitful in the perspective of the individuals of groups that commit them. I argue that the fallacies incorporated in human discussions, dialogues, and deliberations strongly potentiate the establishment and stability of *moral bubbles*, which have to be considered completely homomorphic with the epistemic bubbles. They regard moral/violent aspects and not mere cognitive/epistemological ones. In conclusion: ignorance of our errors is frequently linked to a lack of awareness of the deceptive/aggressive nature of our speech (and behavior).

Moral bubbles, from this perspective, are a great psychological mechanism that allows humans to legitimate and dissimulate violence at the same time. A derived result is a protection of our moral convictions at the individual level and of the moral frameworks that are acting in our collectives.

I introduced above the concept of moral "viscosity", which grants the preservation of a moral framework notwithstanding continuous transgressions. Moral bubbles are also extremely important in the light of moral viscosity because they can grant the stability of moral frameworks exactly permitting to avoid detection of the frequent violent effects that derive from their application. Viscosity describes how a moral actor might break moral laws, such as taking valuable items but still considering theft to be bad. In general, to make an example, he can commit acts of violence while preaching nonviolence as the path to happiness. This is not only the hypocritical effect of moral bubbles, it is one of the many central aspects of human moral bubbles, which provide a solution to the possible contrast and inconsistency between our moral adopted rules and subsequent actions.

In sum, thanks to moral bubbles, we value that, within our moral bubbles, even if we can be easily aware of probable real violent outcomes, this possibility is not activated and disappears from awareness. The actions that descend from serious moral convictions and rules are always endowed with a strong cognitive value because they are firmly tied to us, and we adhere to them without hesitation: possible generated violence disappears, because its cognitive value is terribly secondary and/or it is completely justified, and thus it can be disregarded and the subsequent unawareness legitimized.

6. Conclusions

In this article, I described, while exploiting the concepts of coalition enforcement, cognitive niche, and moral bubble, the important eco-cognitive aspects of moral and violent human behavior, thanks to the naturalization of morality in an evolutionary perspective. I also presented new insights on the illustration of the intertwining between morality and violence, providing a unified point of view substantiated by a naturalistic framework in which physical, biological, and cognitive processes can be concomitantly taken into account and the related naturalized role of violence unveiled. The last part addresses, in detail, the relevant problem of so-called moral bubbles, which prevent human agents from seeing the potential violence generated by their own moral acts.

Funding: Research for this article was supported by the PRIN 2017 Research 20173YP4N3—MIUR, Ministry of University and Research, Rome, Italy.

Acknowledgments: Some themes of this article were already discussed in my book *Understanding Violence. The Intertwining of Morality, Religion and Violence: A Philosophical Stance*, Springer, Cham, Switzerland, 2011. For the informative critiques and interesting exchanges that assisted me to enrich my analysis of the issues treated in this article, I am obligated to the two reviewers, and to my collaborators Selene Arfini and Alger Sans Pinillos.

Conflicts of Interest: The author declares no conflict of interest.

Notes

1 Fox News Channel, Published 28 April 2013—Last Update 9 December 2015: https://www.foxnews.com/world/2-police-officers-shot-outside-italian-premiers-office-in-alleged-plot-to-attack-politicians.

2 The reader interested in more details regarding the relationship between morality, religion, and violence and the related issue of relativism is addressed to chapter six of my book [1].

3 Among the philosophers, Derrida is certainly the only one to distinctly describe and analyze the link between violence and writing, offering considerations of great value on the subject [2–4]. His conclusion echoes, from a general philosophical perspective, the link between morality and violence that is at the basis of this article. The structure of the trace of writing (or difference), described by Derrida, reflects violence in the sense that the common and obvious violence is the vestige of a more fundamental and constitutive "arche-violence". See also chapter three of my book [1]. I add that surely violence is, and has been for a long time, a subject of reflection—for example, in Plato, addressing the relationships between persuasion and violence. Other philosophers, such as Hegel, Sartre, Kant, Weber, and Benjamin, have also discussed various roles of violence: it is not my concern to treat these

results in the present article because, in these cases, the relationship between morality and violence is completely ignored, and violence is standardly considered as the opposite of morality.

4 Those cognitive human behaviors that change the natural environment into a cognitive one are known as delegations of cognitive representations. They are cognitive delegations to the outside world that the mind has created through the construction of so-called "cognitive niches" over the history of culture. Humans have constructed voluminous cognitive niches, hugely endowed with informational, cognitive, and, more recently, computational processes and many kinds of artifacts, as illustrated by recent research in the field of sciences of evolution by Odling-Smee, Laland, and Feldman [9–11].

5 On the coevolution of intelligence, sociality, and language in the perspective of cognitive niches cf. Pinker [13].

6 Cf. Boehm [14].

7 On the moral and, at the same time, violent character of gossiping and fallacious reasoning, see the contributions given by Bertolotti, Bardone, and Magnani [1,15,16].

8 Humans are "docile" in this way, according to Simon [17], in the sense that their fitness is boosted by the inclination to rely on suggestions, recommendations, persuasion, and information gained through social channels as a primary basis for choosing.

9 For much of their evolutionary history, human groups have had an innate moral character, including behavioral prescriptions, social surveillance, and punishment of deviants, cf. Boehm [14] (p. 62) and [20].

10 It is precisely abduction—that is reasoning to hypotheses—that can first and foremost offer the possibility of detecting some appropriate *chances* presented by the environment and that can concurrently produce the possible subsequent efficient *changes* in terms of more sophisticated or innovative niche construction. I have always emphasized the importance of abductive reasoning in human and non-human animal cognition in my research [21–23]. Abductive conjectures can arise via selection from a collection of pre-stored hypotheses (selective abduction—for example, in medical diagnosis) or from the production of new ones (creative abduction—for example, in scientific discovery) [22].

11 On the role of extragenetic information in the evolutionary framework of the cognitive niche theory cf. the recent article of mine [24].

12 Constructing and manipulating visual representations, thought experiments, analogical reasoning, and so on are examples of model-based cognition; however, it also refers to the cognition that animals can obtain through emotions and other experiences. As stated by Peirce, all inference is a form of sign activity, where the word sign encompasses multiple model-based forms of cognition: "feeling, image, conception, and other representation" ([25], 5.283).

13 Cf. Hall [28].

14 Cf. Coolidge and Wynn [29].

15 The authors add in [31] that social approval/disapproval of behavior is adaptive because it tends to homogenize the behaviors, beliefs, and values of groups whose members engage cooperatively and docilely for reciprocal gain. It is hypothesized that a fundamental character of the man is their status of *Homo suadens*: if behavior is lauded it is correct behavior.

16 I note that, with naturalization of the relationship between morality and violence, I do not aim at building a psychological/behavioral theory but rather a new philosophical "stance" on that relationship, taking advantage of various results coming from different areas of science.

17 I specify that, clearly, the moral bubble involves people not seeing their acts as violent as opposed to their seeing their acts as morally justified violence or, alternatively, pleasing violence. The concept of the moral bubble is important because it addresses the common amazing human habit of obliterating potential or actual violence when based on moral concerns; in this case, when violence is activated, it is not seen at all as present, or it is simply disregarded as violence because the "moral" aspect dominates the cognitive scene at stake.

References

1. Magnani, L. *Understanding Violence. The Intertwining of Morality, Religion, and Violence: A Philosophical Stance*; Springer: Berlin/Heidelberg, Germany, 2011.
2. Derrida, J. *Of Grammatology [1967]*; Spivak, G.C., Translator; John Hopkins University Press: Baltimore, MD, USA, 1976.
3. Derrida, J. *Dissemination [1972]*; Johnson, B., Translator; Chicago University Press: Chicago, IL, USA, 1981.
4. Derrida, J. Force of law. "The mystical foundation of authority". *Cardozo Law Rev.* **1990**, *11*, 919–1045.
5. Dennett, D. Kinds of things—Towards a bestiary of the manifest image. In *Scientific Metaphysics*; Ross, D., Ladyman, J., Kincaid, H., Eds.; Oxford University Press: Oxford, UK, 2013; pp. 96–107.
6. Bingham, P.M. Human uniqueness: A general theory. *Q. Rev. Biol.* **1999**, *74*, 133–169. [CrossRef]
7. Bingham, P.M. Human evolution and human history: A complete theory. *Evol. Anthropol.* **2000**, *9*, 248–257. [CrossRef]
8. Boyd, R.; Richerson, P.J. Norms and bounded rationality. In *Bounded Rationality. The Adaptive Toolbox*; Gigerenzer, G., Selten, R., Eds.; The MIT Press: Cambridge, MA, USA, 2001; pp. 281–296.
9. Odling-Smee, F.J.; Laland, K.N.; Feldman, M.W. *Niche Construction. The Neglected Process in Evolution*; Princeton University Press: Princeton, NJ, USA, 2003.
10. Laland, K.N.; Sterelny, K. Perspective: Seven reasons (not) to neglect niche construction. *Evol. Int. J. Org. Evol.* **2006**, *60*, 4757–4779. [CrossRef]

11. Laland, K.N.; Brown, G.R. Niche construction, human behavior, and the adaptive-lag hypothesis. *Evol. Anthropol.* **2006**, *15*, 95–104. [CrossRef]
12. Wilson, D.S.; Timmel, J.J.; Miller, R.R. Cognitive cooperation. When the going gets tough, think as a group. *Hum. Nat.* **2004**, *15*, 1–15.
13. Pinker, S. The cognitive niche: Coevolution of intelligence, sociality, and language. *Proc. Natl. Acad. Sci. USA* **2010**, *107*, 8993–8999. [CrossRef] [PubMed]
14. Boehm, C. *Hierarchy in the Forest*; Harvard University Press: Cambridge, MA, USA, 1999.
15. Bardone, E.; Magnani, L. The appeal of gossiping fallacies and its eco-logical roots. *Pragmat. Cogn.* **2010**, *18*, 365–396. [CrossRef]
16. Bertolotti, T.; Magnani, L. An epistemological analysis of gossip and gossip-based knowledge. *Synthese* **2014**, *191*, 4037–4067. [CrossRef]
17. Simon, H.A. Altruism and Economics. *Am. Econ. Rev.* **1993**, *83*, 157–161.
18. Rohwer, Y. Hierarchy maintenance, coalition formation, and the origin of altruistic punishment. *Philos. Sci.* **2007**, *74*, 802–812. [CrossRef]
19. Lahti, D.C.; Weinstein, B.S. The better angels of our nature: Group stability and the evolution of moral tension. *Evol. Hum. Behav.* **2005**, *2*, 47–63. [CrossRef]
20. Boehm, C. Variance reduction and the evolution of social control. In Proceedings of the Annual Workshop on the Co-Evolution of Behaviors and Institutions, Santa Fe, NM, USA, November 2002.
21. Magnani, L. *Abductive Cognition. The Eco-Cognitive Dimensions of Hypothetical Reasoning*; Springer: Berlin/Heidelberg, Germany, 2009.
22. Magnani, L. *Abduction, Reason, and Science. Processes of Discovery and Explanation*; Kluwer Academic/Plenum Publishers: New York, NY, USA, 2001.
23. Magnani, L. *The Abductive Structure of Scientific Creativity. An Essay on the Ecology of Cognition*; Springer: Cham, Switzerland, 2017.
24. Magnani, L. Cognitive niche construction and extragenetic information: A sense of purposefulness in evolution. *J. Gen. Philos. Sci.* **2021**, *52*, 263–276. [CrossRef]
25. Peirce, C.S. *Collected Papers of Charles Sanders Peirce*; Hartshorne, C., Weiss, P., Burks, A.W., Eds.; Harvard University Press: Cambridge, MA, USA, 1974; Volume 1–6, pp. 1931–1958.
26. Bandura, A. *Social Learning Theory*; Prentice-Hall: Englewoods Cliffs, NJ, USA, 1977.
27. O'Gorman, R.; Wilson, D.S.; Miller, R. An evolved cognitive bias for social norms. *Evol. Hum. Behav.* **2008**, *29*, 71–78. [CrossRef]
28. Hall, K.R.L. Aggression in monkey and ape societies. *The Natural History of Aggression*; Carthy, J., Ebling, F., Eds.; Academic Press: London, UK, 1964; pp. 51–64.
29. Coolidge, F.L.; Wynn, T. Working memory, its executive functions, and the emergence of modern thinking. *Camb. Archealogical J.* **2005**, *5*, 5–26. [CrossRef]
30. Castro, L.; Toro, M.A. The evolution of culture: From primate social learning to human culture. *Proc. Natl. Acad. Sci. USA* **2004**, *101*, 10235–10240. [CrossRef] [PubMed]
31. Castro, L.; Castro-Nogueira, L.; Castro-Nogueira, M.A.; Toro, M.A. Cultural transmission and social control of human behavior. *Biol. Philos.* **2010**, *25*, 347–360. [CrossRef]
32. Kruger, J.; Dunning, D. Unskilled and unaware of it: How difficulties in recognizing one's own incompetence lead to inflated self-assessments. *J. Personal. Soc. Psychol.* **1999**, *77*, 1121–1134. [CrossRef]
33. Woods, J. *Errors of Reasoning. Naturalizing the Logic of Inference*; College Publications: London, UK, 2013.
34. Arfini, S. Ignorant Cognition: A Philosophical Investigation of the Cognitive Features of Not-Knowing. In *Studies in Applied Philosophy, Epistemology and Rational Ethics*; Springer: Cham, Switzerland, 2019; Volume 46.
35. Thom, R. *Esquisse d'Une Sémiophysique [Semio Physics: A Sketch]*; Meyer, V., Translator; InterEditions: Paris, France; Addison Wesley: Redwood City, CA, USA, 1990.

philosophies

Article

Information in Explaining Cognition: How to Evaluate It?

Nir Fresco [1,2]

[1] Department of Cognitive and Brain Sciences, Ben-Gurion University of the Negev, Be'er Sheva 841050, Israel; nfresco@bgu.ac.il
[2] Department of Philosophy, Ben-Gurion University of the Negev, Be'er Sheva 841050, Israel

Abstract: The claims that "The brain processes information" or "Cognition is information processing" are accepted as truisms in cognitive science. However, it is unclear how to evaluate such claims absent a specification of "information" as it is used by neurocognitive theories. The aim of this article is, thus, to identify the key features of information that information-based neurocognitive theories posit. A systematic identification of these features can reveal the explanatory role that information plays in *specific* neurocognitive theories, and can, therefore, be both theoretically and practically important. These features can be used, in turn, as desiderata against which candidate theories of information may be evaluated. After discussing some characteristics of explanation in cognitive science and their implications for "information", three notions are briefly introduced: natural, sensory, and endogenous information. Subsequently, six desiderata are identified and defended based on cognitive scientific practices. The global workspace theory of consciousness is then used as a specific case study that arguably posits either five or six corresponding features of information.

Keywords: cognition; cognitive science; sender; receiver; natural information; endogenous information; sensory information; desiderata; semantic information; scientific explanation

Citation: Fresco, N. Information in Explaining Cognition: How to Evaluate It? *Philosophies* **2022**, *7*, 28. https://doi.org/10.3390/philosophies7020028

Academic Editor: Marcin J. Schroeder

Received: 31 January 2022
Accepted: 22 February 2022
Published: 8 March 2022

1. Introduction

How is "information" used in the cognitive sciences? There is broad agreement that the brain and cognition involve information processing, and many theories explain neural, cognitive, and behavioural phenomena in informational terms [1–4]. However, given that "information" means different things to different people, it is hard to answer this question.

The need to bring conceptual order to the messy domain of existing theories of information was already recognised by Dretske [5]. There are many competing theories, ranging from quantitative theories of information flow [6–8], through evolutionary signalling game theories [9,10], to theories of semantic and biosemantic information [5,11–14]. In order to evaluate which theory of information should be appealed to in cognitive scientific explanations, one should first determine which notion of information is invoked.

The present analysis is motivated by the explanatory work semantic information often does in the cognitive sciences in virtue of its semantic properties. Such information can be found, for example, in animal communication studies. Referential signals exchanged between animals, plus possibly specific features of the sender, allow them to infer environmental states. Paradigmatic cases include the honeybee waggle dance and vervet alarm calls. What enables animals to make predictions based on such signals are correlations between signals and worldly states or events. The waggle dance correlates with the distance to and direction of nectar-rich flowers. Vervets produce acoustically distinct calls in response to (three) different predator types. The semantic content of these signals is supposedly the state or event with which the signal correlates [15].

More generally, semantic information in cognitive explanations often tracks the way changes occurring in the receiver's environment correlate with changes in the receiver. Cognitive explanations typically aim to understand how purposeful behaviour is produced by representations of the environment. Such representations supposedly carry different types of semantic information. Motor representations carry prescriptive information to behave in a specific way, whereas perceptual representations carry descriptive information about environmental states [16]. Semantic information is believed to close the loop between the signal and the environment: the signal is shaped by the environmental state, and the receiver's actions are (eventually) directed toward some environmental state [17].

Insofar as "information" is a theoretical construct that binds together different cognitive systems studied across the cognitive sciences, this concept—even when various subtypes of information are concerned—should be explicated. For it is often unclear which notion of information is used, and whether this notion does explanatory work in a respective neurocognitive theory or model.[1] The scientist would, thus, do well to be explicit about the notion of information that is used, if "information" is to play an explanatory role in their neurocognitive theory. The desiderata below can be used to match neurocognitive theories to suitable theories of information.

A key assumption underlying the motivation for specifying the desiderata is that information does explanatory work in the neurocognitive theory *T* and is not merely an explanatory gloss. Whether or not information does explanatory work is determined by *T*, rather than by the candidate information theory. If information plays an explanatory role in *T*, then the scientist should heed the proposed desiderata. In this sense, information should satisfy an analogue of Ramsey's "job description" challenge for representational explanations [18] (p. 34) by answering two questions affirmatively with respect to *T*. (a) Is there an explanatory benefit in *T* describing some cognitive, neural, or biological processes in informational terms? (b) If *T* uses "information" in describing such processes, do the underlying cognitive/neural/biological states play this sort of informational role, and, if so, how? The ensuing analysis provides a means for evaluating whether a candidate information theory entails the informational features that are posited in *T*.

The aim of the article is, thus, to identify the key features of information that a candidate theory of information should explain to qualify as a potentially good match for the neurocognitive theory *T*, which posits these informational features. The corresponding theory of information should satisfy the desiderata (rather than *T* as a neurocognitive theory). Given that some neurocognitive theories may posit very few features, and other will posit more, the proposed desiderata are not intended as necessary conditions. Moreover, the desiderata are defeasible and may change alongside the evolution of successful practices of the cognitive sciences, for they depend on what the requirements of *T* are with respect to *T*'s predictive accuracy, fruitfulness, and biological plausibility, for example.

We begin (Section 2) by briefly discussing some characteristics of explanation in the cognitive sciences in order to better motivate the proposed desiderata. In Section 3, a conceptual trichotomy of information is proposed as a backdrop for identifying and defending general desiderata for a cognition-friendly theory of information (in Section 4). In Section 5, we briefly evaluate the global workspace theory of consciousness—as a case study—for its appeal to different features of information. Section 6 concludes the article.

2. Unificatory Explanation in the Cognitive Sciences?

The success of the present analysis leans heavily on what qualifies as the "explanatory power" of a given neurocognitive theory, and the "explanatory role" of "information" in that theory. There are many theories of explanation out there with different implications for what is and is not explanatory, under which conditions, and why. For now, we briefly discuss some features of explanation in the cognitive sciences and their implications for the concept of information.

As an interdisciplinary endeavour, the cognitive sciences should supposedly be unified under some general standards of explanation, but current scientific practices may suggest otherwise. It is unclear whether the multiple research programmes ultimately converge together or are only loosely coordinated.[2] The latter alternative leads to some form of explanatory pluralism in which explanations are simply mutually constrained [19]. There exist many cognitive phenomena, often calling for competing explanations, and some approaches and models of these phenomena will likely be more explanatorily useful than others [20]. The upshot is that if multiple information-processing research programmes should be unified, or even only merely coordinated, in virtue of some common concept of "information", then "information" should be consistent across these programmes, thereby placing a greater clarificatory burden on the scientists.

The explanatory power of "information" as a theoretical concept cannot be determined independently of a theory. Does "information" contribute to the neurocognitive theory's predictive accuracy? (How well can the theory explain the available data by appealing to "information"?) Does it contribute to the theory's fruitfulness? (How well can the theory predict "novel" facts and deal with anomalies by appealing to "information"?) Does it contribute to the theory's unifying power? (How well can the theory bring together disparate domains by appealing to "information processing"?) Does it contribute to the theory's logical consistency? (Are the various appeals to "information" in the theory consistent?) [21] (pp. 53–54). As theories develop, new explanations are offered and tested. In that process, concepts, such as "computation", "information", and "representation", are refined. Consequently, the desiderata are not intended to be fixed and will likely change alongside the evolution of successful explanatory practices in the cognitive sciences.

Whether the cognitive sciences are (or should be) unified or simply coordinated, the desiderata can be useful. If neurocognitive theories trade in "information" and "information processing" in describing the entities and processes concerned, then regimenting usage by identifying key features of "information" can contribute to the unificatory effort. Nevertheless, even if cognitive scientists adopt divide-and-conquer strategies, thereby leading to explanatory pluralism, the desiderata can help to chart some mutual constraints on the various theories indicating where coordination is possible and where it is not. Regimenting the usage of "information" can contribute to stressing the explanatory role that this concept plays in information-based neurocognitive theories.

3. A Useful Trichotomy of Information in the Cognitive Sciences

As a backdrop for identifying and defending the desiderata, in this section we propose a trichotomy of information that distinguishes between some very general uses of "information" in the cognitive sciences. One notion refers to the correlation between different worldly events or states that can be exploited by the receiving organism in guiding its actions. Another notion refers to the content of sensory, motor, and cognitive states of an organism. We first briefly discuss a common conceptual distinction in the philosophy of science that is intended to encompass these two notions: natural and non-natural information. We then introduce the proposed trichotomy.

3.1. Natural and Non-Natural Information: The Gricean Path

The distinction between natural and non-natural information follows its Gricean counterpart between natural and non-natural meaning [22]. Natural meaning is both factive and agent-independent (e.g., these spots naturally mean measles). Non-natural meaning is non-factive and agent-dependent (e.g., three rings on the bell non-naturally mean that the bus is full—even when the bus is not full). In recent years, the Gricean analysis has been taken as distinguishing between two different types of information-carrying vehicle. Let us describe how the Gricean distinction is applied to information.

Natural (or correlational) information depends on a correlation between a bearer of natural information and its correlate. For Dretske, this correlation is both factive (i.e., the conditional probability that s is in state F, given that the cue—or signal—r is equal to 1) and lawful (this conditional probability relation is fixed by a law of nature, rather than by mere coincidences). However, these requirements are, arguably, too strong, and should be relaxed (as proposed, e.g., by Scarantino [14]) such that natural information simply increases the probability of the state of affairs it is about. (Still, only reliable correlations qualify as natural information.) As such, the content of natural information is not a fact, but the extent to which a given possibility is more or less likely.[3] Hence, smoke carries natural information about the presence of fire—even in the absence of fire. Similarly, insofar as hormone-mediated maternal effects can adaptively modulate offspring developmental trajectories in variable, yet predictable, environments [23], prenatal hormone levels carry natural information about the environmental state due to parental responses to that very environment.[4] Gricean natural meaning roughly corresponds to a cue or signal.

Non-natural information, however, need not reliably correlate with what it is about, and can be false. The statement "There is smoke" carries non-natural information about smoke whether it correlates with the presence of smoke or not [14] (p. 430). A visual representation of an external object with a 3D shape, *S*, likewise carries non-natural information about *S*′ spatial features that allows a mental rotation of this representation when *S* and another shape call for a comparison [24]. Such representations may, of course, be erroneous. One may misperceive a surface-colour gradient—due to errors in normal sensory processing—as an illumination gradient. A schizophrenic, or someone under the influence of psychedelic drugs, may see, hear, or smell things that do not exist outside their mind, thereby misrepresenting reality. Unlike natural information, non-natural information—just like non-natural meaning—is often understood as representation [14] (p. 430), Refs. [25,26] that is alethically evaluable.

Nonetheless, it is not clear whether non-natural information and representation are extensionally equivalent [13,27]. For one thing, even though some may be "convinced that a naturalistic theory of non-natural information/representation ought to be grounded in natural information" [14] (p. 430), it is unclear how that may be so for resemblance-based representation (recall the mental rotation task above). The answer to the above question also depends on whether one adopts a more restrictive or liberal concept of representation. On a liberal concept, such as Millikan's [28], representation simply requires that a consumer subsystem can fulfil its task normally when the producer subsystem goes into a state (e.g., eye blinking—as a defensive reflex action that protects the corneal surface from potential physical injury) that correlates with a given environmental condition (e.g., a noxious stimulus in the proximal environment).[5] Thus, even very rudimentary non-natural information, by Millikan's and similar views, may qualify as representation.

However, more restrictive concepts of representation may rule out many instances of non-natural information as non-representational. Let us consider just two more restrictive concepts—neither of which, it should be noted, deals explicitly with non-natural information as such. On Lloyd's view [29], for example, it is insufficient for a physical state to yield some behavioural output (e.g., eye blinking in response to a noxious stimulus) for that state to qualify as a representational vehicle (of the noxious stimulus). This state also has to depend—via multiple channels—on the simultaneous conjunction of multiple events (e.g., receptor events) all responding to the same, single environmental condition (e.g., the noxious stimulus). So, *even* if each individual channel carries information about the same environmental condition, they do not each qualify as a representational vehicle: "(w)ithout some further constraint, information provides a bad fit with (a) metatheory of representation" [29] (p. 51).

In the case of perceptual representation, according to Schulte [30], constancy mechanisms are needed in addition to the functions of a given system that track some environmental feature by means of processing information.[6] Constancy mechanisms have the function of producing "representational states that covary reliably with certain objective features in the world" based on vastly variable sensory information [30] (p. 128). "(T)heir presence in a sensory system is necessary and sufficient for the system's being a perceptual system" [31] (p. 413).

These two views suggest that a more prudent approach here would be to leave open the question of whether non-natural information and representation are extensionally equivalent, for additional theoretical constraints and/or mechanisms are required for some neural state to represent X besides just carrying information about X.

It is not always clear that "carrying information" about X suffices for the representation of X. For one thing, some neurocognitive theories simply try to explain whether a neuronal activity detects or responds to a particular environmental feature without claiming that it represents that feature. For example, the question of whether neurones in the striate cortex are sensitive to the presence of a circle or a vertical line need not presuppose that these neurones represent vertical lines. If natural information by itself is indeed insufficient for representation [3,27,32], it remains unclear in virtue of what that non-natural information qualifies as representation. Is it in virtue of functional factors (e.g., the information being exploitable by the processing organism)? Why is non-natural information always alethically evaluable? These questions are still open, and it is far from clear that non-natural information is a more well-understood category than representation.

3.2. Natural, Sensory, and Endogenous Information

To avoid such potential problems about the nature of non-natural information and to help to determine the applicability of the desiderata below in different explanatory contexts, we suggest distinguishing amongst natural, sensory, and endogenous information. To be sure, these three types of information are *not* on an equal ontological footing: sensory and endogenous information do not exist in the absence of a cognitive or sensorimotor system. The present analysis does not attempt to carve out the different types of information that are necessary and, supposedly, sufficient for the naturalisation of cognition. Others have suggested additional notions of information—which *are* on an equal ontological footing with natural information—that are required for such a naturalisation project, including control information [33] and information-as-structural-similarity [34] (p. 76). The present analysis is guided by the types of information posited by neurocognitive explanations. Let us discuss them in turn.

Natural information—as it is typically conceived—is exogenous: it concerns correlations, and often causal relations, between events, features, and states of affairs. "(O)ur brains (...) process exogenous information about the external environment by transducing physical phenomena (e.g., changes in energy, molecular concentrations, etc.) into sensory perceptions that allow us to generate and maintain a sense of what is happening around us" [35].

Photoreceptors, mechanoreceptors, and other receptors (even in more rudimentary organisms, such as molluscs) transduce different physical features of the environment affecting the organism into sensory information that is (typically) used to guide the organism's actions. It should, for starters, be clear that natural and sensory information are different. Why?

Even though sensory information can also be classified as being covariational in nature, it is not exhausted by tracking statistical regularities of the environment. Consider just a few reasons, starting with neural adaptation. The responsiveness of sensory cells tends to gradually decrease over time when facing the same stimulus repetitively. Receptors that are repeatedly exposed to smoke, say, may undergo neural adaptation, and thus their activation state may no longer be in a high correlation with fire. Another reason is the noisy environment: one cannot simply assume an optimal sampling of the environmental stimuli by the sensory apparatus. For example, the stochastic nature of reflection by environmental features and photon emission results in a variability in the number of photons sampled by photoreceptors within a given neural integration time [36]. Besides, even setting aside the possible corruption of sensory information by neural noise, sensory receptors are designed to maximise the signal-to-noise ratio for detecting sensory inputs. The photoreceptor, for example, is designed to intensify photon absorption to extract maximum light information [37]. Smoke is correlated with fire whether the latter is small or big, whereas the photoreceptor intensifies, and does not merely covary with, the absorbed photon. Many of the above differences stem from a more basic dissimilarity: vehicles of sensory information are evolutionarily designed to carry that information, whereas cues, as vehicles of natural information, are not. We may, therefore, conclude that sensory and natural information are not equivalent.

The last type of information in this trichotomy is thoroughly endogenous. Whilst sensory information is stimulus-driven, endogenous information can be roughly described as being expectation and/or goal driven. "(O)ur brains also process endogenous information that reflects our current internal homeostatic states, past experiences, and future goals" [35].

Prior information, which plays a key role in the debate on cognitive penetrability (i.e., is perception informationally encapsulated from cognitive information?) and in Bayesian models of cognition, is endogenous. A relatively recent study in cognitive penetrability has provided evidence that "V1 contains specific color information related to (an observed) object even (when) the sensory bottom-up signal is entirely achromatic" [38] (p. 65).

Similarly, V1 neurones were shown to fire on the apparent motion path as though real motion was present, even in the absence of a bottom-up signal (i.e., sensory information) through the retina or through lateral interactions [38] (p. 66). Additionally, binocular rivalry seems to show that complex, abstract information ("two objects cannot simultaneously be in the same place") may bias low-level visual processes (i.e., switching between the sight of a face and a house in the same visual field). Cognitive penetrability, then, seems to presuppose, under some circumstances, the modulation of sensory information by endogenous information.

Consider another, more radical example of endogenous information that is manifested in hallucinations. Hallucination is sometimes explained in terms of learning, Bayesian inference, and a reliability-based tradeoff between sensory information and prior expectations biased toward high-level priors. Expectation may dominate perception when high-precision prior predictions exert an inordinate influence over perceptual inferences, thereby yielding percepts with no corresponding environmental stimuli [39]. Thus understood, prior beliefs—as vehicles of endogenous information—play an active role in the construction of percepts in the absence of objectively identifiable natural information. It is "(t)he integration of exogenous (i.e., natural) and endogenous information (that) allows us to meaningfully interpret our surroundings, prioritize information that is relevant to our goals, and develop action plans" [35]. In the proposed conceptual trichotomy, natural information is exogenous and supposedly reflects a mind-independent, statistical regularity in the world (but see more on that in Section 4.6) that is the basis for sensory information. Non-natural information is a proper subset of the union of sensory and endogenous information; it refers to representations that can be true or false.

We can now proceed to identifying and defending the desiderata based on this tentative conceptual trichotomy.

4. Desiderata for Cognition-Friendly Theories of Information

In this section, we identify (and defend) six desiderata that are often posited in information-processing neurocognitive theories. These are: (1) the quantifiability of information; (2) the substrate neutrality of information; (3) the sender neutrality of information; (4) information being receiver dependent; (5) some information being non-symbolic, and (6) some information being mistakenly tokened. This list is not supposed to be either fixed—as noted above—or exhaustive: there can be other cognitive-theory-specific desiderata. Our focus is on the more common ones. We describe and elaborate on each of them in turn.

4.1. (D1) Quantifiability

Information should be quantifiable. The shift of psychology, from the stimulus-response paradigm of behaviourism to cognitive science, was partly inspired by information theories that attempted to formalise "information" and provide measures for quantifying it [1] (p. 1415). A very early example can be found in Miller's work [40] evaluating the limits on the amount of information that people can receive, process, and remember, drawing on information theory. There are at least two main reasons for the importance of quantification in psychological research. The first is that—like in the natural sciences—quantification supposedly ensures objectivity, precision, and rigour, thereby removing any biased influence by the scientist on the explanandum. The measurable properties of the explanandum are considered inherent in the phenomenon itself and are not merely imposed by the scientist. The second reason is that correlational analysis is required to identify effects and constructs that underlie the measured phenomena [41]. Thus, the quantifiability of information as an explanatory posit in psychology is desirable.

Quantification is likewise important in cognitive neuroscientific research, which studies and measures neural information processing. Neuroscientific techniques include single-cell recordings—tracking the firing of individual cells, fMRI—looking at neuronal activity at a larger scale, and EEG/MEG—tracking time-locked responses to stimuli, thereby providing insights into the underlying neural representations. Information in the brain is typically taken to be encoded by (a) patterns of activity across a single or many neurones (encoding information by neural populations or clusters), (b) the timing of spike trains (encoding information only in the location of the spike times), (c) the timing or phase of continuous neural activity (encoding information in the signal's amplitude), (d) synchrony across a neural population (encoding information by synchronised groups of neurones), or (e) some combination of the above [1]. The quantification of information is essential for scientists to find regularities amongst these patterns, calculate signal-to-noise ratios, evaluate the optimal neural coding, and identify bounds on information transmission and storage.

Consider two examples in which information is both explanatory and quantified. The first one concerns the sensory information that spike trains convey about the environment [42]. Scientists assert that single neurons convey large amounts of information, on the order of several bits per spike, and signals with more natural temporal correlations are said to be more efficiently coded. Rieke et al. ask "(h)ow do we quantify the notion that the spike train of a single cell "conveys information" about the sensory world?" and add that they are "search(ing) for sharper versions of (such) questions by forcing (themselves) to adopt a more precise and more mathematical language" [42] (p. 13).

Ultimately, they argue that Shannon's information theory provides a suitable mathematical framework for dealing with questions about information transmission by spike trains. Insofar as Rieke et al.'s analysis also encompasses endogenous anticipatory signals[7] or endogenous control signals, both sensory and endogenous information are amenable to quantification.

Another example concerns spatial cognition in foraging ants. Remembering landmarks and estimating distances and directions travelled over a specific period (i.e., also processing endogenous information) are cognitively hard problems. Ants can reduce their cognitive load by conveying information about the spatial coordinates of distant goals to their nestmates. Some ant species (e.g., *Formica cunicularia*) can even switch between different foraging strategies depending on environmental stimuli (e.g., the size of the available food sources) and "internal" stimuli (e.g., colony growth). One paradigm for quantifying such information transfers is based on the "binary tree" model, which requires ants to send (sensory) information about the sequence of turns they have to take to reach the perceived food. The model enables the experimenter to measure the ants' ability to share directional information with nestmates [43] (p. 1151). The experimenters know the amount of information (in bits) that should be sent—based on the high correlation between the food's location and a specific sequence of turns (thereby forming natural information). Thus, they can measure the time ants actually spend to send that information. The upshot is that any notion of information that is not quantifiable is too vague and imprecise to perform real theoretical work in cognitive science.

4.2. (D2) Substrate Neutrality

Information is neutral with respect to the implementing underlying substrate. The same information can be conveyed by various physical means (for a critique, see, e.g., Polger and Shapiro [44]). The proposed trichotomy reveals this desideratum quite straightforwardly: natural information may be conveyed by a variety of physical substrates in nature (e.g., smoke as a product of a material in combustion or light waves reflecting from an object). Sensory information is conveyed by electrochemical signals that result from the transduction of stimuli (e.g., light waves or patterns of air vibrations) by sensory receptors. Endogenous information is conveyed by hormones, neurotransmitters, electrochemical activities, etc. "(W)hen someone shouts "Fire!" (, substrate-neutral) information flows through a series of distinct information-media, each of which instantiates the information in its own unique way: first the ear drum, then the middle ear, cochlea, basilar membrane, (...) and, finally, the auditory nerve" [45] (p. 137).

Nevertheless, substrate neutrality—implying some degrees of freedom for implementing a given message—should not be confused with substrate independence [46] (p. 70). "(I)rrespective of the amount of Shannon information that can be embodied in a particular substrate, what (the signal) can and cannot be about also depends on the specific details of the medium's modifiability and its capacity to modify other systems" [47] (p. 402). The *same* message can be encoded by many *different* signals. In the transmission from a source to a destination, it can be encoded, for example, into an acoustic, electrical, or chemical signal—so long as the underlying medium has sufficient degrees of freedom. Spatial information about a particular object that is conveyed by light waves (which reflect from the object and its surroundings) can sometimes be conveyed, for example, by the reflections of high frequency sound waves (which bounce off that object). (That is, assuming that the receiver is equipped with visual and auditory apparatuses. See also desideratum D3.) However, the information does depend on being implemented by some substrate as its bearer. This idea has led to the inevitable claim about the impossibility of physically disembodied information[8]—particularly in the physics of computation [48,49].

A theory of information that ignores D2 may fall short of providing the theoretical scaffolding for neurocognitive theories that explain, for example, cross-modal sensory integration and combination, or the formation of episodic memories. Let us briefly consider both. Sensory information from different modalities is integrated to influence perception, decision making, and behaviour. "(Q)ualitatively different kinds of information from the various sense organs are put together in the brain to produce a unified, coherent representation of the outside world" [50] (p. 284). Clusters of neurones between sensory-specific areas were found to not only respond to the sensory information of different modalities, but also be capable of integrating these multisensory inputs. Sensory integration enhances and accelerates the detection, localisation, and reaction to biologically noteworthy events. It is also a key asset in signal disambiguation in both animal communication and human speech [51]. This fundamental characteristic of perception and cognition is enabled by the substrate neutrality of information.

Similarly, when we experience the world, our "brain is constantly bombarded by massive amounts of external sensory information which potentially could be encoded and stored into (episodic) memory" [52] (pp. 1198–1199). The binding of different sensory information into a unique, coherent episodic memory likely depends on neuronal activity between the entorhinal cortex and hippocampus. This activity binds temporal information about the sequence of events ("when"), spatial information about the experience ("where"), and the experiential information ("what"). Importantly, "memory retrieval can be cued by all types of sensory stimuli" [52]. The sight, smell, or taste of a particular teacup can be all that is needed to trigger memory retrieval: information about the presence of the teacup can be implemented by different substrates, yet it needs to be physically implemented.

4.3. (D3) Sender Neutrality

A sender should not always be presupposed in the flow of information. The need for this desideratum stems from the classical sender-receiver model in Shannon's information theory and signalling game theories. The common distinction between cues and signals in animal communication studies helps to defend the claim about sender neutrality. A cue, such as smoke or dark clouds, is "a feature of the world, animate or inanimate, which can be used by an animal to guide future actions" [53] (p. 3). A signal is "any act or structure which alters the behavior of other organisms, which evolved because of that effect, and which is effective because the receiver's response has also evolved" [53]. Signal exchange amongst organisms presupposes an actual sender that sends information, whereas cues do not.

A sender should, therefore, be distinguished from a source of information. A source of information can be a physical object (e.g., a fallen tree trunk in the wood) or a physical process (e.g., a wave breaking on the beach) that conveys information (e.g., an obstacle in the path in the wood) without any communicatory goals, and may remain completely unchanged as a result. Smoke and dark clouds—as cues—are correlated with fire and rain, respectively, and are, hence, abiotic sources—rather than senders—of information. Ignoring this desideratum is, typically, the result of focusing too much on symbolic information, which presupposes, at the very least, a potential sender. However, information flow in perceptual and other biological systems does not always require a sender. Identifying information flow with communication is too restrictive [17,54,55] and, thus, explaining how information flow is possible in the absence of a sender—particularly, in organism–environment interactions—is an important desideratum.

4.4. (D4) Receiver Dependence

Information flow ultimately depends on there being a receiver. Whilst it may be sender neutral, being receiver neutral takes the edge off of information informing an informee. Where signals are concerned, a sender is part of a special type of informational exchange that occurs when both it and the receiver have co-evolved to interact with each other on a regular basis [55] (p. 583). Where cues are concerned, however, the reaction to the source may contribute to a type of receiver's response that is beneficial over evolutionary time (ibid, 580–581). "A physical signal has semantic properties only where there is an interest-driven justification for the response it engenders" [16] (p. 96). A receiver is at the centre stage whether information is conveyed by cues or signals.

D4 implies a weaker and a stronger constraint on information. Consider again smoke and dark clouds as cues. They are arguably informative only relative to a receiver that is (a) sensitive to them and (b) can, at least in principle, exploit them [13,27,56–58]. A spatiotemporal correlation between smoke and fire cannot qualify as information to a receiver that is deprived of vision and olfaction. Even if two events are perfectly correlated, but no organism on the planet can detect them (not even humans equipped with cutting-edge technology), why should we say that one event carries information about the other?[9] Thus, the weaker requirement implied by D4 is the sensitivity of the receiver to the physical substrate embodying the information concerned. The stronger requirement, which entails the former, is that the receiver be able, at least in principle, to exploit the information concerned.[10] An amoeba may change its trajectory moving in the direction of a food cue. An adult ape may seek shelter from rain at the sight of dark clouds. A candidate theory of information may adopt either only the weaker constraint or both constraints.

Why is D4 important in the context of cognitive scientific explanatory practices? In short, because in the context of cognitive science, information makes little sense in the absence of an entity to be informed, be that an entire organism or interacting parts in the brain. Where sensory information is concerned, events, objects, and (other) organisms in the world are information sources, and the organism sensing its environment is the receiver. Natural events just unfold in the world and organisms that are sensitive to these events may exploit any extracted information to benefit them presently or in the future.

What about endogenous information exchanged between interacting parts of the brain? Do these parts act alternately as senders and receivers? Hallucination, again, is a paradigmatic example of endogenous information processing that is devoid of any real correlate in the external world. It may be argued, however, that some part of the brain produces that information and transmits it to other parts (and is, thereby, the sender). Despite the many conceptual and technical challenges that Shannon's sender-receiver model raises for neuroscientific techniques, neuroscientists should adopt the "cortex-as-receiver" perspective "to track the causal dynamics from one area to the next to establish whether a measured response is indeed information used by the rest of the brain" [1] (p. 1418). Does the firing of a V1 neurone ever cause activity in the MT or V2 areas, for example? "If it does not, it is a difference that never makes a difference. It is not information, even if it correlates with behavior" [1].

4.5. (D5) Symbolic/Non-Symbolic Information

Some information is non-symbolic, whereas other information is symbolic. Symbolic information includes statements, propositions, and sentences (not only in natural language, but also in programming languages and mathematics). It also includes maps, diagrams, and traffic signs. Because information processing in cognitive systems is arguably not limited to the processing of symbols, D5 seems to be a trivial desideratum. Despite the undeniable importance of language for the higher cognition of humans (e.g., remembering sentences is typically easier than remembering sensorimotor patterns), pre-linguistic babies and nonhuman animals process information that is non-symbolic. Of course, fully developed adult humans who are capable of processing symbolic information also regularly process non-symbolic information when sensing and acting on the natural world around them.

This claim is reminiscent of the heated debate—which culminated in the 1980s and 1990s—about whether cognitive processing is symbolic or sub-symbolic. The firings of neurones in vertebrate brains need not involve symbolic tokens, despite claims to the contrary [59]. On the symbolic view, cognitive processing is essentially symbolic: it consists of computing the consequences of enacting propositional attitudes based on inference rules. On the sub-symbolic view, cognitive processing is essentially associationist: spatiotemporally congruent events are associated in the brain by spreading activation patterns. Nonetheless, insofar as—at the very least—some information processing (e.g., proprioceptive or olfactory information) is non-symbolic, the neurocognitive theory concerned should (also) appeal to a theory of non-symbolic information.

Although a detailed evaluation of symbolic information exceeds the scope of this article, let us make some general observations to enable the ensuing analysis of D5. Symbolic reference is often contrasted with *iconic* and *indexical* reference. Iconic reference—used, e.g., in simple depiction and pantomime—depends on form similarity between the informational vehicle and its referent. Indexical reference—used, e.g., in pointing and innate forms of communication, such as facial expressions—depends on contiguity, correlation, or causal relations. Symbolic reference, however, is independent of any likeness or physical relation between the informational vehicle and its referent [60] (pp. 393–394). To interpret a collection of pebbles (shaped as "SOS") as symbolising a call for help, one must also understand social conventions, for neither the form nor the physical makeup of the pebble collection carries this information intrinsically. Symbolic reference depends on an encompassing system of relations within which the formal similarities and/or correlative aspects of the symbolic vehicle are embedded [60] (p. 399). This reference emerges from reflexive relations that symbols have to one another. For present purposes, we take symbolic information to be carried by spatiotemporal detachable signals that are part of a systematic, rule-governed, signalling system (adapted from Fresco et al. [13], p. 562). It depends on the signal's relations to objects and events in the world, as well as to other signals in the system. Whether or not a signal qualifies as a symbol depends on whether one adopts either a liberal or a restrictive definition of symbolic information.

Whilst in logic, and psycholinguistic theories, for example, the primary focus is on symbolic information, a vast number of research areas in cognitive science study the processing of non-symbolic information. Logic operates on symbolic structures; numbers are added together, and characters are compared and concatenated to form more elaborate symbolic structures. Psycholinguistic theories study the mental processes that are implicated in the acquisition, production, and comprehension of language. The cue-based retrieval theory, for example, accounts for the processing difficulty in language comprehension, and is based on architectures and mechanisms of human memory. The expectation-based parsing theory models classic sentence-processing phenomena using a Bayesian framework to predict which parts of a sentence will be more difficult to process. The central parsing challenge that these two influential (and other) psycholinguistic theories face is how to incorporate incoming, new symbolic information (phonemes, syllables, morphemes, and lexical items) into a dynamically forming complex representation [61]. Any such theory should provide "an account of what constitutes the input to the mental process—that is, what information is operated upon by those processes" [62] (italics added). Psycholinguistic theories, then, should appeal to a theory of symbolic information.

Nevertheless, other neurocognitive theories study phenomena that are underpinned by the processing of non-symbolic information. Models of animal signalling, for example, are relatively clear cases. Chemical communication in ants and the honeybee waggle dance are not typically analysed in terms of symbolic information (but cf. Gallistel [63], p. 145) who argues that the waggle dance symbolically specifies the direction of and distance to the nectar). The distance to the nectar is correlated with several dance components, including the duration of the sound, the number of waggles, the duration of the wagging run, and the duration of the return run [64] (p. 143). However, unlike symbolic information that *may* refer to spatiotemporally distal objects, events, or states of affairs, basic animal signals, such as the waggle dance, refer to the here-and-now, driven by the immediate circumstances of the message production [65] (p. 341). We may consider vervet (and other) alarm calls as functionally referential signals that offer important insights into the evolution of symbolic communication in human language.[11] Nonetheless, the notion of information that is used by animal communication scientists to describe these alarm calls is functional, rather than symbolic.

Moreover, theories that explain more complex neural, and cognitive, phenomena appeal to sensory and/or endogenous information that are not characterised as being symbolic. Let us only consider the appeal to "information" in motor control and learning studies. Motor control theories study the production of controlled, adaptive, and automatic movements, as well as the performance of efficient, coordinated, goal-directed movement patterns spanning multiple levels within the nervous system. The relationship between the task and the environment is critical when the cognitive agent selects and enacts specific motor plans. "(O)ptimal performance is achieved by feedback control law that resolves redundancy moment-by-moment—using all available information to choose the best action under the circumstances" [66] (p. 1227). The relevant information includes both endogenous information (e.g., control signals and goal states) and sensory information (e.g., combined visual and auditory information for estimating the position of a stimulus, or proprioceptive information about the location of one's limb combined with visual information of the limb itself [67] (p. 514)).

Relatedly, motor learning scientists study the complex neural processes during practice or repetition that lead to an improvement in the accuracy and smoothness of movement patterns in task performance. They examine variables that contribute to the formation of motor programmes, the strength of motor schemas, and the sensitivity of error-detection feedback processes. Information available during, and following, each practice or repetition is remembered and forms the basis for motor learning. The main sources of information are the action plan (endogenous information) and feedback (typically, sensory information). Is there any reason to assume that these kinds of information are necessarily symbolic? The following sample of questions—which underlie motor learning research—seems to suggest a negative answer. How are reward signals used to train a strategic process with a nearly infinite action space? How are error signals used to update internal models that map desired goals and the motor responses that are necessary to achieve those goals? How are discrepancy signals used to adjust the relevant movement strategies? [68]. The upshot is that such neurocognitive theories posit the processing of non-symbolic information, and, therefore, require a theory of non-symbolic information.

4.6. (D6) Mistaken Tokening

Finally, information may be mistakenly tokened. The possibility of "mistaken information" clearly exists. Consider illusory and hallucinatory perceptions: they are not veridical, as they do not track the actual state of affairs. There is nothing in the environment corresponding to the mistaken information that the agent processes. Yet, veridical, illusory, and hallucinatory percepts, arguably, stand on an equal cognitive—though not epistemic—footing, insofar as they are all processed in a like manner (for a discussion to that effect see, e.g., Corlett et al. [39], and for an opposing view, see, e.g., Moran [69]). It is, hence, important to consider the production of information that can be mistaken in some cases.

Is it up to the theory of information, then, to specify the conditions under which information is mistaken? Notice, first, that D6 deliberately does not specify a priori whether mistaken tokening amounts to information being false or simply inaccurate. Several philosophers have recently shifted to attributing, at least to perceptual states, accuracy rather than truth conditions [70–73]. Accuracy and truth are related, but have distinct properties, for the latter is binary (at least under classical logic), whereas the former admits of degrees [70] (p. 458). Maps and pictures, for example, are more or less accurate—rather than true or false—depending on their degree of resemblance or isomorphism to what they stand for [71] (fn. 35). Truth conditions, on the other hand, imply that the information concerned asserts that something is the case in a manner akin to propositions or linguistic constructs [72] (pp. 59–60). Hence, whilst it certainly makes sense for some instances of symbolic information (e.g., propositions in logical or natural languages) to be true or false, it is far from clear that non-symbolic information (e.g., an eagle alarm call produced by an infant vervet monkey in response to a bird that is not an eagle) can be alethically evaluable—though it can be more or less accurate.

Three different explanatory approaches to information suggest that a theory of natural information need not specify the conditions under which natural information is mistakenly tokened. On the Dretskean approach, natural information is factive, and, thus, it cannot be erroneous. As Dretske famously put it, "false information, and mis-information are not kinds of information—any more than decoy ducks and rubber ducks are kinds of ducks" [5] (p. 45). It is the representations that fail to carry the information that they are supposed to carry that qualify as misrepresentations [5] (p. 192). Neural informational states that do not properly track the world should, on Dretske's view, be part of a naturalistic explanation of misrepresentation. Nevertheless, explaining such normative factors (i.e., the misalignment between neural states and the world), on his view, falls outside the scope of theories of natural information per se.

According to recent probabilistic approaches to natural information, such as Scarantino's [14], information cannot be mistaken for another reason. There is no error if a cue, or signal, carries information that X is more probable, but X does not occur [14] (pp. 439–440). In other words, natural information is non-factive. Recall that, in Scarantino's theory, the requirement of perfect conditional probability is dropped. The probability of an event occurring can truly be 0.9, even if the correlated event does *not* occur at a given time. Natural information is simply an incremental change in the probability of an underlying event (e.g., there being fire) relative to some prior probability. (This prior probability is fixed by specific background data—as in Bayesian confirmation theory.) An error occurs if the receiver takes the cue/signal to stand for X when X is not the case. This, however, is an instance of either non-natural information or sensory information.

On a third approach, advocated by Baker [74], natural information is factive, but is based on physical necessity, rather than on laws of nature—as in Dretske's approach. "Covariance between smoke and fire, between foxes and rabbit brains, and between fundamental particles all share the feature of being invariant under a range of initial conditions" [74] (p. 14). Whilst there are no laws of nature that apply without exception in cases of information transmission between foxes and rabbit organs, for example, there are relationships of physical necessity that are sensitive to initial conditions. Such initial conditions include the presence of a live fox, patterns of air vibration in the rabbit's ears, the presence of a working rabbit brain, and the presence of a life-sustaining planet.[12] Even on Baker's approach, theoretical resources beyond the theory of information are needed to explain how information may be mistakenly tokened in some neural states.

The interim upshot is that organisms clearly make mistakes, and it remains unclear how such mistakes may be accounted for by natural information. Perhaps, "(s)ome mistakes are due precisely to the reception of probabilistic information about events that fail to obtain" [75] (p. 319). So long as there is a receiver of information, there always exists the possibility of a mistake based on that information. However, natural information—even when it is weakly construed—cannot be either false or mistaken. What may be the cause of such mistakes? If we adopt the classical sender-receiver model, there seem to be three options available: the message may be distorted by a noisy channel, the receiver's decoding procedure may malfunction, or the receiver may misinterpret the (decoded) message. If mistaken information is unaccounted for, the explanatory value of information is unclear.

Sensory and endogenous information are often described in neurocognitive theories as incorrect or inaccurate. The ubiquity of cognitive phenomena in which there is a mismatch between the world and how it is perceived by an organism calls for an explanation of that mismatch. A neurocognitive theory that attributes the mismatch to mistaken information can (and should) explain how that information may have been distorted by noise. The McGurk effect is a clear example of sensory integration in which a perceptual error occurs when one misperceives sounds due to a mismatch between the audio and visual parts of speech. Even motion sickness is sometimes explained by a discrepancy amongst current visual information, vestibular information, and proprioceptive information, based on a temporal comparison with prior information from the immediate past. Nevertheless, it is the relevant theories of information that should specify the accuracy (or truth) conditions of information. The burden of explaining how sensory and endogenous information can be mistakenly tokened lies both with the corresponding theories of information and the neurocognitive theories that appeal to the underlying types of information.

This concludes our outline of the desiderata for cognition-friendly theories of information (see Table 1 below).

Table 1. A list of identified desiderata and their applicability to three types of information.

Desideratum	Natural Information	Sensory Information	Endogenous Information
1. *Quantifiability*	✔	✔	✔
2. *Substrate-Neutrality*	✔	✔	✔
3. *Sender-Neutrality*	✔	✔	✔
4. *Receiver-Dependence*	✔	✔	✔
5. *Symbolic/Non-Symbolic*	n/a	✔	✔
6. *Mistaken Tokening*	n/a	✔	✔

5. Global Workspace Theory: Which Features of Information Are Posited?

In this penultimate section, we briefly evaluate how one influential theory in cognitive science—the global workspace theory (GWT) of consciousness—presupposes the above desiderata (at least partially; see the discussion below concerning D1). According to GWT, what one experiences as a conscious state,[13] at any given moment, is the global broadcasting of information across an interconnected network of prefrontal-parietal areas and other distant high-level sensory areas [76] (p. 911). Several sensory and other specialised modular circuits compete and cooperate for access to the limited-capacity global workspace, and only the more salient inputs are those that are eventually selected and broadcasted. This processing remains unconscious, until some underlying activity exceeds a certain relevance threshold and ignites the global workspace. When ignition occurs, the salient information is broadcasted and sustained until it decays and remains silent [76] (p. 912). That information is made consciously accessible to many local processes, including memory, attention, motor planning, and verbal reporting [77]. Despite GWT still being a controversial theory [76], a large body of empirical findings seems to be consistent with this theory [78]. Nonetheless, we simply focus here on the features of information posited by a neurobiologically informed version of GWT—the global neuronal workspace (GNW) theory (GNWT).

How do the localised modular cortical areas and the GNW interact? The localised modules all connect to the GNW, and process specific perceptual, motor, memory, and evaluative information preconsciously [78] (p. 777). The GNW is formed by a large, interconnected network of long-range cortical neurones distributed over the prefrontal, cingulate, and parietal regions with reciprocal horizontal projections to neurones in other cortical areas through long-range excitatory axons [79] (p. 210). GNW neurones typically accumulate information through recurrent top-down/bottom-up loops, in a competitive manner such that only a single representation ultimately achieves a global conscious status. There is, thus, no single brain centre where conscious information is collected and dispatched, but rather a synthesis of multiple processes converging to a cohesive metastable state [80] (pp. 56–58).

Let us evaluate, in order, which features of information GNWT posits, starting with D1. GNWT predicts that consciously available information can be identified with a deeper and more prolonged propagation of information through long-distance connections, as compared with information that remains unconscious. Dehaene et al. report, for example, that some "paradigms afforded a precise measurement of the timing of information progression and conscious access in the visual system" [80] (p. 71). They add that various neuroimaging "data suggest that conscious access causes a major change in the availability of information that is easily detected by a variety of subjective and objective measures" [80] (p. 67). A recent EEG study in humans indexed levels of consciousness using mid-range and long-range weighted mutual information as a measure of information sharing [78] (p. 787). Nevertheless, it seems that though GNWT presupposes that information *is* quantifiable, its present measurements are mostly *indirect*. That is, rather than applying standard information-theoretic measures, such as differential entropy, mutual information, or Kullback-Leibler divergence, indirect measures of brain activity are used. Consider, for example, an ongoing international adversarial collaboration that aims to reveal the footprints of consciousness [76]. Participating scientists claim that GNWT predicts that the ignition of the global workspace (i.e., information sharing) is measurable by long-range synchrony between the prefrontal and sensory cortices. Patterns of this information sharing can be *approximated* by measures of (gamma/beta) synchronisation between the sensory response/intrinsic activity and the evoked stimulus [81]. The upshot is that *if* the explanatory role of information in GNWT is contingent on that information being measured *directly*, then this role may be less central then it would otherwise appear to be.

The fact that different types of information are processed in the localised modular cortical areas prior to ever reaching the GNW implies that D2 is likewise assumed by GNWT. The relevant cortical areas are said to process perceptual, motor, memory, and evaluative information. GNWT aims to explain how information across distributed cortical processes can be integrated despite their difference. The hypothesised binding mechanisms are believed to co-select distributed feature representations that are part of a single object, thereby explaining why conscious object representation is usually coherent and integrated [78] (p. 783).

Whilst D4 is conspicuously posited by GNWT, D3 is only trivially assumed by the theory insofar as some sensory information received originates in the abiotic world. GNWT explains that when one is conscious of some information, different brain circuits have access to that information. Hence, there is an organism that not only receives the information but is conscious of that information. Furthermore, distinct specialised modular cortical areas first receive specific perceptual, motor, memory, and evaluative information from the respective neural subsystems. Then, when competing for access to the GNW, they become information senders. Likewise, the GNW acts as a receiver when it receives information from the winning modules, and then as a sender that broadcasts that information to other cognitive, sensory, and motor subsystems. Information thus plays an explanatory role in that it is part of a causal chain that may start either in the external world (sensory information) or in the organism itself (endogenous information), culminating in a successful receipt by the GNW. In sum, the receiver concerned is either the GNW that receives bottom-up information from localised modular cortical areas, or any specific modular cortical area that receives top-down information from the GNW.

D5 is posited in various aspects of GNWT. First, the empirical paradigms that are used to test conscious and unconscious information-processing appeal to both symbolic and non-symbolic information. Language masking experiments, for example, test how words that are presented in close spatial and temporal proximity with other visual stimuli sometimes become unconscious. An early such study by Dehaene et al. [82] (p. 757) showed that unmasking words (i.e., symbolic information) enables the propagation of the activation and ignition of a large-scale correlated cerebral assembly. Focusing on non-symbolic, sensory information, a study by King et al. [83] tested how brains encode many features of a visual stimulus, and showed that only task-relevant features (in that case, presence, angle, and visibility) are later maintained during the delay period, even when the stimulus is reported as unseen. Subjects were asked to detect and mentally maintain the orientation of a masked grating, which is clearly non-symbolic. Second, the precise format of information in the GNW remains an open question [84] (p. 166). One possibility is that it reflects the underlying structure of the winning sensorimotor system. Another is that *symbolic* information is the suitable medium for integrating the output of heterogeneous processes. Last, understanding self-consciousness, in GNWT, may be based on the capacity for recursive thought, which may require, in turn, *symbolic* capacities [78] (p. 791). The upshot is that GNWT posits the processing of both symbolic and non-symbolic information.

Finally, GNWT also posits D6: effortful conscious processing is error-prone and the communication between brain circuits and GNW is susceptible to noise. First, some hypothesise that a salience network in the GNW monitors the relevance and/or salience of sensory and endogenous information entering the GNW relative to unimportant background noise [85]. Dysfunctional gatekeeping may result in the inadvertent broadcasting of information, thereby leading to abnormal conscious perceptions—as in hallucination. Furthermore, theoretical constructs of signal detection theory are claimed to map onto specific pre-stimulus and post-stimulus states of GNW neuronal activity [78] (p. 791): a noisy internal representation of the stimulus, and a decision threshold corresponding to the ignition threshold in the GNW. Mistaken information that might become conscious is a live possibility according to GNWT. More generally, insofar as any *specific* global workspace model appeals to the very distinction between message and signal for explaining the possibility of noise, distortion, or mismatch between encoding and decoding, information plays an explanatory role in that model in describing disturbances of consciousness.

By way of concluding this section, we briefly reply to a possible objection pertaining to GNWT as a particular case study. Above we have claimed that "information" is highly dependent on context with respect to the *specific* neurocognitive theory that uses it. Given that GNWT seems to satisfy all six desiderata, *possibly* except for D1, it might seem to have been cherry-picked to provide further evidence for these desiderata. However, examining every information-based neurocognitive theory is clearly impractical. Having chosen a different case study that shows that only four (or three) features of information are posited would not have invalidated the remaining desiderata. Rather, it could have shown that "information" plays a lesser explanatory role in *that* theory as compared with GNWT. Agreeing upon *all* the desiderata is not an easy matter, as "information" is used differently and *possibly* inconsistently by extant neurocognitive theories. Nonetheless, dissent about the importance and relevance of the desiderata is a step in the right direction in determining the explanatory work "information" does in specific information-based neurocognitive theories.

6. Conclusions

Information is sometimes only an explanatory gloss rather than a key explanatory construct in understanding cognition; but when information does play an important explanatory role, it is methodologically useful to explicate that role. To that aim, having distinguished amongst natural, sensory, and endogenous information, we have identified and defended six desiderata that information-based neurocognitive theories often posit: (a) *quantifiability*, (b) *substrate neutrality*, (c) *sender neutrality*, (d) *receiver dependence*, (e) *symbolic/non-symbolic*, and (f) *mistaken tokening*. These desiderata can be used to evaluate (1) how competing theories of information fare as a foundation for such neurocognitive theories, and (2) the explanatory role that information plays in a given neurocognitive theory. For example, a perceptual theory that does not appeal to the possibility of information being false or inaccurate in explaining misperceptions or hallucinations is likely less reliant on information as an explanatory construct. To show how central the notion of information might be in some information-based neurocognitive theories, we have briefly evaluated the global workspace theory of consciousness for its appeal to different features of information. Of course, other neurocognitive theories may posit fewer (or, alternatively, even more) features of information. In such cases, the neurocognitive theory concerned would place fewer (or, alternatively, more) constraints on suitable theories of information. Understanding the explanatory role and the precise notion(s) of information used in cognitive science is an important steppingstone in explaining representation.

Funding: This research was funded by the Israel Science Foundation Grant 386/20.

Acknowledgments: My gratitude goes to all those who contributed to this article both through lively discussions and helpful comments on earlier versions. They include Dimitri Coelho Mollo, Matteo Colombo, Lotem Elber-Dorozko, Tomer Fekete, Eva Jablonka, Gil Lipkin-Shahak, Marcin Miłkowski, Jonathan Najenson, Itay Navon, Andrea Scarantino, Oron Shagrir, Nicholas Shea, Adam Singer, Aaron Sloman, Phillip Staines, Gal Vishne, Marty Wolf and Rotem Yahel. Thanks are also due to several anonymous reviewers and anyone else whom I may have inadvertently omitted. This research was also partly supported by stipendiary fellowships from both Sidney M. Edelstein Centre for History and Philosophy of Science Technology and Medicine, and the Kreitman School of Advanced Graduate Studies at Ben-Gurion University of the Negev.

Conflicts of Interest: The author declares no conflict of interest.

Notes

1 Despite the differences between them, cognitive "theories" and "models" are used interchangeably hereafter.

2 For a critique of the prospects of an integrative neuroscience, see, e.g., Sullivan [86]. She argues that the multiplicity of distinct experimental protocols used to examine the same supposed phenomenon provide evidence against this integrative endeavour.

3 We discuss the factivity of natural information further in Section 4.

4 Therefore, maternal hormones provide a mechanism for transferring environmental cues from parents to offspring.

5 An odd consequence of this view is that an activation of the reflex without the stimulus counts as a misrepresentation.

6 One may also argue that other mechanisms (e.g., a simultaneity constancy mechanism) are required when combining multisensory information (e.g., visual and tactile) about a single environmental event (because different kinds of information take varying amounts of time to be processed in the brain).

7 A recent study of V1 and V2 activity in macaque monkeys reports that the "anticipatory signal reflects a nonsensory component of cortical activity that is (. . .) not related to stimulus coding or choice behavior" [87] (p. 5199).

8 "Silence may be very informative. This is a peculiarity of information: its absence may also be informative" [12] (p. 88). If so, then silence supposedly qualifies as *disembodied* information. However, it does not qualify as *information* per se, but rather as being *informative* as part of an inferential process that includes other background information [88].

9 For a similar reason, Scarantino, for example, deems natural information "an objective commodity" that is nonetheless "mind-dependent" [14] (p. 432) relativised to potential receivers.

10 Rathkopf similarly argues that if an organism cannot exploit an XY correlation, even in principle, for some biological end, then that correlation cannot be legitimately used to compute the mutual information between events X and Y [58] (p. 324).

11 This view has been contested [89,90].

[12] It remains unclear, though, (a) why cognitive systems should be in tune with such physical necessities and initial conditions, and (b) how physical necessity is grounded if not by a law of nature.

[13] The question of whether consciousness is the cause or rather the outcome of access to the global workspace does not affect our main argument here.

References

1. De-Wit, L.; Alexander, D.; Ekroll, V.; Wagemans, J. Is neuroimaging measuring information in the brain? *Psychon. Bull. Rev.* **2016**, *23*, 1415–1428. [CrossRef] [PubMed]
2. Koch, C. *Biophysics of Computation: Information Processing in Single Neurons*; Oxford University Press: New York, NY, USA, 1999.
3. Ramos, R.T. The concepts of representation and information in explanatory theories of human behavior. *Front. Psychol* **2014**, *5*, 1034. [CrossRef] [PubMed]
4. Tononi, G.; Boly, M.; Massimini, M.; Koch, C. Integrated information theory: From consciousness to its physical substrate. *Nat. Rev. Neurosci.* **2016**, *17*, 450–461. [CrossRef] [PubMed]
5. Dretske, F.I. *Knowledge & The Flow of Information*; The MIT Press: Cambridge, MA, USA, 1981.
6. Fisher, R.A. On the Mathematical Foundations of Theoretical Statistics. *Philos. Trans. R. Soc. A Math. Phys. Eng. Sci.* **1922**, *222*, 309–368.
7. Hartley, R.V.L. Transmission of Information. *Bell Syst. Tech. J.* **1928**, *7*, 535–563. [CrossRef]
8. Shannon, C.E. A mathematical theory of communication. *Bell Syst. Tech. J.* **1948**, *27*, 379–423. [CrossRef]
9. Lewis, D. *Convention: A Philosophical Study*; Wiley-Blackwell: Cambridge, MA, USA, 1969.
10. Skyrms, B. *Signals: Evolution, learning, & information*; Oxford University Press: Oxford, UK, 2010.
11. Bar-Hillel, Y.; Carnap, R. Semantic Information. *Br. J. Philos. Sci.* **1953**, *4*, 147–157. [CrossRef]
12. Floridi, L. *The Philosophy of Information*; Oxford University Press: Oxford, UK, 2011.
13. Fresco, N.; Ginsburg, S.; Jablonka, E. Functional Information: A Graded Taxonomy of Difference Makers. *Rev. Philos. Psychol.* **2020**, *11*, 547–567. [CrossRef]
14. Scarantino, A. Information as a Probabilistic Difference Maker. *Australas. J. Philos.* **2015**, *93*, 419–443. [CrossRef]
15. Stegmann, U.E. A primer on information and influence in animal communication. In *Animal Communication Theory: Information and Influence*; Stegmann, U.E., Ed.; Cambridge University Press: Cambridge, UK, 2013; pp. 1–39.
16. Rathkopf, C. What Kind of Information is Brain Information? *Topoi* **2017**, *39*, 95–102. [CrossRef]
17. Cao, R. A teleosemantic approach to information in the brain. *Biol. Philos.* **2012**, *27*, 49–71. [CrossRef]
18. Ramsey, W. *Representation Reconsidered*; Cambridge University Press: Cambridge, UK, 2007.
19. Miłkowski, M.; Hohol, M. Explanations in cognitive science: Unification versus pluralism. *Synthese* **2020**, *199*, 1–17. [CrossRef]
20. Colombo, M.; Hartmann, S. Bayesian Cognitive Science, Unification, and Explanation. *Br. J. Philos. Sci.* **2017**, *68*, 451–484. [CrossRef]
21. McMullin, E. Explanatory Success and the Truth of Theory. In *Scientific Inquiry in Philosophical Perspective*; Rescher, N., Ed.; University Press of America: Lanham, MD, USA, 1987; pp. 51–73.
22. Grice, H.P. Meaning. *Philos. Rev.* **1957**, *66*, 377–388. [CrossRef]
23. Groothuis, T.G.G.; Hsu, B.-Y.; Kumar, N.; Tschirren, B. Revisiting mechanisms and functions of prenatal hormone-mediated maternal effects using avian species as a model. *Philos. Trans. R. Soc. B Biol. Sci.* **2019**, *374*, 20180115. [CrossRef]
24. Hyun, J.-S.; Luck, S.J. Visual working memory as the substrate for mental rotation. *Psychon. Bull. Rev.* **2007**, *14*, 154–158. [CrossRef]
25. Birch, J. Propositional content in signalling systems. *Philos. Stud.* **2014**, *171*, 493–512. [CrossRef]
26. Ramstead, M.J.D.; Veissière, S.P.L.; Kirmayer, L.J. Cultural Affordances: Scaffolding Local Worlds through Shared Intentionality and Regimes of Attention. *Front. Psychol.* **2016**, *7*, 1090. [CrossRef]
27. Shea, N. *Representation in Cognitive Science*; Oxford University Press: Oxford, UK, 2018.
28. Millikan, R.G. Biosemantics. *J. Philos.* **1989**, *86*, 281–297. [CrossRef]
29. Lloyd, D.E. *Simple Minds*; The MIT Press: Cambridge, MA, USA, 1989.
30. Schulte, P. Perceptual representations: A teleosemantic answer to the breadth-of-application problem. *Biol. Philos.* **2015**, *30*, 119–136. [CrossRef]
31. Burge, T. *Origins of Objectivity*; Oxford University Press: Oxford, UK, 2010.
32. Artiga, M.; Sebastián, M.Á. Informational Theories of Content and Mental Representation. *Rev. Philos. Psychol.* **2020**, *11*, 613–627. [CrossRef]
33. Corning, P.A. Control information theory: The "missing link" in the science of cybernetics. *Syst. Res. Behav. Sci.* **2007**, *24*, 297–311. [CrossRef]
34. Miłkowski, M. The Hard Problem of Content: Solved (Long Ago). *Stud. Log. Gramm. Rhetor.* **2015**, *41*, 73–88. [CrossRef]
35. Chang, L.J.; Jolly, E.; Cheong, J.H.; Rapuano, K.M.; Greenstein, N.; Chen, P.-H.A.; Manning, J.R. Endogenous variation in ventromedial prefrontal cortex state dynamics during naturalistic viewing reflects affective experience. *Sci. Adv.* **2021**, *7*, eabf7129. [CrossRef]
36. O"Carroll, D.C.; Wiederman, S.D. Contrast sensitivity and the detection of moving patterns and features. *Philos. Trans. R. Soc. B Biol. Sci.* **2014**, *369*. [CrossRef]

37. Okawa, H.; Miyagishima, K.J.; Arman, A.C.; Hurley, J.B.; Field, G.D.; Sampath, A.P. Optimal processing of photoreceptor signals is required to maximize behavioural sensitivity: Rod signal-to-noise and behavioural threshold. *J. Physiol.* **2010**, *588*, 1947–1960. [CrossRef]
38. Vetter, P.; Newen, A. Varieties of cognitive penetration in visual perception. *Conscious. Cogn.* **2014**, *27*, 62–75. [CrossRef]
39. Corlett, P.R.; Horga, G.; Fletcher, P.C.; Alderson-Day, B.; Schmack, K.; Powers, A.R. Hallucinations and Strong Priors. *Trends Cogn. Sci.* **2019**, *23*, 114–127. [CrossRef]
40. Miller, G.A. The magical number seven, plus or minus two: Some limits on our capacity for processing information. *Psychol. Rev.* **1956**, *63*, 81–97. [CrossRef]
41. Tafreshi, D.; Slaney, K.L.; Neufeld, S.D. Quantification in psychology: Critical analysis of an unreflective practice. *J. Theor. Philos. Psychol.* **2016**, *36*, 233–249. [CrossRef]
42. Rieke, F.; Warland, D.; de Ruyter van Steveninck, R.; Bialek, W. *Spikes: Exploring the Neural Code*; The MIT Press: Cambridge, MA, USA, 1996.
43. Reznikova, Z. Spatial cognition in the context of foraging styles and information transfer in ants. *Anim. Cogn.* **2020**, *23*, 1143–1159. [CrossRef] [PubMed]
44. Polger, T.W.; Shapiro, L.A. *The Multiple Realization Book*; Oxford University Press: Oxford, UK, 2016.
45. Mangan, B. Against functionalism: Consciousness as an information-bearing medium. In *Toward a Science of Consciousness II: The Second Tucson Discussions and Debates*; Hameroff, S.R., Kaszniak, A.W., Scott, A., Eds.; The MIT Press: Cambridge, MA, USA, 1998; pp. 135–141.
46. Miłkowski, M. *Explaining the Computational Mind*; The MIT Press: Cambridge, MA, USA, 2013.
47. Deacon, T.W. *Incomplete Nature: How Mind Emerged from Matter*; W. W. Norton & Company: New York, NY, USA, 2011.
48. Bennett, C.H. Logical Reversibility of Computation. *IBM J. Res. Dev.* **1973**, *17*, 525–532. [CrossRef]
49. Landauer, R. Information is Physical. *Phys. Today* **1991**, *44*, 23–29. [CrossRef]
50. Ghazanfar, A.; Schroeder, C. Is neocortex essentially multisensory? *Trends Cogn. Sci.* **2006**, *10*, 278–285. [CrossRef]
51. Stein, B.E.; Stanford, T.R.; Rowland, B.A. The neural basis of multisensory integration in the midbrain: Its organization and maturation. *Hear. Res.* **2009**, *258*, 4–15. [CrossRef]
52. Sugar, J.; Moser, M. Episodic memory: Neuronal codes for what, where, and when. *Hippocampus* **2019**, *29*, 1190–1205. [CrossRef]
53. Maynard Smith, J.; Harper, D. *Animal signals*; Oxford University Press: Oxford, UK, 2003.
54. Godfrey-Smith, P. Sender-Receiver Systems within and between Organisms. *Philos. Sci.* **2014**, *81*, 866–878. [CrossRef]
55. Jablonka, E. Information: Its Interpretation, Its Inheritance, and Its Sharing. *Philos. Sci.* **2002**, *69*, 578–605. [CrossRef]
56. Fresco, N. Information, Cognition, and Objectivity. *Am. Philos. Q.* **2021**, *58*, 251–268. [CrossRef]
57. Mann, S.F. Consequences of a Functional Account of Information. *Rev. Philos. Psychol.* **2018**, *11*, 1–19. [CrossRef]
58. Rathkopf, C. Neural information and the problem of objectivity. *Biol. Philos.* **2017**, *32*, 321–336. [CrossRef]
59. Newell, A. *Unified Theories of Cognition*; Harvard University Press: Cambridge, MA, USA, 1990.
60. Deacon, T.W. The symbol concept. In *The Oxford Handbook of Language Evolution*; Gibson, K.R., Tallerman, M., Eds.; Oxford University Press: Oxford, UK, 2011; pp. 393–405.
61. Martin, A.E. Language Processing as Cue Integration: Grounding the Psychology of Language in Perception and Neurophysiology. *Front. Psychol.* **2016**, *7*, 120. [CrossRef] [PubMed]
62. Altmann, G. *Psycholinguistics: History in Encyclopedia of Language & Linguistics*; Elsevier: Amsterdam, The Netherlands, 2006; pp. 257–265.
63. Gallistel, C.R. Learning and Representation. In *Learning and Memory*; Bryne, J.H., Ed.; Elsevier Science: Amsterdam, The Netherlands, 2017; pp. 141–154.
64. Michelsen, A.; Andersen, B.B.; Jesper, S.; Kirchner, W.H.; Lindauer, M. How Honeybees Perceive Communication Dances, Studied by Means of a Mechanical Model. *Behav. Ecol. Sociobiol.* **1992**, *30*, 143–150. [CrossRef]
65. Anderson, S.R. The Place of Human Language in the Animal World. In *Formal Models in the Study of Language*; J. Blochowiak, J., Grisot, C., Laenzlinger, C., Eds.; Springer: Berlin, Germany, 2017; pp. 339–351.
66. Todorov, E.; Jordan, M.I. Optimal feedback control as a theory of motor coordination. *Nat. Neurosci.* **2002**, *5*, 1226–1235. [CrossRef] [PubMed]
67. Wolpert, D.M. Probabilistic models in human sensorimotor control. *Hum. Mov. Sci.* **2007**, *26*, 511–524. [CrossRef]
68. Taylor, J.A.; Ivry, R.B. The role of strategies in motor learning. *Ann. N. Y. Acad. Sci.* **2012**, *1251*, 1–12. [CrossRef] [PubMed]
69. Moran, A. Naïve Realism, Hallucination, and Causation: A New Response to the Screening Off Problem. *Australas. J. Philos.* **2019**, *97*, 368–382. [CrossRef]
70. Crane, T. Is perception a propositional attitude? *Philos. Q.* **2009**, *59*, 452–469. [CrossRef]
71. Orlandi, N. *The Innocent Eye: Why Vision Is Not a Cognitive Process*; Oxford University Press: Oxford, UK, 2014.
72. Siegel, S. *The Rationality of Perception*; Oxford University Press: Oxford, UK, 2017.
73. Zahnoun, F. Truth or Accuracy? *Theoria* **2020**, *86*, 643–650. [CrossRef]
74. Baker, B. Natural information, factivity and nomicity. *Biol. Philos.* **2021**, *36*, 1–21. [CrossRef]
75. Scarantino, A.; Piccinini, G. Information without truth. *Metaphilosophy* **2010**, *41*, 313–330. [CrossRef]
76. Melloni, L.; Mudrik, L.; Pitts, M.; Koch, C. Making the hard problem of consciousness easier. *Science* **2021**, *372*, 911–912. [CrossRef] [PubMed]

77. Baars, B.J. *A Cognitive Theory of Consciousness*; Cambridge University Press: Cambridge, UK, 1988.
78. Mashour, G.A.; Roelfsema, P.; Changeux, J.-P.; Dehaene, S. Conscious Processing and the Global Neuronal Workspace Hypothesis. *Neuron* **2020**, *105*, 776–798. [CrossRef] [PubMed]
79. Dehaene, S.; Changeux, J.-P. Experimental and Theoretical Approaches to Conscious Processing. *Neuron* **2011**, *70*, 200–227. [CrossRef]
80. Dehaene, S.; Changeux, J.-P.; Naccache, L. The Global Neuronal Workspace Model of Conscious Access: From Neuronal Architectures to Clinical Applications. In *Characterizing Consciousness: From Cognition to the Clinic?* Dehaene, S., Christen, Y., Eds.; Springer: Berlin, Germany, 2011; pp. 55–84.
81. Mudrik, L.; Melloni, L.; Pitts, M.; Blumenfeld, H.; de Lange, F.; Jensen, O.; Kreiman, G.; Luo, H.; Chalmers, D.; Devore, S.; et al. Preregistration File-v2-May 2020. Available online: osf.io/f87xq (accessed on 20 February 2022).
82. Dehaene, S.; Naccache, L.; Cohen, L.; Bihan, D.L.; Mangin, J.-F.; Poline, J.-B.; Rivière, D. Cerebral mechanisms of word masking and unconscious repetition priming. *Nat. Neurosci.* **2001**, *4*, 752–758. [CrossRef]
83. King, J.-R.; Pescetelli, N.; Dehaene, S. Brain Mechanisms Underlying the Brief Maintenance of Seen and Unseen Sensory Information. *Neuron* **2016**, *92*, 1122–1134. [CrossRef]
84. Shanahan, M.; Baars, B. Applying global workspace theory to the frame problem. *Cognition* **2005**, *98*, 157–176. [CrossRef]
85. Hare, S.M. Hallucinations: A Functional Network Model of How Sensory Representations Become Selected for Conscious Awareness in Schizophrenia. *Front. Neurosci.* **2021**, *15*. [CrossRef]
86. Sullivan, J.A. The multiplicity of experimental protocols: A challenge to reductionist and non-reductionist models of the unity of neuroscience. *Synthese* **2009**, *167*, 511–539. [CrossRef]
87. Goris, R.L.T.; Ziemba, C.M.; Stine, G.M.; Simoncelli, E.P.; Movshon, J.A. Dissociation of Choice Formation and Choice-Correlated Activity in Macaque Visual Cortex. *J. Neurosci.* **2017**, *37*, 5195–5203. [CrossRef]
88. Fresco, N.; Michael, M. Information and Veridicality: Information Processing and the Bar-Hillel/Carnap Paradox. *Philos. Sci.* **2016**, *83*, 131–151. [CrossRef]
89. Scarantino, A. Rethinking Functional Reference. *Philos. Sci.* **2013**, *80*, 1006–1018. [CrossRef]
90. Wheeler, B.C.; Fischer, J. Functionally referential signals: A promising paradigm whose time has passed. *Evol. Anthropol. Issues News Rev.* **2012**, *21*, 195–205. [CrossRef] [PubMed]

philosophies

Article

The Simulative Role of Neural Language Models in Brain Language Processing

Nicola Angius *, Pietro Perconti, Alessio Plebe and Alessandro Acciai

Department of Cognitive Science, University of Messina, Via Concezione 8, 98121 Messina, Italy; pietro.perconti@unime.it (P.P.); alessio.plebe@unime.it (A.P.); alessandro.acciai@studenti.unime.it (A.A.)
* Correspondence: nicola.angius@unime.it

Abstract: This paper provides an epistemological and methodological analysis of the recent practice of using neural language models to simulate brain language processing. It is argued that, on the one hand, this practice can be understood as an instance of the traditional simulative method in artificial intelligence, following a mechanistic understanding of the mind; on the other hand, that it modifies the simulative method significantly. Firstly, neural language models are introduced; a study case showing how neural language models are being applied in cognitive neuroscience for simulative purposes is then presented; after recalling the main epistemological features of the simulative method in artificial intelligence, it is finally highlighted how the epistemic opacity of neural language models is tackled by using the brain itself to simulate the neural language model and to test hypotheses about it, in what is called here a co-simulation.

Keywords: simulative artificial intelligence; synthetic method; mechanism; neural language models; brain language processing; deep learning

Citation: Angius, N.; Perconti, P.; Plebe, A.; Acciai, A. The Simulative Role of Neural Language Models in Brain Language Processing. *Philosophies* **2024**, *9*, 137. https://doi.org/10.3390/philosophies9050137

Academic Editors: Marcin J. Schroeder and Gordana Dodig-Crnkovic

Received: 10 July 2024
Revised: 23 August 2024
Accepted: 28 August 2024
Published: 29 August 2024

1. Introduction

The use of machines to predict and explain the intelligent and adaptive behaviours of biological systems traces back to the birth, in the middle of the twentieth century, of cybernetics, due to the groundbreaking work of Norbert Wiener [1]. Cybernetics was also conceived as an attempt to promote a *mechanistic* view of living systems in apparent contrast with the vitalism of Henri Bergson and the use of the "vital force" principle to explain natural evolution and adaptation [2]. The epistemological setting of cybernetics has been fully inherited by Artificial Intelligence (AI), especially in the simulative approach of the pioneers Hallen Newell and Herbert Symon. The so-called simulative, or *synthetic*, method in AI amounts to using computational systems to test cognitive hypotheses about some natural cognitive system [3]. The synthetic method influenced research in AI, under both the symbolic and sub-symbolic paradigm, and in robotics.[1]

AI is now living what has been a called a *Renaissance* era [4], thanks to the unexpected success of *Deep Learning* (DL). Roughly speaking, two main paths can be identified along which the resurgence of AI has unfolded in the last ten years. In the first five years, the most successful path was vision, leading for the first time to artificial systems with a visual recognition ability similar to that of humans [5–9], arousing surprise and interest in the science of vision [10–12]. Five years later, it was the turn of language, a path opened by the Transformer model [13], quickly followed by various evolutions and variants [14–17], generically called here Neural Language Models (NLMs). In this case too, the sudden and unexpected availability of artificial systems with linguistic performances not so far from human ones has deeply shaken the scientific community of language scholars [18–22].

The success of DL in crucial cognitive tasks such as vision and language has prompted different reactions from the cognitive neuroscience community, ranging from acknowledgment [11], to curiosity [12], to refusal [23]. One main reason for such different attitudes towards DL is that whereas traditional Artificial Neural Networks (ANNs) were explicitly inspired by the functioning of the brain, the development of the Transformer architecture has not been influenced by the functional or structural organization of the brain. And nonetheless, a new line of research in cognitive neuroscience uses Transformer-based models to simulate brain activities. More specifically, NLMs are being used to predict cortex activations while processing language [24–26].

This paper intends to show how the application of DL networks in the study of brain language processing can be understood, from an epistemological and methodological point of view, as an instance of the simulative method as considered in [3], in continuity with the mechanistic approaches in the philosophy of cognitive science. In particular, it is examined how NLMs are used to simulate human agents involved in linguistic tasks, providing predictions about the human cognitive system.

The main aim of this paper is, nonetheless, highlighting significant methodological differences that arise when DL is involved in simulative tasks. In traditional simulative AI, cognitive hypotheses are tested by experimenting on the simulative system, as long as one cannot directly experiment on the simulated system, due to ethical concerns or when the simulated system is *epistemically opaque*. However, epistemic opacity and non-interpretability is one essential feature of DL models as well [27]; this marks a significant difference between NLMs and the simulative programs of symbolic AI or the ANN of the connectionist approach. It is argued here that, in order to overcome the limited intepretability of NLMs when used to simulate brain language processing, the brain itself is used as a model of the NLM in what is called here a *co-simulation*. The idea of using a natural cognitive system to simulate an artificial computational one strengthens even more the mechanistic view of the human mind.

This paper is organized as follows. Section 2 introduces NLMs and the Transformer architecture; Section 3 shows how NLMs are being used in cognitive neuroscience for simulative purposes in the context of brain language processing; Section 4 underlines the main epistemological features of the simulative method in AI and bio-robotics; Section 5 analyses how the simulative method is applied and modified in NLM simulations; finally, Section 6 concludes the paper.

2. Neural Language Models

The conquest of natural language has been one of the most difficult challenges for AI, and for a long time, ANNs have played a secondary role compared to conventional Natural Language Processing. The first attempt to integrate ANNs into natural language processing was undertaken by [28], concentrating on inflectional morphology. Their aim was to show, through an artificial model, that learning the morphology of the past tense of English verbs does not necessitate explicit or innate rules, but it is instead acquired from experience. Their model succeeded and was able to replicate the typical learning curves observed in young children. However, Rumelhart and McClelland faced a significant challenge in employing ANNs for language processing due to a seemingly irreconcilable discrepancy between the two formats. Language is an ordered sequence of auditory signals (in the case of spoken language) or symbols (in the case of written language), whereas a neural layer is a real vector with a fixed dimension. This creates a problem in encoding an arbitrary length datum (the word) with a fixed-dimension vector (the neural layer), even for models restricted to the processing of single words.

A second challenge in applying ANNs to natural language processing is that representing words with neural vectors becomes more problematic when moving from single-word morphology to syntax. Feedforward ANNs are static, making it difficult to establish a sense of order for multiple words in a sentence.

An additional challenge for traditional ANNs arises from the very technique that determined their success in the '90s: backpropagation learning [29]. Efficient backpropagation requires tasks where inputs and outputs are clearly identifiable, and examples of these input-output pairs must be available, i.e., supervised training. However, the ability to understand language, and even more so to produce it, extends beyond tasks where the necessary inputs and outputs for supervised training can be distinctly identified.

Fueling the confidence in those who, despite these negative premises, have persevered, is the fact that the symbolic nature of language seems antithetical even to the neurons of our brain, which apparently have solved these problems very well. This confidence was well placed, and finally crowned by the Transformer architecture [13] combining several effective strategies to cope with the symbolic nature of natural language. The first strategy is *word embedding*, which learns from examples to optimally convert words into vectors of neural activity. Introduced by [30], its key feature is that the vector representation is semantically meaningful. These numerical vectors can be manipulated in ways that respect lexical semantics. For instance, let vector $\vec{w}(\cdot)$ represent the word embedding transformation and let $\vec{w}(\texttt{king})$ be the vector for the word 'king'; by subtracting from it the vector $\vec{w}(\texttt{male})$ for 'male' and adding the vector $\vec{w}(\texttt{female})$ for 'female', one obtains vector $\vec{q}$:

$$\vec{q} = \vec{w}(\texttt{king}) - \vec{w}(\texttt{male}) + \vec{w}(\texttt{female})$$

which is closer to the vector $\vec{w}(\texttt{queen})$ for the word 'queen' than to any other word embedding vector.

The second strategy is the *attention* mechanism, firstly introduced by [31] in the framework of patter recognition and later on, in the context of language generation, by [13]. This method dynamically identifies relevant information and relationships among words in a sentence. The Transformer employs these strategies in an innovative way. Firstly, word embedding is learned as the entire neural model processes corpora. Secondly, the attention mechanism completely replaces recursion, allowing all words, along with their vector embeddings, to be simultaneously presented as input.

Furthermore, the Transformer incorporates an elegant solution to bypass supervised learning, as introduced by [32]: the concept of the *autoencoder*. This deceptively simple idea involves assigning the ANN the task of reproducing its own input as output. The architecture implementing this concept is typically organized into two components. The encoder generates an internal representation of the input, while the decoder reproduces the output from this representation, which coincides with the input. The popular term "stochastic parrots" [33] for Transformer models originates from this autoencoder structure. Although the term accurately reflects the training technique, it becomes irrelevant when used derogatorily towards NLMs. This exemplifies what [34] have termed a *Redescription Fallacy*, where a NLM's skills and abilities are judged based on irrelevant characteristics, such as the training strategy in this case.

The remarkable efficiency of the Transformer has led to many variations, including ViT *Vision Transformer* [35] and BERT (*Bidirectional Encoder Representations from Transformers*), where attention is applied to both the left and right side of the current word [14]. The original Transformer was designed for translation, so it includes an encoder for the input text and a decoder for the text generated in a different language. A simplification was later adopted by GPT (Generative Pre-trained Transformer), which consists only of a decoder part, primarily for generating text by completing a given prompt [15]. The popular public interface ChatGPT is based on later models of the GPT family [16]. The autoencoding strategy during learning is the task of just predicting the next token in a text. In a strictly mathematical sense, the output of the Transformer is the probability of tokens being generated at the next time step. It is important to note that often this interpretation—although entirely correct in itself—is mistakenly regarded as the overall task performed by the Transformer, thus leading to a misleading underestimation of it. Similarly, it would not be incorrect to assert that when a person writes a word, it corresponds to the highest probability in a space of brain neural activations of the entire vocabulary. But if one were to limit oneself to this to account, for example, for the words we authors put one after the other in this sentence, it would be a truly disappointing explanation.

The subsequent description pertains to the streamlined GPT architecture, with an overall scheme shown in Figure 1. The input text consists of *tokenst*$_i$, where each token is an integer index into the vocabulary, which comprises words, punctuation marks, and parts of words. The vocabulary size N typically includes several tens of thousands of entries. A crucial operation on the input token is embedding, performed with the embedding matrix $W_E \in \mathbb{R}^{D\times N}$, where D is the embedding dimension. For a token t_i in the input stream, the embedded vector is computed as follows:

$$\vec{x}_i = W_E^{(t_i)} + p(i) \tag{1}$$

where $W_E^{(j)}$ is the j-th column of W_E and $p(\cdot) : \mathbb{N} \to \mathbb{R}^d$ is a function that encodes the position of the token inside the stream of text.

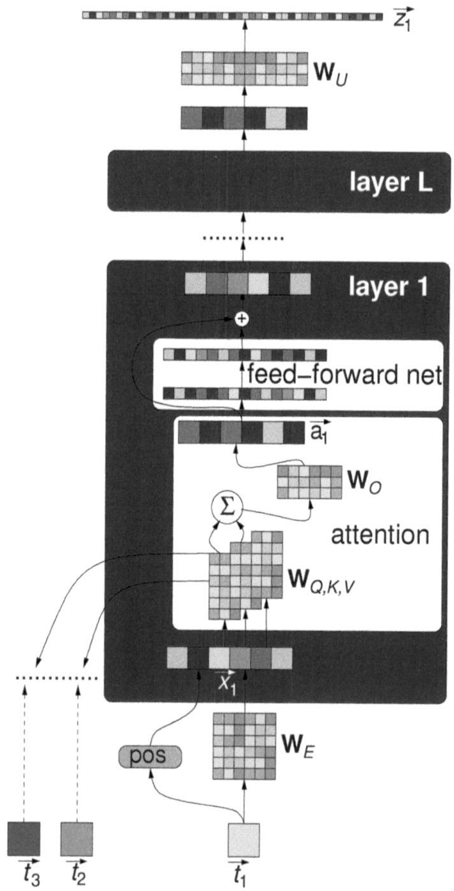

Figure 1. A simplified scheme of the overall Transformer architecture. All components are described in the text.

The model consists of a chain of L layers, with each layer comprising an attention block followed by a feedforward neural network, and each block reading from and writing to the same residual stream. Figure 1 details only one layer for a single token, although all tokens are processed in parallel. The output of the last layer is mapped back to the vocabulary space by the unembedding matrix $W_U \in \mathbb{R}^{N\times D}$ and then fed into a softmax layer. Each element in the output vector $\vec{z}_i$ represents the probability of a token being the successor to $\vec{t}_i$.

A zoom into the attention mechanism is provided in Figure 2. It is based on linear algebra operations using the following matrices:

- $W_K \in \mathbb{R}^{A \times D}$—the "key" matrix;
- $W_Q \in \mathbb{R}^{A \times D}$—the "query" matrix;
- $W_V \in \mathbb{R}^{A \times D}$—the "value" matrix;
- $W_O \in \mathbb{R}^{D \times A}$—the "output" matrix.

A is the dimension of the vector used in the attention computation, in most current NLMs is equal to D. The matrices $W_{K,Q,V}$ map an embedded token into the vectors "query" $\vec{q}$; "key" $\vec{k}$; and "value" $\vec{v}$. The scalars s_i in Figure 2, called "score", result from the multiplication of the "query" and "key" vectors, and modulate the amount of the "value" vectors.

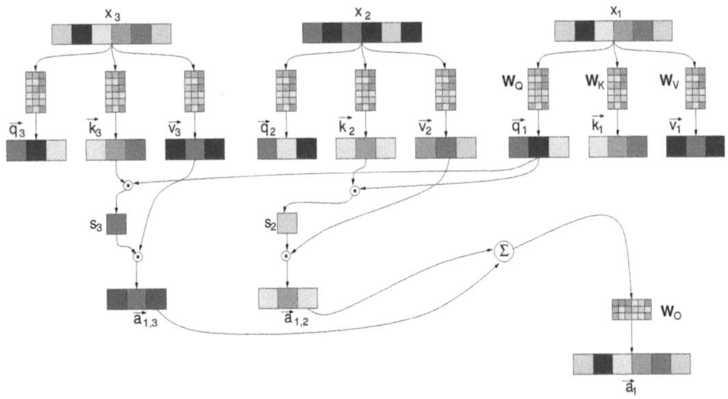

Figure 2. Detail of the attention mechanism, for the current embedded token $\vec{x}_1$ with respect to the previous tokens $\vec{x}_2$ and $\vec{x}_3$.

In a discursive manner, the attention mechanism generates a vector where information from all preceding words is combined, weighted by the relevance of each previous word to the current one. This mechanism synergizes with the other fundamental component of the Transformer: word embedding. The ability to encapsulate all relevant information of a word into a numerical vector for any context of use enables simple linear algebra operations to effectively capture the syntactic and semantic relationships within a text. Now here is the mathematical expression of the operations carried out by the attention:

$$\vec{a}_i = W_O W_V \begin{bmatrix} \vec{x}_i \\ \vec{x}_{i+1} \\ \cdots \\ \vec{x}_{i+T} \end{bmatrix} \left(\frac{1}{\sqrt{D}} \begin{bmatrix} \vec{x}_i \\ \vec{x}_{i+1} \\ \cdots \\ \vec{x}_{i+T} \end{bmatrix}^{\top} W_K^{\top} W_Q \vec{x}_i \right) \tag{2}$$

where T is the span of tokens preceding the current token $\vec{x}_i$.

The scientific community has been profoundly impacted by the sudden and unforeseen emergence of artificial systems, enabled by Transformer-based models, which exhibit linguistic performances approaching those of humans [20–22,36,37]. For sure, no Transformer-based system matches humans in mastering language in all its possible uses,[2] but the leap made in approaching human performance has been extraordinary. Currently, NLMs continue to progress, whether this means surpassing humans in the near future, or continuing to approach them at an increasingly slower pace [38], is not a matter addressed in this article.

The crucial philosophical issue has become that of providing explanations for the kind of mind that emerges in NLMs and allows its performance, its "alien intelligence" using the words of [39]. Explanations that are currently largely lacking, although some initial attempts can be seen. The almost total absence of explanations for the linguistic abilities of the NLMs contrasts with the relative simplicity of their computational architecture and their way of learning. Again, there is a vast technical literature that computationally illustrates the implementations of the various NLMs [40,41], but there is a huge gap from here to identifying what in these implementations gives language faculty. One of the best illustrative texts on Transformer architectures ([42], p. 71) underscores the issue well: "It has to be emphasized again that there's no ultimate theoretical reason why anything like this should work. And in fact, as we'll discuss, I think we have to view this as a—potentially surprising—scientific discovery: that somehow in a neural net like ChatGPT it's possible to capture the essence of what human brains manage to do in generating language".

Such an explanatory request concerns how the relatively simple algorithmic components of the Transformer provide it with the ability to express itself linguistically and to reason at a level comparable to humans. It's worth noting that while linguistics has generated highly sophisticated and detailed descriptions of language, how it is understood and generated by the brain remains essentially a mystery, much like in NLMs. At the same time, one of the ambitions of simulative AI has been to explain aspects of natural cognition by designing their equivalents. However, the presupposition was that these artificial equivalents would be understandable, which is not the case with NLMs.

Before examining how this challenges the traditional epistemology of simulative AI, let us preliminarily see how NLMs are being used in simulative studies of the brain.

3. Using NLMs to Simulate the Brain

There is a current line of research which investigates the relationships between NLM structures and brain structures, through functional magnetic resonance imaging (fMRI), when engaged in the same linguistic task. It is a surprising inquiry, unexpected even for its own protagonists. Indeed, apart from the generic inspiration from biological neurons for artificial neurons, there is nothing specific in the Transformer mechanisms that has been designed with the brain language processing in mind. However, early results show surprising correlations between activation patterns measured in the models and in the brain, and some analogies in the hierarchical organizations in models and cortex.

Ref. [24] aim at explaining one main difference occurring between NLMs and brain language processing, namely that while NLMs are trained to guess the most probable next word, the brain is able to predict sensibly longer-range words.

Ref. [24], in collaboration with Meta AI, did several experiments to examine correlations between NLMs and brain activities using a collection of fMRI recordings of 304 subjects listening to short stories, and prompting the GPT-2 model with the same stories. Individuals were tested using 27 stories between 7 and 56 min, on average 26 min for each subject, and a total of 4.6 brain recording hours for the 304 subjects. The GPT-2 model involved a pre-trained, 12 layer, Transformer, trained using the Narratives dataset [43].

The first experiment was turned to correlate activations in the Transformer to fMRI brain activation signals for each brain voxel and each individual. Correlations were quantified in terms of a "brain score", determined through a linear ridge regression. In particular, GPT-2 activations linearly mapped on such brain areas as the auditory cortex, the anterior temporal area, and the superior temporal area.[3]

In a second set of experiments, the authors evaluated whether considering longer-range word predictions in the Transformer produces higher brain scores. Longer-range predictions were obtained by concatenating the Transformer activation for the current word with what the authors named a *"forecast window"*, that is, a set of w embedded future words, where w is called the width of the window, and where each word is parameterised by a number d, designating the distance of the word in the window with the current word. The experiment yielded higher predictions scores, in this case called "forecast score" (on average +23%) for a range of up to 10 words ($w = 10$), with a peak for a 8 word-range ($d = 8$). Again, forecast score picks correlate model activations with brain activation in cortex areas that are associated with language processing.

In the third, most revealing, experiment, ref. [24] started by the consideration that the cortex is structured into anatomical hierarchies and asked whether different layers in the cortex predict different forecast windows w. In particular, they aimed at evaluating the hypothesis that the prefrontal area is involved in longer-range word predictions than temporal areas. Similarly, the authors considered the different Transformer layers and looked for correlations between activations of the cortex layer and activations of GPT-2 layers. Subsequently, they computed, for each layer and each brain voxel, the highest forecast score, that is, the highest prediction from Transformer layer activations to brain activations. The experiment results were in support of the initial hypothesis.[4]

As stated at the beginning of this section, the work of [24] belongs to a whole line of research looking for correlations between brain structures and NLM structures. To quickly give another example, Kumar and coworkers at the Princeton Neuroscience Institute [26] investigated possible correlations between the individual attention heads[5] in the Transformer, and brain areas when listening to stories. They used the simple model BERT, with 12 layers and 12 attention heads, and applied Principle Component Analysis to the 144 model activations along the story, correlating them with brain areas obtained through fMRI.

What emerges from this line of research, is that Transformer based NLMs are used to model and predict activation patterns in the brain, usually observed through fMRI, in order to collect additional evidence on the brain areas involved in specific linguistic tasks. Schematically, both systems, the NLM and the brain, are given the same task, namely elaborating acoustic signals (the listened story) to process language understanding. The artificial system is then used to predict behaviours (brain activations) of the natural one. This method can be preliminarily considered an instance of the simulative method in AI, that we now turn to analyse.

4. The Simulative Method in Cognitive Science

The *simulative method* in science [45,46] consists in representing a target, natural, system by a means of a mathematical model, usually a set of differential equations, implementing the model in a computational one, typically a simulative program, and executing the latter to provide predictions of the target system behaviours. One characterising feature of computer simulations in science is that they are required to mimic the evolution of the target system in order to provide faithful predictions.

In the realm of cognitive science, the simulative method amounts to implementing an artificial system, either a robot or a computer program, aimed at testing some given hypothesis on a natural cognitive system [47,48]. That is, the main aim of simulations in cognitive science is epistemological: their characterising feature is that they are involved in advancing and testing cognitive hypotheses over the simulated system by building an artificial system and experimenting on it. Experimental strategies are thus performed on the artificial system in place of the natural one. Given a cognitive *function*, hypotheses usually concern the *mechanism* implementing that function in the natural cognitive system.[6] The simulative or, as it is often called, the "*synthetic*" method in cognitive science develops an artificial cognitive system implementing that mechanism for the given function and compares the behaviours of artificial and natural systems. Hypothesised mechanisms play the epistemic role of program *specifications* for artificial computational systems.[7] In case the displayed function of the simulative system matches with the behaviours of the simulated system, the initial hypothesis concerning how the function under interest is realised in terms of the implemented mechanisms is corroborated. Once corroboration is achieved, simulations on the artificial system are used to predict, and explain, the future behaviours of the natural system. Additionally, new mechanisms identified in the artificial system for some displayed function are used as hypotheses for explaining similar behaviours in the natural system.

The synthetic method in cognitive science finds in the *Information Processing Psychology* (IPP) of [52] one important pioneering application. In the approach of Newell and Symon, a human agent is given a problem solving task, typically a logic exercise or the choice of moves in a chess game, asking him to think aloud, thus obtaining a verbal account of her mental processes while carrying out the task. Verbal reports are analyzed in order to identify the solution strategies adopted by the agent and the specific operations performed while carrying out the task. The analysed verbal reports are then used to develop a program that simulates the behaviour of the human agent. Subsequently, new problem solving tasks are given to both the program and the human agent, and verbal reports of the latter are compared with the execution traces of the simulative program to ascertain that the two systems use the same solution strategies. Finally, the program execution traces for new tasks are used for predicting the strategies and mental operations that the human agent performs when given the same tasks.

In the IPP approach, human agents' verbal reports are used to hypothesise the mechanism used by the agents to profitably solve the administered cognitive task. The solution strategies hypothesised by Newell and Symon typically consisted in research mechanisms in decision trees. Research mechanisms of this sort are used as program specifications to develop computer programs, using such programming languages as *Information Processing Language* and *List Processor* (LISP), being able to realise those solution strategies. The *Logic Theorist* and the *General Problem Solver* are well-known examples of such programs. Computer programs are then used to test the initial hypothesis, namely the solution strategy advanced on the basis of the verbal reports. The hypothesis is tested by administering new cognitive task to the program, such as proving logic theorems from Russel and Whitehead's *Principia Mathematica*. In case the solution strategies adopted by the simulative program are the same used by the tested human agent, the initial hypothesis is considered as corroborated.

The synthetic method has been also, and more recently applied, to *biorobotics*. For instance, ref. [53] argue that the synthetic method in simulative AI is the method applied, among others, to the robotic simulation of chemiotaxis in lobsters [54].[8] Ref. [54] hypothesise the biological mechanism implementing lobster chemiotaxis, namely the ability to trace back the source of food, leaving chemical traces in the sea, through chemical receptors put on the two antennae. The very simple advanced mechanism is that the receptor stimulation activates, in a proportional manner, the motor organs of the side opposite to that of the antenna. In other words, the stimulation of receptors of the right antenna activates the left motor organs and the stimulation of receptors of the left antenna activates the right motor organs. The higher the receptor stimulus, the higher the motor organ activation. This simple mechanism would, according to [54], allow lobsters to constantly steer towards the food source following the chemical trail.

Such a hypothesis is tested by building a small robot lobster, named RoboLobster, provided with two chemical receptors, put on the left and right side, and wheels in place of legs. RoboLobster implements the hypothesised mechanism: the left artificial receptor causes, upon stimulation, a directly proportional activation of the right wheel, the right receptor activates the left wheel. RoboLobster was tested in an aquarium containing a pipe releasing a chemical trail. However, the robot was able to trace back the pipe only when put within a 60 cm distance from the pipe; while when put 100 cm away from the chemical source the robot was unable to locate the pipe. The synthetic experiments led the authors to falsify and reject the hypothesis.

Ref. [53] are very careful to notice that when the initial hypothesis gets falsified while testing the artificial system, researchers still use the simulation to understand why the hypothesis was falsified and whether the problem was the hypothesis itself or rather other side phenomena. In other words, they look for an explanation concerning why the supposed mechanism is not able to implement the interested cognitive function. Researchers usually evaluate whether the developed artificial system is a faithful implementation of the hypothesised mechanism. Another source of mistake may be that the mechanism implemented by the developed system is not a faithful description of the biological mechanism.[9]

Ref. [54] suppose that RoboLobster was unable to trace the chemical source because of a wrong distance between the two receptors or of the initial orientation of the robot in the aquarium. However, even modifying the receptor distance and the robot orientation, RoboLobster is still unable to find the pipe when put 100 cm away from it. The authors conclusion is that RoboLobster fails since from a certain distance the chemical trail is scattered and is not informative enough for the robot about the direction to take.

In this third case, the artificial system is used to *discover* new hypothesis about the natural cognitive system and its environment. It is indeed hypothesised that chemical trails are informative with respect to the food location for real lobsters only at a certain distance, the reason being that lobster receptors at a certain distance are not able to detect a difference in chemical concentrations.

To sum up, the synthetic method in cognitive science is a simulative approach applied in all those cases in which testing a cognitive hypothesis directly on the natural system is not feasible. An artificial system is built, in the form of a computer program or robot, and the hypothesis is tested on the artificial system instead. This is done by implementing the hypothesis, in the form of a mechanism for the given cognitive function, in the artificial system and comparing the behaviours of the simulative system with those of the simulated one. In case the artificial system performs the same cognitive function of the natural simulated system, the initial hypothesis is corroborated, otherwise the hypothesis is falsified. In both cases, artificial systems can be used to advance new hypotheses about the behaviours of artificial and natural systems which are tested again on the artificial one. The epistemological relations entertained by the natural cognitive system and the simuative model are depicted in Figure 3.

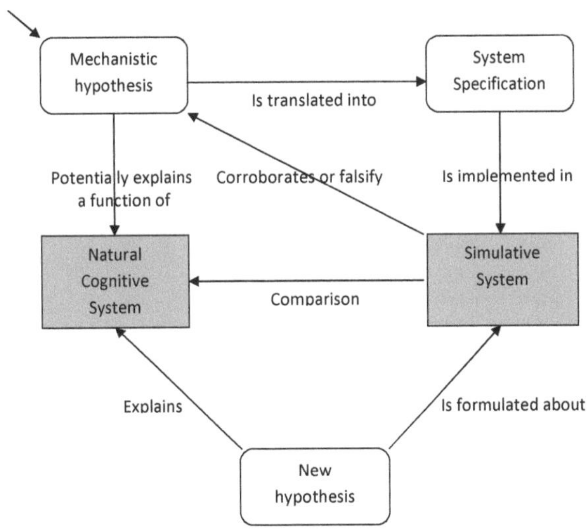

Figure 3. The epistemological framework of simulative AI. The incoming arrow indicates where the process starts.

5. Co-Simulations of Neural Activations Using NLMs

Even though NLMs have been developed with engineering purposes only, namely for developing language processing systems, the early work of [24] and of [26] shows how they are being fruitfully applied to simulative AI as well.[10] However, the way NLMs are used to predict and explain brain activations in the cortex puts significant methodological challenges for the synthetic method in simulative AI.

One first main difference between the simulative method in AI and the application of NLMs in neuroscience is that NLMs are not developed so as to implement mechanisms corresponding to hypotheses about linguistic functions of the brain. The aim of NLMs is not that of corroborating any such hypotheses, as it happens with the simulative method in traditional AI. From an epistemological and methodological point of view, NLMs seem not to be simulative models. And nonetheless, NLMs are used to simulate the brain, that is, to obtain predictions of cortex activations. It is astonishing how, as the work of [24] shows, even though NLMs were developed without considering structural properties of the cortex, once trained they bear structural similarities with language processing areas of brain. An astonishment one also feels while considering DL models involved in vision.[11]

In the synthetic method, hypothesised mechanisms are used as specifications to develop simulative systems and, as stated above, it is required that simulative programs or robots be correct implementations of those mechanisms. As it is in software development, the specification set determines a blueprint of the system to be developed and both correct and incorrect behaviours of the implemented system are defined and evaluated by looking at the specifications [61]. In the case of a correctly implemented system, the specification set provides a means to represent and explain the behaviours of the systems [62]. The opportunity to understand and explain machine behaviours allows scientists to use computational artificial systems for simulating natural ones which, by contrast, are not known and explained.

ANNs in general, and DL models in particular, do not fall under this epistemological framework. DL systems are not developed so as to comply with a set of specifications, that is, functions are not declared and then implemented in a DL network, as it is for traditional software. Functions do not depend only from the network architectural choices, but they rather emerge from the model during training and depend much more on the training dataset [63]. Again NLMs are not developed as implementing neurological mechanisms one supposes realise linguistic functions. The absence of a specification set for NLMs is at the basis of the known *epistemic opacity* of those models: except from some architectural choices (i.e., kind of DL models or the number of models) and hyper-parameters (such as the number of neuron layers or the size of the layers) one is unaware of the inner structure of a trained model. In particular, one cannot come to know how the model parameters are updated at each backpropagation of the network.

In the synthetic method, simulative systems are used as some sort of *proxy* for the simulated cognitive system: since one cannot directly experiment on the cognitive system, as long as it is opaque to the scientist, an artificial system is built and hypotheses are evaluated over it. In the case of Newell and Symon's IPP, since one does not know whether the hypothesised solution strategies for a given task are the ones actually implemented in the brain, the identified research mechanisms for decision trees are implemented in a computer program, the program is subsequently executed to test the hypothesised solution strategies.

The second main epistemological difference of simulations using NLMs is that that they are opaque systems as well and cannot play the epistemic role of proxies for the simulated systems. As what concerns the language function, one is in the difficult situation in which both the natural and the AI system need to be explained. Our knowledge about how the brain processes language is limited in the same way as it is our knowledge about why NLMs show linguistic abilities close to those of humans. As stated in Section 2, such an explanatory gap has been recognized and theorised in one of the most recent technical introduction to NLMs [42].

What the work of [24] shows is that, in front of two opaque systems, they are used to understand each other. As already noted, the simulation starts with no initial hypothesis, being the NLM developed independently from any previous study of brain language processing. Subsequently, and in accordance with the standard synthetic method, both the natural cognitive systems (the 304 tested subjects) and the NLM (GPT-2) are given the same task, namely listening, and processing, 27 short stories, and it is evaluated whether behaviours of the artificial system cope with behaviours of the natural system. In this case, it is tested whether activations in the Transformer can be correlated with fMRI brain activation signals.

Once obtained a positive answer, new experiments are performed to test whether considering longer-range word predictions would decrease the correlation score. One should notice that a hypothesis is involved here, namely that the Transformer differs from the brain while processing language in that the former is able to predict only short-range words, typically the next word in a context. The outcome of the experiment is that the Transformer correlates to the brain more than expected, viz. while predicting up-to-10-range words.

The third experiment is devoted to understand why this is the case, that is, why the initial hypothesis was partly falsified. Notice that this is what happens with the synthetic method too: in case the initial hypothesis gets falsified, further experiments on the simulative system are carried out to understand why this happened. In the case of RoboLobster, once the initial hypothesis concerning the mechanism allowing chemiotaxis was falsified, researches supposed that the inability of the robot to trace back the chemical source, when put on a 100 cm distance, was due to the distance between the two receptors or to the initial orientation of the robot, rather than to the falsity of the hypothesis per se. The robot was tested at different orientations in the aquarium and changing the distance between antennae: experiments were still carried over the artificial system.

Getting back to the GPT-2 experiment, ref. [24] try to evaluate whether the fact that the artificial system and the natural one are both able to predict long-range words can be related to structural similarities between the cortex and the Transformer. This is achieved by considering the cortex *as a model of* the Transformer! In particular, it is hypothesised that the hierarchical organization of the cortex resembles, both structurally and functionally, the hierarchical organization of the Transformer. The hypothesis is tested by administering again the same task to both systems and computing the forecast score, obtaining positive evidence.

When NLMs are used for simulation purposes, one is dealing with a system which is at least as opaque as the natural system about which she would like to acquire knowledge. In the work of [24] the problem is tackled by modifying the simulative approach in such a way that the two opaque systems are used to *simulate each other*, and thus to acquire knowledge about both in the form of corroborated, or falsified, hypotheses. In what can be called a *co-simulation*, the NLM is initially used to simulate the brain by looking for correlations while involved in the same task. In this case, hypotheses to be tested relate to the brain (its ability to predict longer-range words) and correlations are Transformer predictions of brain activations. In case one needs additional information concerning why a certain hypothesis was corroborated or falsified, the natural system is used to simulate the artificial one. Hypotheses now concern the Transformer (its hierarchical organization) and simulations involve brain predictions of Transformer activations. The simulative relations entertained by the brain and NLM are depicted in Figure 4.

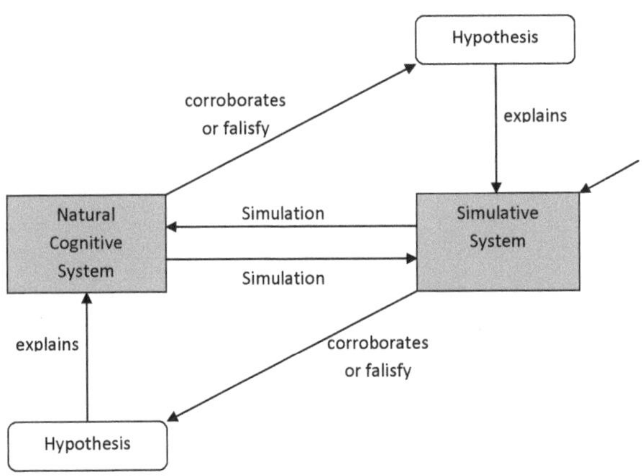

Figure 4. The epistemological framework of NLM simulations.

6. Conclusions

Contemporary DL applications often feature simulation-based scenarios where a model exposed to data from a natural system develops internal structures that correspond to aspects of that system. For instance, ref. [64] utilized a convolutional DL model to simulate parton showers, with each layer representing a different angular scale for emissions. Similarly, in the neural model by [65], which simulates the Hénon-Heiles potential, the autoencoder's internal layer with four neurons captures the four dimensions of the Hénon-Heiles system.

This paper examined another crucial field wherein DL simulations are being applied, namely cognitive neuroscience. NLMs, initially engineered to automatise language translation and generation, are now applied to the simulative investigations of brain language processing. Whereas using artificial computational systems to simulate natural ones is a well-affirmed practice in AI, this paper showed how the applications of NLNs in brain simulations involves significant epistemological and methodological modifications of the synthetic method in cognitive science. The epistemic opacity of NLMs implies that, while they are used to simulate the brain, knowledge is attained about the model as well. This is achieved by a co-simulation wherein the brain is used as a model of the NLM, providing predictions of the Transformer behaviours, and corroborating hypotheses about the latter.

Author Contributions: All authors contributed equally to the conceptualization, methodology, investigation, writing, and reviewing of this work. All authors have read and agreed to the published version of the manuscript.

Funding: Nicola Angius was supported by the PRIN project PRIN_20223E8Y4X_002–CUP J53D23019 490006, "SMARTEST–Simulation of Probabilistic Systems for the Age of the Digital Twin", funded by the Italian Ministry of University and Research. Pietro Perconti, Alessio Plebe, and Alessandro acciai were supported by the PRIN project PRIN_2022MM8KM_003–CUP J53D23007150006, "ALTEREGO–How to Emulate Intentionality and Awareness in Remote Communication by Means of Software Surrogates", funded by the Italian Ministry of University and Research.

Institutional Review Board Statement: Not applicable.

Informed Consent Statement: Not applicable.

Data Availability Statement: The original contributions presented in the study are included in the article, further inquiries can be directed to the corresponding author/s.

Conflicts of Interest: The authors declare no conflicts of interest. The funders had no role in the design of the study; in the collection, analyses, or interpretation of data; in the writing of the manuscript; or in the decision to publish the results.

Abbreviations

The following abbreviations are used in this manuscript:

AI	Artificial Intelligence
ANN	Artificial Neural Networks
DL	Deep Learning
NLM	Neural Language Models

Notes

1 This will be extensively illustrated in Section 4 below.

2 A notable case is that no NLM is able to simulate or explain language acquisition by children.

3 More specifically, the brain score was quantified in the following way. First, a sequence M of words w corresponding to the short stories in the Narratives dataset was defined. The corresponding fMRI recordings from the Narratives Dataset were then sampled with time samples t = 1.5 s and preprocessed using the fMRIprep tool [44] to analyse the cortical voxels; the latter were then projected and morphed onto a brain model, obtaining brain activations Y for each w and having size $T \times V$ (where T is the total number of fMRI samples t and V is the total number of voxels). NLM activations were obtained by tokenising words w in M for being inputted to the network; each activation X corresponded to a vector of size $M \times U$ where U is the number of neurons per layer (768 for the used GPT2 model); activations were mostly extracted from the eighth layer. Finally, for each individual s, each word sequence M, and each voxel v, it was evaluated the mapping between Y and X. The brain score $R^{(s,v)}$ was obtained by using a linear ridge regression to predict a brain activation Y for a given network activation X; the obtained mappings were evaluated using a Pearson correlation between predicted Y and actual activations Y^*. For further technical details the reader should refer to [24].

4 For technical details the reader should refer to [24].

5 Embedded vectors in the Transformer are actually divided into portions, called *heads*, and the attention mechanism is applied separately to each head, and only in the end are the various portions re-joined. The idea is that an embedded vector combines different properties of a word, and that certain categories–for example, the tense of verbs or the gender and number of nouns and adjectives–always occupy the same portions of the vector, and therefore it is convenient to process separately the network of relationships between the separate characteristics of the various words in the text.

6 By mechanism it is referred here to *biological* mechanism as intended in [49], namely as a set of "entities and activities organized such that they are productive of regular changes from start or set-up to finish or termination condition" (p. 3). See [50] for how mechanisms of this sort are able to implement cognitive functions.

7 Program specifications in computer science express the behavioural properties that the system to be developed must realise [51], and their formulation is the first step of most software development methods.

8 Other biorobotic applications of the synthetic method can be found in the simulation of phonotaxis in crickets [55], ants homing [56], or rats navigation [57].

9 In the context of the epistemology of computer simulations in science, the two problems are known as the *verification* and *validation* problem for simulative models. Verification is about ascertaining that the simulative system is a correct implementation of the simulative model; validation is about evaluating whether, and the extent to which, the simulative model is a faithful representation of the target simulated system.

10 It should be indeed recalled that AI has been historically characterised by two main research traditions, an engineering one, concerning the development of artificial systems showing intelligent behaviour, and a simulative one, using artificial intelligent systems to study cognition.

11 The neuroscience of vision is another field wherein neural architectures keep some feature of the natural system, and important similarities have been found between DL models and the visual cortex [58,59]. DL models have been even found to reproduce structural hallmarks of the visual face network in the inferior temporal cortex [60].

References

1. Wiener, N. *Cybernetics or Control and Communication in the Animal and the Machine*; MIT Press: Cambridge, MA, USA, 1948.
2. Bergson, H. *Creative Evolution*; Dover: New York, NY, USA, 1911.
3. Simon, H.A. *The Sciences of the Artificial*, 3rd ed.; MIT Press: Cambridge, MA, USA, 1996.
4. Tan, K.H.; Lim, B.P. The artificial intelligence renaissance: Deep learning and the road to human-Level machine intelligence. *APSIPA Trans. Signal Inf. Process.* **2018**, *7*, e6. [CrossRef]
5. Krizhevsky, A.; Sutskever, I.; Hinton, G.E. ImageNet Classification with Deep Convolutional Neural Networks. In Proceedings of the Advances in Neural Information Processing Systems, Lake Tahoe, NV, USA, 3–6 December 2012; pp. 1090–1098.
6. Simonyan, K.; Zisserman, A. Very deep convolutional networks for large-scale image recognition. In Proceedings of the International Conference on Learning Representations, San Diego, CA, USA, 7–9 May 2015.
7. Szegedy, C.; Liu, W.; Jia, Y.; Sermanet, P.; Reed, S.; Anguelov, D.; Erhan, D.; Vanhoucke, V.; Rabinovich, A. Going Deeper with Convolutions. In Proceedings of the IEEE International Conference on Computer Vision and Pattern Recognition, Boston, MA, USA, 7–12 June 2015; pp. 1–9.
8. Land, E.H.; McCann, J.J. Lightness and retinex theory. *Josa* **1971**, *61*, 1–11. [CrossRef]
9. McCann, J.J. Retinex at 50: Color theory and spatial algorithms, a review. *J. Electron. Imaging* **2017**, *26*, 031204. [CrossRef]
10. Gauthier, I.; Tarr, M.J. Visual Object Recognition: Do We (Finally) Know More Now Than We Did? *Annu. Rev. Vis. Sci.* **2016**, *2*, 16.1–16.20. [CrossRef]
11. VanRullen, R. Perception Science in the Age of Deep Neural Networks. *Front. Psychol.* **2017**, *8*, 142. [CrossRef] [PubMed]
12. Grill-Spector, K.; Weiner, K.S.; Gomez, J.; Stigliani, A.; Natu, V.S. The functional neuroanatomy of face perception: From brain measurements to deep neural networks. *Interface Focus* **2018**, *8*, 20180013. [CrossRef] [PubMed]
13. Vaswani, A.; Shazeer, N.; Parmar, N.; Uszkoreit, J.; Jones, L.; Gomez, A.N.; Kaiser, Ł.; Polosukhin, I. Attention Is All You Need. In Proceedings of the Advances in Neural Information Processing Systems, Long Beach, CA, USA, 4–9 December 2017; pp. 6000–6010.
14. Devlin, J.; Chang, M.W.; Lee, K.; Toutanova, K. BERT: Pre-training of Deep Bidirectional Transformers for Language Understanding. In Proceedings of the North American Chapter of the Association for Computational Linguistics: Human Language Technologies, Minneapolis, MN, USA, 2–7 June 2019; Association for Computational Linguistics: Kerrville, TX, USA, 2019; pp. 4171–4186.
15. Brown, T.B.; Mann, B.; Ryder, N.; Subbiah, M.; Kaplan, J.; Dhariwal, P.; Neelakantan, A.; Shyam, S.; Sastry, G.; Askell, A.; et al. Language Models are Few-Shot Learners. *arXiv* **2020**, arXiv:2005.14165.
16. Ouyang, L.; Wu, J.; Jiang, X.; Almeida, D.; Wainwright, C.; Mishkin, P.; Zhang, C.; Agarwal, S.; Slama, K.; Ray, A.; et al. Training language models to follow instructions with human feedback. In Proceedings of the Advances in Neural Information Processing Systems, New Orleans, LA, USA, 28 November–9 December 2022; pp. 27730–27744.
17. Touvron, H.; Lavril, T.; Izacard, G.; Martinet, X.; Lachaux, M.A.; Lacroix, T.; Rozière, B.; Hambro, N.G.E.; Azhar, F.; Rodriguez, A.; et al. LLaMA: Open and efficient foundation language models. *arXiv* **2023**, arXiv:2302.13971.

18. Alishahi, A.; Chrupała, G.; Linzen, T. Analyzing and interpreting neural networks for NLP: A report on the first BlackboxNLP workshop. *Nat. Lang. Eng.* **2019**, *25*, 543–557. [CrossRef]
19. Baroni, M. Linguistic generalization and compositionality in modern artificial neural networks. *Philos. Trans. R. Soc. B* **2019**, *375*, 20190307. [CrossRef]
20. Boleda, G. Distributional Semantics and Linguistic Theory. *Annu. Rev. Linguist.* **2020**, *6*, 213–234. [CrossRef]
21. Green, M.; Michel, J.G. What Might Machines Mean? *Minds Mach.* 2022, *forthcoming*. [CrossRef]
22. Pavlick, E. Symbols and grounding in large language models. *Philos. Trans. R. Soc. A* **2023**, *381*, 20220041. [CrossRef] [PubMed]
23. Robinson, L.; Rolls, E.T. Invariant visual object recognition: Biologically plausible approaches. *Biol. Cybern.* **2015**, *109*, 505–535. [CrossRef] [PubMed]
24. Caucheteux, C.; Gramfort, A.; King, J. Evidence of a predictive coding hierarchy in the human brain listening to speech. *Nat. Hum. Behav.* **2023**, *7*, 430–441. [CrossRef] [PubMed]
25. Caulfield, J.; Johnson, J.L.; Schamschula, M.P.; Inguva, R. A general model of primitive consciousness. *J. Cogn. Syst. Res.* **2001**, *2*, 263–272. [CrossRef]
26. Kumar, S.; Sumers, T.R.; Yamakoshi, T.; Goldstein, A.; Hasson, U.; Norman, K.A.; Griffiths, T.L.; Hawkins, R.D.; Nastase, S.A. Reconstructing the cascade of language processing in the brain using the internal computations of a transformer-based language model. *bioRxiv* **2023**. [CrossRef]
27. Lipton, Z.C. The mythos of model interpretability: In machine learning, the concept of interpretability is both important and slippery. *Queue* **2018**, *16*, 31–57. [CrossRef]
28. Rumelhart, D.E.; McClelland, J.L. On learning the past tenses of English verbs. In *Parallel Distributed Processing: Explorations in the Microstructure of Cognition*; Rumelhart, D.E., McClelland, J.L., Eds.; MIT Press: Cambridge, MA, USA, 1986; Volume 2, pp. 216–271.
29. Rumelhart, D.E.; Hinton, G.E.; Williams, R.J. Learning Representations by Back-Propagating Errors. *Nature* **1986**, *323*, 533–536. [CrossRef]
30. Mikolov, T.; Sutskever, I.; Chen, K.; Corrado, G.; Dean, J. Distributed representations of words and phrases and their compositionality. In Proceedings of the Advances in Neural Information Processing Systems, Lake Tahoe, NV, USA, 5–10 December 2013; pp. 3111–3119.
31. Carpenter, G.A.; Grossberg, S. ART 2: Self-organization of stable category recognition codes for analog input patterns. *Appl. Opt.* **1987**, *26*, 4919–4930. [CrossRef] [PubMed]
32. Hinton, G.; Zemel, R.S. Autoencoders, minimum description length and Helmholtz free energy. In Proceedings of the Advances in Neural Information Processing Systems, Denver, CO, USA, 28 November–December 1994; pp. 3–10.
33. Bender, E.M.; Gebru, T.; McMillan-Major, A.; Shmitchell, S. On the Dangers of Stochastic Parrots: Can Language Models Be Too Big? In Proceedings of the 2021 ACM Conference on Fairness, Accountability, and Transparency, Virtual, 3–10 March 2021; ACM: New York, NY, USA, 2021; pp. 610–623.
34. Milligan, K.; Astington, J.W.; Dack, L.A. Language and theory of mind: Meta-analysis of the relation between language ability and false-belief understanding. *Child Dev.* **2007**, *78*, 622–646. [CrossRef] [PubMed]
35. Dosovitskiy, A.; Beyer, L.; Kolesnikov, A.; Weissenborn, D.; Zhai, X.; Unterthiner, T.; Dehghani, M.; Minderer, M.; Heigold, G.; Gelly, S.; et al. An image is worth 16x16 words: Transformers for image recognition at scale. In Proceedings of the International Conference on Learning Representations, Virtual, 3–7 May 2021.
36. Søgaard, A. Understanding models understanding language. *Synthese* **2022**, *200*, 443. [CrossRef]
37. Perconti, P.; Plebe, A. Do Machines Really Understand Meaning? (Again). *J. Artif. Intell. Conscious.* **2023**, *10*, 181–206. [CrossRef]
38. Plebe, A.; Perconti, P. The slowdown hypothesis. In *Singularity Hypotheses: A Scientific and Philosophical Assessment*; Springer: Berlin/Heidelberg, Germany, 2013; pp. 349–365.
39. Frank, M.C. Baby steps in evaluating the capacities of large language models. *Nat. Rev. Psychol.* **2023**, *2*, 451–452. [CrossRef]
40. Tingiris, S. *Exploring GPT-3*; Packt Publishing: Birmingham, UK, 2022.
41. Rothman, D. *Transformers for Natural Language Processing*; Packt Publishing: Birmingham, UK, 2022.
42. Wolfram, S. *What Is ChatGPT Doing ... and Why Does It Work*; Wolfram Media: Champaign, IL, USA, 2023.
43. Nastase, S.A.; Liu, Y.F.; Hillman, H.; Zadbood, A.; Hasenfratz, L.; Keshavarzian, N.; Chen, J.; Honey, C.J.; Yeshurun, Y.; Regev, M.; et al. The "Narratives" fMRI dataset for evaluating models of naturalistic language comprehension. *Sci. Data* **2021**, *8*, 250. [CrossRef] [PubMed]
44. Esteban, O.; Markiewicz, C.J.; Blair, R.W.; Moodie, C.A.; Isik, A.I.; Erramuzpe, A.; Kent, J.D.; Goncalves, M.; DuPre, E.; Snyder, M.; et al. fMRIPrep: A robust preprocessing pipeline for functional MRI. *Nat. Methods* **2019**, *16*, 111–116. [CrossRef]
45. Winsberg, E. *Science in the Age of Computer Simulation*; Chicago University Press: Chicago, IL, USA, 2010.
46. Durán, J.M. *Computer Simulations in Science and Engineering: Concepts-Practices-Perspectives*; Springer Nature: Cham, Switzerland, 2018.
47. Boden, M.A. *Mind as Machine: A History of Cognitive Science*; Oxford University Press: Oxford, UK, 2008.
48. Datteri, E. Biorobotics. In *Agent-Based Modelling in Population Studies: Concepts, Methods, and Applications*; Magnani, L., Bertolotti, T., Eds.; Springer: Berlin/Heidelberg, Germany, 2017; pp. 817–837.
49. Machamer, P.; Darden, L.; Craver, C.F. Thinking about Mechanisms. *Philos. Sci.* **2000**, *67*, 1–84. [CrossRef]

50. Piccinini, G.; Craver, C.F. Integrating psychology and neuroscience: Functional analyses as mechanism sketches. *Synthese* **2011**, *183*, 283–311. [CrossRef]
51. Turner, R. Specification. *Minds Mach.* **2011**, *21*, 135–152. [CrossRef]
52. Newell, A.; Simon, H.A. *Human Problem Solving*; Englewood Cliffs: Prentice Hall, NJ, USA, 1972.
53. Datteri, E.; Tamburrini, G. Biorobotic experiments for the discovery of biological mechanisms. *Philos. Sci.* **2007**, *74*, 409–430. [CrossRef]
54. Grasso, F.W.; Consi, T.R.; Mountain, D.C.; Atema, J. Biomimetic robot lobster performs chemo-orientation in turbulence using a pair of spatially separated sensors: Progress and challenges. *Robot. Auton. Syst.* **2000**, *30*, 115–131. [CrossRef]
55. Webb, B. Robots in invertebrate neuroscience. *Nature* **2002**, *417*, 359–363. [CrossRef] [PubMed]
56. Lambrinos, D.; Möller, R.; Labhart, T.; Pfeifer, R.; Wehner, R. A mobile robot employing insect strategies for navigation. *Robot. Auton. Syst.* **2000**, *30*, 39–64. [CrossRef]
57. Burgess, N.; Donnett, J.G.; Jeffery, K.J.; O-keefe, J. Robotic and neuronal simulation of the hippocampus and rat navigation. *Philos. Trans. R. Soc. Lond. Ser. B Biol. Sci.* **1997**, *352*, 1535–1543. [CrossRef] [PubMed]
58. Güçlü, U.; van Gerven, M.A. Deep neural networks reveal a gradient in the complexity of neural representations across the ventral stream. *J. Neurosci.* **2015**, *35*, 10005–10014. [CrossRef]
59. Khaligh-Razavi, S.M.; Kriegeskorte, N. Deep supervised, but not unsupervised, models may explain IT cortical representation. *PLoS Comput. Biol.* **2014**, *10*, e1003915. [CrossRef]
60. Lee, H.; Margalit, E.; Jozwik, K.M.; Cohen, M.A.; Kanwisher, N.; Yamins, D.L.; DiCarlo, J.J. Topographic deep artificial neural networks reproduce the hallmarks of the primate inferior temporal cortex face processing network. *bioRxiv* **2020**. [CrossRef]
61. Turner, R. *Computational Artefacts: Towards a Philosophy of Computer Science*; Springer: Berlin/Heidelberg, Germany, 2018.
62. Angius, N.; Tamburrini, G. Explaining engineered computing systems' behaviour: The role of abstraction and idealization. *Philos. Technol.* **2017**, *30*, 239–258. [CrossRef]
63. Angius, N.; Plebe, A. From Coding To Curing. Functions, Implementations, and Correctness in Deep Learning. *Philos. Technol.* **2023**, *36*, 47. [CrossRef]
64. Monk, J.W. Deep learning as a parton shower. *J. High Energy Phys.* **2018**, *2018*, 21. [CrossRef]
65. Choudhary, A.; Lindner, J.F.; Holliday, E.G.; Miller, S.T.; Sinha, S.; Ditto, W.L. Physics-enhanced neural networks learn order and chaos. *Psychon. Bull. Rev.* **2020**, *27*, 217–236. [CrossRef] [PubMed]

philosophies

MDPI

Article

Aesthetic Gadgets: Rethinking Universalism in Evolutionary Aesthetics

Onerva Kiianlinna

Department of Philosophy, History, and Art, University of Helsinki, 00100 Helsinki, Finland; onerva.kiianlinna@helsinki.fi

Abstract: There is a growing appetite for the inclusion of outcomes of empirical research into philosophical aesthetics. At the same time, evolutionary aesthetics remains in the margins with little mutual discussion with the various strands of philosophical aesthetics. This is surprising, because the evolutionary framework has the power to bring these two approaches together. This article demonstrates that the evolutionary approach builds a biocultural bridge between our philosophical and empirical understanding of humans as aesthetic agents who share the preconditions for aesthetic experience, but are not determined by them. Sometimes, philosophers are wary of the evolutionary framework. Does the research program of evolutionary aesthetics presuppose an intrinsic aesthetic instinct that would determine the way we form aesthetic judgments, regardless of the environment with which we interact? I argue that it does not. Imitation and mindreading are considered to be central features of the aesthetic module. Recently, and contrary to the prior view, it has been shown that imitation and mindreading are not likely to be innate instincts but socially learned, yet evolved patterns of behavior. Hence, I offer grounds for the idea that the cognitive aesthetic module(s) is socially learned, too. This outcome questions the need for the traditional differentiation between empirical and philosophical aesthetics.

Keywords: aesthetic judgment; bioculturalism; cognitive gadgets; cultural evolutionary psychology; evolutionary aesthetics; global aesthetics; innateness; instincts; modularity; social learning

Citation: Kiianlinna, O. Aesthetic Gadgets: Rethinking Universalism in Evolutionary Aesthetics. *Philosophies* **2022**, *7*, 71. https://doi.org/10.3390/philosophies7040071

Academic Editors: Gordana Dodig-Crnkovic and Marcin J. Schroeder

Received: 15 May 2022
Accepted: 17 June 2022
Published: 21 June 2022

1. Introduction

In this article, a conception of an aesthetic (or art) instinct (or impulse) is at stake. From Yrjö Hirn (1900) to Denis Dutton (2009), it has appeared in the tool kit of philosophers talking about our evolved capacities at play in aesthetic judgment [1,2].[1] Many later, often more sophisticated, evolutionary aesthetics works no longer use the term [3,4]. However, occasionally aesthetic instinct still appears in research today [5] (pp. 86–87). This alongside with the history of theorizing about an aesthetic instinct has suggested reliance on innateness in evolutionary aesthetics. Consequently, one of the common worries about evolutionary aesthetics I have encountered both in conversations and in writing is that its appeal to species-typical traits seems to without warrant entail an essentialist and deterministic picture of human behavior [6] (pp. 41–42). I show that this worry is misguided and that it should not be attributed to all versions of evolutionary aesthetics today.

Empirical evidence of what has been interpreted as aesthetic aspects of life emerges from all known human cultures. Moreover, philosophical aesthetics has traditionally held—perhaps most influentially by Kant—that all humans share the potential capacity to form aesthetic judgments [7] (pp. 123–125). Yet, although it is common to speak about aesthetic judgments, it is controversial to hold that all people make aesthetic judgments, either in the sense that everybody would experience the same objects similarly or that there could be universal standards of taste.

Social environments affect the perceptions of sensory data in a unique way for each individual. Along these lines, some scholars have argued that we should not try to look for universal features common to all aesthetic judgments. For example, Bence Nanay emphasizes the role of top-down influences on perception and uses it as grounds for abandoning cultural universalism in aesthetics [8] (p. 87). However, Nanay still tries to locate some universally relevant features of aesthetic artifacts in order to be able to speak about global aesthetics. The global—culture-dependent—approach calls for "... a conceptual framework that can talk about any artifact, no matter where and when it was made" [8] (p. 93).[2] If one takes up this challenge, one falls back into advocating a form of universalism.

The usefulness of an evolutionary viewpoint is in that it offers tools for tackling this discrepancy. It can be used to clarify how aesthetic judgments can be both universal and individually or culturally unique. To be more specific, the evolutionary framework clarifies what kind of a relationship there is between the culture-dependent dimension of aesthetic judgment and the underlying universalist claim that we should nevertheless be able to speak about aesthetic judgments that vary culturally. Having read this paper, the reader will have an understanding of how universalism of aesthetic judgment could be true, but at the same time, nativism (innateness) false.

The issue being addressed in this paper is thus that universalism in evolutionary aesthetics should not be considered as equal to innateness of aesthetic judging. It is important, because even though it has always been clear that many singular aesthetic judgements are not innate, the history of talking about an aesthetic instinct and equivalents has painted evolutionary aesthetics in a light in which, sometimes more and sometimes less justifiably, evolutionary aestheticians would be expected to hold that aesthetic judgment as a behavior is an innate trait.

2. Modules, Gadgets, Judgments

Since I am bringing together ideas from both empirical and philosophical approaches to perception, I start by introducing my core terminology in this chapter. In order to examine how aesthetic judging can be universal yet not innate, I provide a clarification of why and what kind of reference to modularity is needed if one wants to hold on to a conception of aesthetic judgement as a functional entity even in the loosest possible sense.

Cognitive or mental *modules* generate patterns of behavior also known as behavioral traits or skills.[3] Unlike some of my sources, such as Helen Longino [9], Fabrizio Desideri [12], Cecilia Heyes [13], and Tomi Kokkonen [10], I do not talk about mechanisms. In the case of aesthetic judgments, the exact mechanisms of function are unknown, so I employ the looser term "modules". Modules are part of flexible functional wholes, organisms, and are often constituted by a variety of other modules. Moreover, although modules are functional entities, they are not completely independent from each other. As Philip Robbins notes, modularity of mind appears across philosophy—to be precise, "[—]in philosophy of science, epistemology, ethics, and philosophy of language[...]" [14]. The general biological definition of modularity is, according to Hugo Mercier and Dan Sperber, the following:

> "All these mechanisms on the instinct-expertise continuum are what in biology (or in engineering) might typically be called *modules:* they are autonomous mechanisms with a history, a function, and procedures appropriate to this function. They should be viewed as components of larger systems to which they each make a distinct contribution. Conversely, the capacities of a modular system cannot be well explained without identifying its modular components and the way they work together". [15] (p. 73)

I have adopted the more recent Carruthersian, instead of the stricter Fodorian, outlook on modularity. Fodor modules are more segregated and inflexible.[4] The debate over if the mind should be seen as massively modular in the Carruthersian sense is not settled, but the reason I use the loosest possible definition of "module" is that it allows virtual domain-generality. Otherwise, the domain of aesthetic judgment would have to be too restricted. In other words, domain-generality is needed because aesthetic judgments apply to a variety of objects, such as nature, the everyday, and systems. By positing an aesthetic module, we gain an understanding of how aesthetic judgement can be a domain-general functional entity. This understanding is needed if one wants to obtain information about aesthetic judgement on the explanatory level of dealing with functional mechanisms while claiming that aesthetic judgment is an entity.

Modules can be innate instincts, individually constructed or socially acquired. "*Cognitive gadgets*" are types of modules that are learned from other people [13] (pp. 146–147). This does not mean one can say genetic evolution plays no part in the explanations concerning cognitive gadgets and rely solely on social inheritance. Cultural evolution is not totally independent from genetic inheritance. We need our genetically inherited domain-general cognitive mechanisms, such as central processors and certain attentional biases, to be able to construct mechanisms via social learning [13] (pp. 52–54).

The benefit of the modular approach in aesthetics is that it helps to move beyond what we have direct introspective access to.[5] This is needed in order to form a conception of aesthetic cognition. The question if there is an aesthetic module or not is misleading, because it is a matter of perspective. Modules in complex systems, such as humans, are not clear-cut. Whether there is an aesthetic module depends on what function we attach to it and not on whether the module itself would be a sufficient cause for the pattern of behavior under scrutiny [9] (p. 144), [10] (p. 62).

My scope is *aesthetic judgment*, with which in this article I mean observable behavior, empirically perceived changes that can be mental, verbal, neural, or bodily. Although some of my sources—for example, John Dewey [16]—talk about aesthetic experience, I will go on to aesthetic experience only to the extent necessary to shed light on aesthetic judgment. This is because aesthetic judgement can be more easily grounded in empirical observation. The reason I use aesthetic judgement in this meaning is that materialism forms the basis of naturalism. Robert Stecker persuasively defends a philosophical view according to which aesthetic experience is a valuable experience and aesthetic judgment, in turn, requires second-order processing: acknowledging the value of what we perceive as something that can evoke aesthetic experience [17] (p. 5). Unlike aesthetic experience, Stecker continues, aesthetic judgements are about instrumental value of forms, qualities, and meanings—the judgements concern their interactive force as providers of aesthetic experience for the subject.

I hold that aesthetic experience can only be empirically accessed by looking at judgments, for example via ratings of horror, amusement, etc.[6] Because aesthetic experience does not take place in a specific brain region, measuring neural activity would not be directly assessing the experience, either, but rather some specific component of it, which tells us even less about the experience in its entirety than measuring judgments.[7] The same holds true for measuring arousal in response to aesthetic stimuli via, for example, skin conductance—although these are valuable contributions to our understanding of the reactions, they only give us indirect and partial information about the experience as a whole. This being said, for the sake of clarity, it is appropriate to state what kind of existing philosophical theories concerning aesthetic experience my argument allows and contradicts. Since metarepresentationality and domain-generality are central for my conception of aesthetic judgement, I must rule out views on aesthetic experience emphasizing either properties of the object or a specific attitude of the subject. This leads me to embrace theories treating aesthetic experience as a relational organization of attitudes (for Perceptual, Attitudinal, and Adverbialist Models of aesthetic experience, see [18] (pp. 71–73)).

3. Level of the Explanation

The importance of clarifying how aesthetic judging can be universal yet not innate is that it marks a paradigm shift in evolutionary aesthetics. Framing the research question on a different explanatory level than before allows shifting the focus from *why* we, as biocultural beings, form aesthetic judgements to *how* it is possible that we, as biocultural beings, form aesthetic judgements. In order to make this shift from the evolutionary level to the proximate level in evolutionary aesthetics, a modular treatment of aesthetic judgment is needed.

Evolutionary aesthetics scholars have a common understanding on little more than that there inevitably are some evolutionary aspects to aesthetic judgment, because the aesthetic subject is a bodily entity. The question is about explicating what the evolutionary aspects are, although these aspects are probably not very simple and uniform—just as aesthetic experience and aesthetic judgments are not. The evolutionary explanation in the wide sense can be done at different levels, not all of which are evolutionary in the narrow sense of the word.[8]

I will take a closer look at the levels of explanation in evolutionary aesthetics by contrasting the evolutionary and proximate levels. I am preoccupied with the proximate level ("how?") of explanation looking at modules at play in aesthetic experience. I am not operating on the evolutionary level ("why?"). With this article, I am responding to Eveline Seghers' call for proximate-level explanations in evolutionary aesthetics [19] (p. 55).

The aim of traditional evolutionary aesthetics, at least since aesthetics and literature professor Hirn [1], has been to hypothesize on the evolutionary level of explanation whether the ability to form aesthetic judgments is adaptive, beneficial for an individual and its potential offspring. In other words, does it have an evolutionary purpose?

Despite various attempts, this ultimate question might prove to be impossible to answer with the methodology and theories we currently have, as has been frequently noted in evolutionary aesthetics. For the time being, I leave it as a black box and modify the research focus towards proximate questions by concentrating on the module itself rather than the possible evolutionary functions of aesthetic perception or some specific aesthetic judgments. The level of adaptations describes the scope of traditional evolutionary aesthetics. If aesthetic judgment is seen as a proximate-level module, the focus shifts to how and in what contexts the behavior takes place—what its domain is. Clarification of the difference between these two levels shows that contemporary evolutionary aesthetics can move beyond the traditional and problematic evolutionary level of explanation.

Talking about one module may seem overly simplistic, as neuroaesthetics has already established that there is no part of the brain solely for the aesthetic and that there is no reason to assume exclusively aesthetic emotions [20] (p. 471), [21]. Aesthetic judgments concern a very wide group of phenomena, and it would seem more likely that there would be many modules that may work together on a case-by-case basis. This is not a problem, because modules are properties of the mind, not the brain.

Aesthetic judgment is most likely not a uniform behavioral system present in all cases of aesthetic judgment, so all attempts to identify strict necessary and sufficient conditions for an aesthetic module are partial. Different capacities are needed in order to form aesthetic judgments, and it is case-specific which skills are employed on each occasion. What is common for all aesthetic judgments, however, is their metarepresentational nature. I hold (without taking a stance on experience here) that "the aesthetic module" concerning aesthetic judgments is a metarepresentational module, or rather, a functional collection of metarepresentational modules.[9] If one wants to explore the potential heuristic value of seeing aesthetic judgment as modular in the first place, the purpose for which the module is 'designed' is forming aesthetic judgments per se [22].

4. Aesthetic Metacognition

Exploring how aesthetic judgement can be universal yet not innate calls for an explanation on what kind of cognitive process is at stake. In this section, let us look at why aesthetic judgement should be seen as part of metacognition and how, if at all, this view supports the claim that it could be culturally transmitted.

Heyes et al. define metacognition as "[—]representation or evaluation of a cognitive state or process[...]" [23] (p. 350). Metacognition has many forms. Imitation is one example with aesthetic judgment being another. The aesthetic module is part of metacognition because it needs to equip us with abilities to evaluate and process sensory inputs.[10]

Deirdre Wilson provides a helpful general definition of metarepresentation: "A metarepresentation is a representation of a representation: a higher-order representation with a lower-order representation embedded within it" [25] (p. 411). The metarepresentational nature of the process of forming aesthetic judgments means, in a nutshell, that aesthetic judgments need to be subjectively justified. Treating aesthetic judgement as a metarepresentational module, as intuitive inference analogical to reasoning, has been examined in another paper, so to avoid overlap, I go through the argument only briefly [22].

The more often used concept in philosophical aesthetics is 'reflection' or 'contemplation'. Yet, I prefer to use metarepresentation because it better captures that the process may be very fast and intuitive. The notion of metarepresentation also acts as a conceptual bridge between philosophical aesthetics and empirical approaches and thus, aids naturalist argumentation in aesthetics. We do not need to make aesthetic judgments demonstrating some predetermined high standard of taste. Cultural conceptions on well-justified aesthetic judgments no doubt shape our aesthetic tastes, but I am not saying that (all) others need to agree with our judgments or that we need to feel we have particularly "good" or sophisticated taste in order to make subjectively well-justified aesthetic judgments. We do not even have to be certain of them, but we can hold degrees of certainty. For example, I can make the judgment "*Valse Triste* is beautiful" or a more modest "I am not sure if I find *Valse Triste* beautiful (please try to convince me if you think it is or is not beautiful)". I am saying that when we make an aesthetic judgment, be it confident or uncertain, it comes from a place where we are able to make the judgment, in other words, where we have found what we intuitively deem 'sufficient reasons' for holding the view—even if we cannot explain what they are. Forming aesthetic judgments is a metacognitive process that requires intuition—which could also be called representation, model, hypothesis, or expectation—of a justified aesthetic judgment in that particular context. The judgment thus forms in a loop of top-down and bottom-up processes as the hypothesis is tested in inference.[11]

Heyes et al. argue that at least some of metacognition is culturally transmitted, and maybe even formed adaptive—"refined for purpose"—by culture:

"While metacognition is adaptive, and found in other animals, we should not assume that all human forms of metacognition are gene-based adaptations. Instead, some forms may have a social origin, including the discrimination, interpretation, and broadcasting of metacognitive representations" [23] (p. 349).

If this is so, it has implications for evolutionary aesthetics, which now has to take this possibility into account.[12]

Heyes et al. challenge some paradigmatic views on innateness of certain traits drawing evidence for their renowned "cultural origins hypothesis" from previous research on education and metacognitive training that point to social learning enhancing metacognitive sensitivity [23] (pp. 356–357). It is of interest for the article at hand to see if aesthetic judgment fulfills the three empirical predictions or implications of the hypothesis. If so, this would suggest that it is a strong candidate for being mostly culturally—rather than mostly genetically—inherited. I am using the epithet "mostly" because in practice, almost all behavioral traits are some combination of both—at the end of the day, there is no "either or" between nurture and nature, as for example Evelyn Fox Keller clarifies [29].[13]

The first point concerns variation as a condition for selection. Heyes et al. predict cultural variation in metacognitive sensitivity [23] (p. 357). The equivalent here would be "aesthetic sensibility" as described by John Bender. There is variation when people "identify certain features, properties, or relations of a work as being aesthetically significant, i.e., as either being value-making or value-lowering" even when their "perceptual or phenomenal experience" is similar [30] (p. 74).[14] However, metacognitive sensitivity and metacognitive bias cannot be measured in aesthetic judgment, because there is no compelling measure of the alleged accuracy—and even less, correctness—of aesthetic judgments. Yet, there seems to be a belief that there is room for aesthetic education, such as professional art criticism, which implies differences exist.

Second, the cultural origins hypothesis indicates that individuals who most effectively transmit metacognitive skills are themselves exceptionally sensitive in metacognition [23] (p. 358). It is no secret that scholars of aesthetics as well as art critics have (at least in the past) praised themselves—and sometimes even each other—for great aesthetic sensibility and considered cultivating their aesthetic sensibility crucial for their profession. Although we could identify their role in shaping other people's aesthetic judgment, we cannot rely on their authoritative testimony on the accuracy and correctness of aesthetic judgments. The same empirical problem arises as above, but interestingly it may not be as paramount as it seems.

It is not clear whether reliable decision accuracy is a necessary component of metacognitive activity and furthermore, that accuracy is a necessary component of forming meta-level judgments of first-order computations. First-order computations here refer to lower-level aesthetic properties in the judgment at hand.[15] There is empirical evidence to indicate that people may rightly experience the confidence of judgment even when they do not have first-order accuracy [31]. This does not give us a reason to believe that there are correct aesthetic judgments. It only points to that aesthetic judgments in general can be formed metacognitively. We feel levels of confidence in our aesthetic judgments as if they were correct or incorrect, even though there is no first-order accuracy. It has also been empirically indicated that expertise does not make aesthetic judgments based on lower-aesthetic properties more uniform so that absolute correctness of judgments could be standardized [32].

The third prediction is that compared to other animals, our species demonstrates stronger links between sociality and metacognition [23] (p. 358). Although we know little about the aesthetic lives of other animals, there seems to be a consensus in evolutionary aesthetics that aesthetic and social behaviors in humans are closely linked together, and that at least humans, as a species, are prone to patterns of behavior we deem to be linked with the aesthetic [33,34] (pp. 1–2). There is also evidence that aesthetic behavior is not restricted to humans but may have analogs in other animals, and Seghers holds it may have evolved utilizing capabilities shared with chimpanzees [35] (p. 270). However, it is another matter if human aesthetic behavior is *more* connected to sociality than that of other animals. To the best of my knowledge, we do not know for sure.[16]

It would not be a strong claim for giving up the analogy that one prediction out of three remains with only modest supporting evidence. Hence, I will continue the thought experiment of treating aesthetic judgment as potentially culturally acquired and transmitted further to other people via social learning.

5. Aesthetic Gadget in Transmission

Finally, I can proceed to elaborate on how, then, aesthetic judging transmits if it is not innate—what cognitive mechanisms give rise to it, and how they are transmitted.

Desideri has studied what he calls "the aesthetic mechanism" from the viewpoint of coevolutionary aesthetics. I base my argumentation on his views and use them as the null hypothesis, the theoretical starting point that sketches out at least some of the sub-modules of the aesthetic module.[17] Desideri states:

"Properly by growing from the soil of perceptual experience (of the «aesthesis»), the aesthetic mechanism cannot be seen as something innate or genetically predisposed. On the other hand, it is not even conceivable that such a mechanism derives only from socio-historical contexts or is transmitted by a cultural tradition" [12] (p. 36).

Although Desideri says that the mechanism is not innate and that his theory resists contrasting innativism (naturism) with historicism (nurturism), and universalism with relativism, he does not enough explain to what extent exactly the mechanism could be both innate and cultural, or universal and relative [12] (pp. 31, 36). I shed more light on these issues. Based on the previous empirical studies, I examine if the factors of Desideri's mechanism are most likely to be learned gadgets, adaptations that have been preserved in cultural selection, not natural selection. If I am correct, this indicates that the module as a functional whole is likely to be socially transmitted, or in Desideri's words, "transmitted by a cultural tradition".[18]

Desideri argues that mimesis, seeking, preference, and play are important for the aesthetic mechanism [12] (p. 31). I proceed to show that at least mimesis—that can more easily than the rest be treated as a module—is largely culturally refined for purpose and culturally transmitted in the context of the aesthetic. I add mindreading, also known as theory of mind, to the list, because it too, more or less implicitly, appears in characterizations of abilities contributing to aesthetic judgment, especially in the case of artifacts. For example, in Gregory Tague's coevolutionary treatment, art is a space for never-ending mindreading [38]. I claim that if art is seen as he does, the central role of mindreading would point away from innateness.

The aim of all of this is to show that the aesthetic module as a functional device facilitating aesthetic judgments and thus, on its part, "material/art culture," as Tague puts it, cannot be a "hardwired" instinct. This is so, although in different instances of aesthetic judgment, the module functions with different combinations of sub-modules, all of which are, in turn, more or less innate. I start with Desideri's factors and then spend more time justifying why I think mindreading should be added.

Mimesis is linked to modelling and learning new things: in Desideri's words, "the expansion of the circle of what is familiar" [12] (p. 31). Heyes argues that in light of empirical evidence, imitation is not an instinct but a socially learned mechanism, a cognitive gadget, that develops in the course of acquiring matching vertical associations; we observe another person doing something and then do it ourselves, learning what it feels like and how it can be activated. Both representations enhance each other, and experience builds up "a repertoire of matching vertical associations", which leads to perceptual sequence learning and motor sequence learning working together. Finally, this forms an imitation mechanism [13] (pp. 122, 142–143). That being said, though, are Desideri and Heyes speaking about the same behavior? I think it is feasible to talk about imitation mechanism in the case of aesthetic judgments. Although Heyes talks about imitating other people, and Desideri about the more general capacity to represent the world, they both refer to learning through producing representations.

With "seeking", Desideri means curiosity, "the pleasure of exploration" [12] (p. 31). Treating as multifaceted behavior employing lots of different modules as seeking as a module induces the common objection to the functional outlook on the mind: should it be treated as a module at all? In other words, would it be too bold a blanket statement, an easy solution for an explanation of what induces the behavior, to assume that it is a functional entity? For this reason, I will restrain from including seeking here as a module. This being said, I do not oppose Desideri in that curiosity is central for making aesthetic judgments.[19]

Moving on to preference, Desideri describes it as "the ability to choose as a degree of freedom and an advantage in the conduct of life" [12] (p. 31). If curiosity was a controversial candidate for being a somewhat unified module, seeking should be, too.[20] The remaining one of Desideri's features, play, is equally controversial. Desideri defines it as "the intra specific and cooperative practice of learning through the exercise and the simulation reinforced by the pleasure" [12] (p. 31). It often goes together with imitation and social learning, but it is not clear if it should be a module.[21]

All in all, it seems plausible that if we stay true to Desideri's aesthetic mechanism as a functional whole, as a module, it is culturally transmitted. The same holds when, for the sake of the argument, mindreading is added as one more central feature of it.

Mindreading is a means of knowing that other organisms and oneself have inner lives: "In prototypical examples of mindreading, an agent works out what another agent is thinking or feeling *right now* [13] (p. 144)." Mindreading is an example of metacognition because it requires first forming a representation about the world that is then ascribed to the mind in question. Moreover, mindreading cannot be ignored when speaking about aesthetic judgements, no matter how privately or intuitively they are formed.

Throughout the history of aesthetics, aesthetic objects that are artifacts—artworks, adornments, rituals, and so on—have been seen as forms of nonverbal communication. Even when there is no external audience but only the artist, the process of art making itself can be argued to be a communicative feedback-loop. For example, Dewey claims that taking the position of the audience, the artist must have reflective or introspective distance from their emotions to be able to produce an artwork that by definition induces emotional responses [16] (pp. 70–73). This translates into mindreading of the audience's mental states; the artist needs to exercise mindreading in order to act as the observer in the artistic process itself as well as to be able to claim a piece should be looked at as art. Namely, in order to claim art status, one is required to make the assumption that others are capable of making the same claim. The same holds even if one disagrees with Dewey that art always has an emotional component, or if one thinks that the artist does not always or constantly assimilate the attitude of an audience during the material working process. In order to hold that the poem I just created is meaningful not only for me but potentially for others, although not necessarily in the same way as it is for me personally, I am already engaged in mindreading. I am attaching to others as vivid, at least to some extent, inner worlds as I have.

For Dewey, experience in itself, not only in the cases of art, is communication:

> "Experience is the result, the sign, and the reward of that interaction of organism and environment which, when it is carried to the full, is a transformation of interaction into participation and communication". [16] (p. 22)

Mindreading does not only refer to attaching mental states to other conscious beings but also to oneself [13] (p. 144). If aesthetic experience is treated as communication where the subject is aware of their mental state (that they are having a certain experience, for example that of beauty, even if forming an aesthetic judgment was intuitive), mindreading could be a feature of the aesthetic module regardless of what one is judging. Here, however, I am talking about cases of social artworks.

According to some current views, even the earliest indications of aesthetic value, such as bodily adornments, geometric rock patterns, and other images, often carry with them a reference to human sociality and thus, mindreading—and mindsharing [33] (p. 229], [38],. According to the hypothesis of Gianluca Consoli, mindreading is a condition for aesthetic experience, because these two coevolved [36] (p. 37).[22] An artifact would be incomprehensible as an aesthetic object without recursively attaching intentional mental state of the maker to it [36] (p. 48).[23] This, in turn, leads to the development of self-reflection, self-constitution, and self-invention, all of which leads into the production of more aesthetic artifacts [36] (pp. 49–50). Tague goes as far as to claim that "[a]rt is a way to test ideas we form via theory of mind" [38] (p. 178). Although Tague thinks that we have what he calls an *innate* impulse for art-making [38] (p. 1), I think the claim has insufficient evidence.[24] I will go into this next to show that innateness should not be taken to be a property of aesthetic judgement in evolutionary aesthetics, and therefore, should not be confused with universalism of aesthetic judgement.

According to Heyes, mindreading is a cognitive gadget. She builds her view on that children gradually learn mental states and their meanings from observing and interacting with adults who are already experts in exercising the mechanism [13] (pp. 153–154, 204). How we read minds does not merely require activation of an instinct by social interaction nor is it constructed by the mindreader herself from environmental cues, but the mechanism is largely defined by the previous generations of mindreaders that teach it to the next one [13] (pp. 146–147). The empirical evidence to back this theory up is that mindreading develops slowly, is cognitively demanding, is linked to specific cortical circuits, and varies from culture to culture [13] (pp. 148–151). In sum, similarly to imitation, although mindreading is adaptive and selected for during evolution for its fitness value, it cannot be assumed to be entirely innate.[25]

Module refers to provisions that allow a certain type of input to be processed rather than a defined machinery that can be reproduced:

> "A cognitive mechanism certainly is not a pellet of information that can be copied inside your head, sent through the air, and planted wholesale in my head. [...] Instead, we can recognize that certain kinds of social interaction, sometimes with many agents over a protracted period of time, gradually shape a child's cognitive mechanisms so that they resemble those of the people around them". [13] (p. 44)

As priors become shared this way, aesthetic agents emerge, aesthetic experience and judgments of individuals may overlap, and we start seeing an aesthetic gadget in action. What bottom-up lower-aesthetic features we take into account and what top-down intuitions about good aesthetic explanations we have depends on social transmission and thus cultural evolution as well as forming habits.[26] This is not to rule out that it also depends on individual qualities that are genetically inherited. The core message of contemporary evolutionary aesthetics is that aesthetic objects and preferences have led to the evolution of aesthetic judgment (here, as a cognitive module), and not only that aesthetic minds would have unilaterally or completely freely produced aesthetic objects.[27] In traditional evolutionary aesthetics, in contrast, aesthetic objects are often treated as developed to match our innate cognitive abilities [50,51] (pp. 416–420).

6. Conclusions

It is a generally accepted notion that aesthetic judgments—for example, what we find beautiful—are largely culturally transmitted. What about the potential mental module for how we form these judgments? My point has been that also the module that determines how the mind works to realize these judgments is culturally transmitted. It is likely that there is no module for aesthetic judgment that people are born with. Rather, there is only an unrealized potential for the ability for aesthetic judgment to develop. In other words, there is a universal—evolutionary—prerequisite for it to be constructable. It is not present at birth nor does it realize itself inevitably, but only in the particular social environments that ensure its transmittance. This statement may sound obvious, but it cannot be directly derived from the mere observation that we have culturally transmitted aesthetic judgements. The value of the article lies within the argumentation itself. To my knowledge, there is no previous account on how exactly the intuitive notion that aesthetic judging is both universal and culture dependent is possible. At present, universalism and culture dependence are often contrasted with each other.

My contribution has been showing how it is possible to agree with both global aesthetics and evolutionary aesthetics. I have argued that an aesthetic gadget is a plausible alternative for an aesthetic instinct. Acknowledging the module(s) for aesthetic judgment as a cognitive gadget(s) rather than innate instinct(s) does not make the evolutionary framework pointless. It shifts the explanation towards the proximate level, alongside the existing developmental level. Most importantly, this also moves the explanation away from the evolutionary level, what has previously been the focus in evolutionary aesthetics emphasizing instincts. This is the new tenet of evolutionary aesthetics in the 2020s.

Funding: Finnish Academy of Science and Letters. Funding number 210007. Open access funding provided by University of Helsinki.

Institutional Review Board Statement: Not applicable.

Informed Consent Statement: Not applicable.

Acknowledgments: I created this article with the help of many exceptionally observant people. I would like to thank Arto Haapala, Petri Ylikoski, Emma Vitikainen, Hanne Appelqvist, Henry Bacon, Javier Gomez-Lavin, Alice Harberd, Tomi Kokkonen, Oiva Kuisma, Joonas Kurjenmiekka, João Lemos, Alexis Makin, Tamra Mendelson, Federico Pianzola, Kalle Puolakka, Bradley Reynolds, Maarten Steenhagen, Gregory Tague, and Jan Verpooten for their comments. Previous versions of this paper have been discussed at the working seminar of aesthetics at the University of Helsinki, the

American Society for Aesthetics Eastern Division Meeting 2022, and the Nordic Society for Aesthetics Annual Conference 2022. I am grateful for all the comments and questions at these events. I would also like to thank the anonymous peer reviewers and the editors for their helpful comments and edits as well as the Finnish Academy of Science and Letters for funding this research.

Conflicts of Interest: The authors declare no conflict of interest. The funders had no role in the design of the study; in the collection, analyses, or interpretation of data; in the writing of the manuscript, or in the decision to publish the results.

Notes

1 Hirn thinks that an artist has an "art-impulse" and a member of the audience has an "art-sense" that both have developed from other traits during the course of evolution [1] (p. 16). Hirn's theory and terminology are in many parts outdated, but I do not think he explicitly claims that the art impulse and art sense would be innate. Dutton, in turn, uses the terminology of innateness in his book *The Art Instinct: Beauty, Pleasure & Human Evolution* and summarizes his point by saying that "[a]rt may seem largely cultural, but the art instinct that conditions it is not" [2] (p. 206). Hirn uses the *Einfühlung* theories of his era. By leaving the role of emotions and feelings untouched here, I do not wish to indicate they were irrelevant for metacognition, aesthetic judgement, its sub-modules, or its transmission. Since dealing with this important and large issue would mean, for instance, delving in length and detail into the vast Cognitivism vs. Emotionalism debates going on in analytic aesthetics at the moment, I leave it for another paper.

2 In the case of aesthetic judgment, this would translate into trying to find some loose features of aesthetic objects, not just aesthetic artifacts.

3 However, defining a cognitive mechanism and a behavioral trait as well as differentiating between psychological and behavioral traits is ambiguous [9] (p. 151), [10] (pp. 192–193, 200–202, 254–256, 262–263). Furthermore, choosing a focus between behavior and psychology contributes to what questions can be answered. I do not see it relevant to differentiate between sociobiology, behavioral ecology, evolutionary psychology, cultural evolution, and gene-culture co-evolution here, because the dividing lines—if there are any—tell more about the history of evolutionary humanities than contribute to understanding the evolutionary aspects of aesthetic judgment today. The approaches are not fixed or mutually exclusive, and scholars can conduct research under several of these labels at the same time [11] (p. 195). For an analysis on how these approaches intertwine and how their research questions differ, see [11] (pp. 210–213).

4 The central features of Fodor modules are domain specificity, mandatory operation, limited central accessibility, fast processing, informational encapsulation, "shallow" outputs, fixed neural architecture, characteristic and specific breakdown patterns, and characteristic ontogenetic pace and sequencing. Carruthers modules' central features are dissociability, weak neural localizability, and central inaccessibility [14].

5 However, it does not allow direct or empirical access to reality but serves as a heuristic device.

6 For example, via phenomenology, we can obtain complementary information about aesthetic experience.

7 I am not saying that empirical aesthetics would give the field of aesthetics no relevant information whatsoever. My point concerns only how we understand the object of the explanation, what we are receiving information about.

8 "... the ethologist Niko Tinbergen (1963) stressed that, when we ask why an animal exhibits a particular behavioural pattern, we could potentially be asking one of four different questions. First, we can ask questions about the *function* of the behavior pattern implying the role that the trait plays in enhancing reproductive success. Second, we can ask about the *evolutionary history* of the behavior pattern, including an account of its original ancestral state and the selective pressures in the evolutionary history of the lineage that led to the species possessing this derived behaviour. Third, we can ask what *proximate* causes leads the individual to express the behaviour pattern, for instance, by looking at the sensory input, neural mechanisms, and effector systems that produce the behaviour. Finally, we can ask what factors during *development* have played a role in directing the appearance of the behaviour at the relevant stage in its lifetime" [11] (p. 205). Answering all of the Tinbergen's dimensions separately is a requirement for the evolutionary understanding of a behavioral trait [11] (p. 7). When it comes to aesthetic judgment, researchers are unanimous about none of them. It is also worth noting that it is controversial how many explanatory levels there are. In this article, I am only considering the proximate and evolutionary levels without paying much attention to the rest.

9 Note that Jérôme Dokic suggests aesthetic experience is not necessarily meta-representational [18] (p. 75).

10 Aesthetic judgment requires inference. The process may be fast, but it includes interpreting evidence—aesthetic properties—for conclusions. As we have to process several aesthetic features and often also data from several senses, we have to employ our working memory. If burdened enough, we experience "aesthetic fatigue" [24]. Further support for treating aesthetic judgment as metacognitive is that we communicate our aesthetic judgments with each other, and they become shared even to the point where they are agreed or disagreed upon. Additionally, so-called external (for the judgment) "second-order" factors, such as previous experience and homeostasis, hunger for example, influence our aesthetic judgments—they are not external to it, strictly speaking [20].Discrimination means the ability to draw apart different signals so that one can build confidence on the correctness, subjective justification, of a judgment. As already touched upon above, we have an intuition about good aesthetic judgments not as an abstract category but on a case-by-case basis, even when we do not consciously go through or are not able to go through the

exact process of forming an aesthetic judgment. This is a precondition for interpersonal discussion on aesthetic judgments, as well as agreement and disagreement with other people. Heyes et al. state:"Explicit metacognition uses conscious representations in working memory to monitor or evaluate–and often to control–cognitive states and processes. Explicit metacognition (here metacognition, when not qualified) is sensitive to cognitive load, and is typically slow, deliberate, and verbally reportable" [23] (p. 350).It can operate either in the level of first-order (lower-level aesthetic properties), but more commonly, second-order (aesthetic properties) computations (or confidence). I speak only about metacognition and leave it open here whether aesthetic judgment is explicit metacognition.

11 Terms 'representation' and 'model' have several usages. Here, representations are embodied, although not necessarily internal to the brain. They are also far from complete and stagnant. Models, too, refer to a state of the organism:The generative model should therefore be interpreted as instantiated by the agent as a whole. In other words, it is not something that one can abstract away from the phenotypic traits of an organism, because it is those traits, including states of its local niche, that instantiate such a model" [26] (p. 57).The loop of top-down and bottom-up processes refers to the predictive processing framework gaining popularity in philosophy of mind at the moment. The central idea of predictive processing is unconscious prediction error minimization. We receive bottom-up messages from the world (including ourselves) concerning effects, but rather than just passively registering, we process them inferentially. This means that our prior beliefs, accumulated during a longer period of time, shape what causes we end up taking to be most likely for a given effect. We position ourselves in the world so that our expectations or hypotheses and sensory feedback match the best way possible. Understood in this way, perception is action, and action is perception in the sense that perceptual inference works to optimize the mental models about the world to fit the data from the senses, and active inference, in turn, tests the hypotheses and changes the sensory input to fit them [27] (p. 183), [28] (pp. 75, 81, 96).

12 Heyes' hypothesis concerns specifically human metacognition. It is not my aim here to study if aesthetic module is human-specific, or if other species have it as well. For example, Desideri thinks the cognitive mechanism is species-typical for humans [12] (p. 32). This would be in line with the idea that the factors characteristic of his aesthetic mechanism are Heyesian gadgets. For my purposes in this article, however, it suffices to say that at least humans have aesthetic metacognition.

13 In a general sense, a scenario where functioning genetic traits would not be realized in culture/nurture is only an abstract thought experiment. An exception could be for example reflexive blinking, but those cases are not relevant for this article.

14 Bender talks about "sensibility" whereas Heyes et al. talk about "sensitivity". For Bender, aesthetic sensitivity refers to differences in the intensity of perceptual experience, sensitivity of the sense organs [30] (p. 76). My argument concerns aesthetic judgments as metarepresentations rather than immediate perceptiveness of the senses, in which case Benderian aesthetic sensibility is analogical to metacognitive sensitivity.

15 For the metarepresentational process of forming aesthetic judgments, see [22] (p. 87).

16 For an opposite stance, see [36] (p. 48).

17 Stephen Davies and Seghers, in turn, talk about "aesthetic sense" [34] (p. 18), [35] (p. 270).

18 My argumentation here still leaves open to what extent the module would be innate. I do not by any means rule out that there could be statistically universal aesthetic preferences, such as certain odors. For an empirical study, see [37]. If there are some innate aesthetic preferences, the module utilizes them.

19 For the sake of the argument, if one still insisted on treating seeking in the case of the aesthetic as a mechanism, it would point to social learning. It is linked to perceiving symmetry or invariance (and thus, asymmetry and variance) [39]. Nanay argues that people learn socially to direct their attention [8] (p. 92). He refers to attention that can be: "i. Distributed with regards to objects and focused with regards to properties ii. Distributed with regards to objects and distributed with regards to properties iii. Focused with regards to objects and focused with regards to properties iv. Focused with regards to objects and distributed with regards to properties" [40] (p. 24). Attention applies here only in the context of the aesthetic, but talking about a functioning aesthetic module, it suffices for the purposes of this article. If Nanay is correct, how we guide our attention, properties that can catch our attention and hold meaning for us in the first place—things we are curious about and that form the soil of our aesthetic judgments—depend on social learning that molds our "mental imagery", or horizon of expectation that affects interpretation of signals [8] (p. 90). This means that the way we seek, what generates this behavior, would not be an instinct.

20 Again, I will make a detour into the wilder speculations of whether preferring, in the case of aesthetic judgment, is mostly learned. At large, we have the capacity to prefer since day one in our lives—for example, over if we eat or refuse milk, or sleep or demand attention—similar to how we have consciousness since day one. Similarly to consciousness at large, the ability to prefer aesthetically forms fully in the course of social life. For the social development of consciousness, see [41] (pp. 229, 249). This is a rather common notion in philosophical aesthetics and can be derived from what was said about directing attention in the previous note. Besides the abundance of philosophical theorizing, there is also empirical evidence that taste varies according to whether a person is interacting in artworlds or not [42] (p. 32).

21 Play, broadly construed, has been considered crucial for the cognitive development of humans [43]. Even if play was treated as a module, it does not entail that we know the evolutionary function and ontogeny of play in humans or in other animals, and it also remains unknown if it is a mostly genetically inherited instinct or not. Neither possibility is ruled out. [44] (pp. 551, 555–556), [45] (p. 1), [46] (pp. 9, 12).

22 Consoli also holds that "[—]aesthetic experience is supported by a multiple set of preexisting mental properties, evolved for other reasons, and then exapted to a new and original adaptive function" [36] (p. 39). Although I use Consoli's stance that mindreading has a role in aesthetic experience, I do not take a stance on the level of adaptations here.

23 The view is not challenged by empirical evidence of the appreciation of AI-created art, because the appreciator (not the AI) is in this case the maker or artist as the appreciator is looking at the object as art, as part of the historical continuum of other artworks, no matter how much the algorithm used previous artworks as reference.

24 Tague says: "In terms of biology, there clearly are striking benefits to making art over the costs, and the behavior is not only passed on by instruction and learning but the impulse is innate and heritable" [38] (p. 1). Although I agree that at least some of the sub-modules of the aesthetic module are heritable, and that some may also be innate, in this article I argue why it is a bit misleading to talk about *the impulse* for artistic behavior.

25 For more on mindreading and imitating other people's mental states ("embodied simulation") as different from each other but both present when looking at movies, see [47].

26 For habits, see [48].

27 See for example, [48,49] (p. 6).

References

1. Hirn, Y. *The Origins of Art: A Psychological & Sociological Inquiry*; MacMillan and Co.: London, UK, 1900.
2. Dutton, D. *The Art Instinct: Beauty, Pleasure, & Human Evolution*; Oxford University Press: Oxford, UK, 2009.
3. Davies, S. *The Artful Species: Aesthetics, Art, and Evolution*; Oxford University Press: Oxford, UK, 2012.
4. Richards, R. *The Biology of Art*; Cambridge University Press: Cambridge, UK, 2019.
5. Luty, J. Is Art an Adaptation? The Timeless Controversy over the Existence of Aesthetic Universals (by the Lens of Evolutionary Informed Aesthetics). *Rocz. Kult.* **2021**, *12*, 75–91. [CrossRef]
6. Rampley, M. *The Seductions of Darwin: Art, Evolution, Neuroscience*; The Pennsylvania State University Press: University Park, PA, USA, 2017.
7. Kant, I. *Critique of Judgment*; Meredith, J.C., Translator; Walker, N., Ed.; Oxford University Press: Oxford, UK, 2007.
8. Nanay, B. *Aesthetics: A Very Short Introduction*; Oxford University Press: Oxford, UK, 2019.
9. Longino, H.E. *Studying Human Behavior: How Scientists Investigate Aggression and Sexuality*; The University of Chicago Press: Chicago, LI, USA, 2013.
10. Kokkonen, T. Evolving in Groups: Individualism and Holism in Evolutionary Explanation of Human Social Behaviour. Ph.D. Thesis, University of Helsinki, Helsinki, Finland, 10 September 2021.
11. Laland, K.N.; Brown, G.R. *Sense and Nonsense: Evolutionary Perspectives on Human Behaviour*, 2nd ed.; Oxford University Press: Oxford, UK, 2011.
12. Desideri, F. Epigenesis and Coherence of the Aesthetic Mechanism. *Aisthesis. Prat. Ling. E Saperi Dell'estetico* **2015**, *8*, 25–40.
13. Heyes, C. *Cognitive Gadgets: The Cultural Evolution of Thinking*; Harvard University Press: Cambridge, MA, USA, 2018.
14. Robbins, P. Modularity of Mind. In *The Stanford Encyclopedia of Philosophy*; Zalta, E.N., Ed.; Winter Edition; 2017; Available online: https://plato.stanford.edu/archives/win2017/entries/modularity-mind/ (accessed on 14 May 2022).
15. Mercier, H.; Sperber, D. *The Enigma of Reason: A New Theory of Human Understanding*; Penguin Books: London, UK, 2018.
16. Dewey, J. *Art as Experience*; Perigee: New York, NY, USA, 2005.
17. Stecker, R. Aesthetic Experience and Aesthetic Value. *Philos. Compass* **2006**, *1*, 1–10. [CrossRef]
18. Dokic, J. Aesthetic Experience as a Metacognitive Feeling? A Dual-Aspect View. *Proc. Aristot. Soc.* **2016**, *116*, 69–88. [CrossRef]
19. Seghers, E. Proximate and Ultimate Factors in Evolutionary Thinking on Art. *Hum. Ethol. Bull.* **2016**, *31*, 47–59. [CrossRef]
20. Skov, M.; Nadal, M. The Nature of Perception and Emotion in Aesthetic Appreciation: A Response to Makin's Challenge to Empirical Aesthetics. *Psychol. Aesthet. Creat Arts* **2021**, *15*, 470–483. [CrossRef]
21. Chuan-Peng, H.; Yi Huang, Y.; Eikchoff, S.B.; Peng, K.; Sui, J. Seeking the "Beauty Center" in the Brain: A Meta-Analysis of fMRI Studies of Beautiful Human Faces and Visual Art. *Cogn. Affect. Behav Neurosci.* **2020**, *20*, 1200–1215. [CrossRef] [PubMed]
22. Kiianlinna, O. Contradiction That Never Was. Epigenesis versus Modularity in Evolutionary Aesthetics. *Aisthesis. Prat. Ling. E Saperi Dell'estetico* **2021**, *14*, 79–91. [CrossRef]
23. Heyes, C.; Bang, D.; Shea, N.; Frith, C.D.; Fleming, S. Knowing Ourselves Together: The Cultural Origins of Metacognition. *Trends Cogn. Sci.* **2020**, *24*, 349–362. [CrossRef]
24. Mikuni, J.; Specker, E.; Pelowski, M.; Leder, H.; Kawabata, H. Is There a General "Art Fatigue" effect? A Cross-Paradigm, Cross-Cultural Study of Repeated Art Viewing in the Laboratory. *Psychol. Aesthet. Create. Arts* **2021**, *16*, 343–360. [CrossRef]
25. Wilson, D. Metarepresentation in Linguistic Communication. In *Metarepresentations: A Multidisciplinary Perspective*; Sperber, D., Ed.; Oxford University Press: Oxford, UK, 2000.
26. Kirchhoff, M.D.; Kiverstein, J. *Extended Consciousness and Predictive Processing: A Third-Wave View*, 1st ed.; Routledge: London, UK, 2019; Taylor & Francis eBooks Complete.
27. Clark, A. Whatever next? Predictive brains, situated agents, and the future of cognitive science. *Behav. Brain Sci.* **2013**, *36*, 181–253. [CrossRef] [PubMed]
28. Hohwy, J. *The Predictive Mind*, 1st ed.; Oxford University Press: Oxford, UK, 2013; EBSCOhost.

29. Keller, E.F. *The Mirage of a Space Between Nature and Nurture*; Duke University Press: Durham, NC, USA, 2010; EBSCOhost.
30. Bender, J.W. Sensitivity, Sensibility, and Aesthetic Realism. *J. Aesthet. Art Crit.* **2001**, *59*, 73–83. [CrossRef]
31. Scott, R.B.; Dienes, Z.; Barrett, A.B.; Bor, D.; Seth, A.K. Blind Insight: Metacognitive Discrimination Despite Chance Task Performance. *Psychol. Sci.* **2014**, *25*, 2199–2208. [CrossRef] [PubMed]
32. Specker, E.; Forster, M.; Brinkmann, H.; Boddy, J.; Immelmann, B.; Goller, J.; Pelowski, M.; Rosenberg, R.; Leder, H. Warm, lively, rough? Assessing agreement on aesthetic effects of artworks. *PLoS ONE* **2020**, *15*, e0232083. [CrossRef] [PubMed]
33. Malotki, E.; Dissanayake, E. *Early Rock Art of the American West: The Geometric Enigma*; University of Washington Press: Seattle, WA, USA, 2018.
34. Davies, S. *Adornment: What Self-Decoration Tells Us about Who We Are*; Bloomsbury Publishing Plc: London, UK, 2020.
35. Seghers, E. Cross-Species Comparison in the Evolutionary Study of Art: A Cognitive Approach to the Ape Art Debate. *Rev. Gen. Psychol.* **2014**, *18*, 263–272. [CrossRef]
36. Consoli, G. The Emergence of the Modern Mind: An Evolutionary Perspective on Aesthetic Experience. *J. Aesthet. Art Crit.* **2014**, *72*, 37–55. [CrossRef]
37. Arshamian, A.; Gerkin, R.C.; Kruspe, N.; Wnuk, E.; Floyd, S.; O'Meara, C.; Rodriguez, G.; Lundström, J.N.; Mainland, J.D.; Majid, A. The perception of odor pleasantness is shared across cultures. *Curr. Biol.* **2022**, *32*, 2061–2066. [CrossRef]
38. Tague, G. *Art and Adaptability: Consciousness and Cognitive Culture*; Brill: Leiden, The Netherlands, 2017; ProQuest Ebook Central.
39. Friston, K.J.; Lin, M.; Frith, C.D.; Pezzulo, G.; Hobson, J.A.; Ondobaka, S. Active Inference, Curiosity and Insight. *Neural. Comput.* **2017**, *29*, 2633–2683. [CrossRef]
40. Nanay, B. *Aesthetics as Philosophy of Perception*; Oxford University Press: Oxford, UK, 2016.
41. Donald, M. *A Mind So Rare: The Evolution of Human Consciousness*, 1st ed.; W.W. Norton & Company: New York, NY, USA, 2001.
42. Verpooten, J.; Dewitte, S. The Conundrum of Modern Art: Prestige-Driven Coevolutionary Aesthetics Trumps Evolutionary Aesthetics among Art Experts. *Hum. Nat.* **2017**, *28*, 16–38. [CrossRef]
43. Ginsburg, K.R.; The Committee on Communications; The Committee on Psychosocial Aspects of Child and Family Health. The Importance of Play in Promoting Healthy Child Development and Maintaining Strong Parent-Child Bonds. *Pediatrics* **2007**, *119*, 182–191. [CrossRef]
44. Sharpe, L.L. Play does not enhance social cohesion in a cooperative mammal. *Anim. Behav.* **2005**, *70*, 551–558. [CrossRef]
45. Lillard, A.S.; Lerner, M.D.; Hopkins, E.J.; Dore, R.A.; Smith, E.D.; Palmquist, C.M. The Impact of Pretend Play on Children's Development: A Review of the Evidence. *Psychol. Bull.* **2013**, *139*, 661–686. [CrossRef] [PubMed]
46. Allison, M.L.; Reed, R.; Michels, E.; Boogert, N.J. The drivers and functions of rock juggling in otters. *R. Soc. Open Sci.* **2020**, *7*, 200141. [CrossRef] [PubMed]
47. Raz, G.; Hendler, T. Forking Cinematic Paths to the Self: Neurocinematically Informed Model of Empathy in Motion Pictures. *Projections* **2014**, *8*, 89–114. [CrossRef]
48. Portera, M. Aesthetics as a Habit: Between Constraints and Freedom, Nudges and Creativity. *Philosophies* **2022**, *7*, 24. [CrossRef]
49. Prum, R. Coevolutionary aesthetics in human and biotic artworlds. *Biol. Philos.* **2013**, *28*, 811–832. [CrossRef]
50. Dissanayake, E. Komar and Melamid Discover Pleistocene Taste. *Philos. Lit.* **1998**, *22*, 486–496. [CrossRef]
51. Pinker, S. *The Blank Slate: The Modern Denial of Human Nature*; Penguin Books: London, UK, 2002.

philosophies

Article

The Cognitive Philosophy of Communication

Trond A. Tjøstheim [1], Andreas Stephens [2,*], Andrey Anikin [1,3] and Arthur Schwaninger [4]

[1] Department of Philosophy, Cognitive Science, Lund University, 221 00 Lund, Sweden; trond_arild.tjostheim@lucs.lu.se (T.A.T.); andrey.anikin@lucs.lu.se (A.A.)
[2] Department of Philosophy, Theoretical Philosophy, Lund University, 221 00 Lund, Sweden
[3] Equipe de Neuro-Ethologie Sensorielle (ENES)/Centre de Recherche en Neurosciences de Lyon (CRNL), University of Lyon/Saint-Etienne, CNRS UMR5292, INSERM UMR_S 1028, 42100 Saint-Etienne, France
[4] Department of Philosophy, University of Zurich, 8044 Zurich, Switzerland; arthur.schwaninger@philos.uzh.ch
* Correspondence: andreas.stephens@fil.lu.se

Received: 15 October 2020; Accepted: 17 November 2020; Published: 19 November 2020

Abstract: Numerous species use different forms of communication in order to successfully interact in their respective environment. This article seeks to elucidate limitations of the classical conduit metaphor by investigating communication from the perspectives of biology and artificial neural networks. First, communication is a biological natural phenomenon, found to be fruitfully grounded in an organism's embodied structures and memory system, where specific abilities are tied to procedural, semantic, and episodic long-term memory as well as to working memory. Second, the account explicates differences between non-verbal and verbal communication and shows how artificial neural networks can communicate by means of ontologically non-committal modelling. This approach enables new perspectives of communication to emerge regarding both sender and receiver. It is further shown that communication features gradient properties that are plausibly divided into a reflexive and a reflective form, parallel to knowledge and reflection.

Keywords: communication; reflexive communication; reflective communication; knowledge; memory; artificial intelligence

1. Introduction

Despite its centrality to fields such as linguistics and ethology, the concept of communication has no generally accepted definition. If the main focus is on language, it seems intuitive to use the so-called "conduit metaphor" [1,2] and to describe communication as a transfer of information from sender to receiver, the purpose being to reconstruct the message as accurately as possible on the receiver's side. Successful communication, according to this classical view, constitutes lossless transfer of information from the sender's to the receiver's mind, where the concept of information is derived from semantics and pragmatics—that is, from the dictionary meaning of words and the way these words are deployed by intentional actors in ongoing social interaction [3].

In contrast, biological accounts of communication emphasize the evolved, adaptive nature of communication. According to this view, communication can be defined as "the process of conveying information from senders to receivers by means of signals, and signals as the behaviors or structures that senders evolved in order to convey information" [4] (p. 2). Now, animals interact in many ways, including patently non-communicational encounters such as predation, accidental eavesdropping across species, and so on. Hence, treating communication as an evolved feature, adaptive to the sender and receiver, focuses the inquiry on those aspects of social interaction that are communicative "by design."

To situate the current state of play in the philosophy of biology, and more specifically, the philosophy of biological communication, we will introduce two main approaches to communication: the informational, and the influential. The former builds on classical formal theories of communication in terms of a sender, a message, and a receiver, initiated by Shannon [5]. The latter prefers the terms signaler, signal, and perceiver, and originates in work by Dawkins and Krebs [6].

The original work by Shannon was developed in the context of optimizing the legibility of human speech communication over telephone lines [5]. This work conceptualized information as independent of content, and focused on stable transmittable differences, quantifiable in terms of bi-valued bits. Later work by Lakoff [2] and others resulted in the conduit metaphor which tend to presuppose human-like communicators informing each other about states of the world. These states may then include the states of the communicator's mind.

While the informational approach works well in the context of humans and machines, Dawkins and Krebs [6] criticized its use in biological settings. They maintained that organisms tend to reflexively optimize energy use, and that this can be achieved by means of signaling and the perception of those signals by other organisms. Hence the point of communication in this sense is to influence the behavior of the organism's environment to avoid excessive energy expenditure, or to gain energy [7] (p. 176) (see also [8]). This is in contrast with organisms trying to inform each other about something, or maintaining desires or beliefs that need to be communicated somehow [9].

We interpret the conduit metaphor to sort under the informational approach, and by highlighting its limitations in the context of biological communication, we also argue for the influential approach. However, by treating communication as a continuum that can include both biological forms of communication, as well as human, and artificial ones, we aim to show that it may be possible to unify the two approaches. Further, this unification is mediated by differences in cognitive capabilities which specifically have to do with the degree to which reflective processing is supported.

More specifically, we will investigate communication from two perspectives. First, as a biological natural phenomenon, focusing on its connection to an organism's embodied structures and memory system. Such a perspective can elucidate how communicative abilities have evolved, consisting of a gradient set distributed on procedural, semantic, and episodic long-term memory as well as on working memory. Moreover, this approach lets us investigate parallels between communication and knowledge—which also map to the memory systems [10–12]. In particular, a reflexive and a reflective form of communication will emerge. Second, from the perspective of artificial neural networks. This will show that communication is possible independent of ontological commitments, only requiring similarity of experience. The article will present arguments for the following three theses:

- Communication can fruitfully be grounded in an organism's embodied structure and memory system.
- Communication features gradient properties that are plausibly divided into a reflexive and a reflective form.
- The conduit metaphor of communication is limited by not taking into account reflexive reward and aversion inducing processes that motivate approach or avoidance.

This teleological view of communication has the additional benefit of being easily extended to artificial systems if we replace evolutionary adaptations with actual intelligent design, in this case by human engineers. To take a very simple example, a red light on a console that alerts the user of memory overload or low battery power can be viewed as communicative if it was designed for this specific purpose. In Section 2, a short background to our biological account of communication is presented, which Section 3 links to embodied structures, memory systems, and knowledge. Section 4 focuses on the communicative sender, both from a reflexive and a reflective perspective, and Section 5 then deals with the communicative receiver in a similar manner. In Section 6, the discussion is connected to the development of biologically grounded communicative features in AI systems.

2. Background

An important point of contention in biological theories of communication concerns the notion that signals carry information, which is then processed by the receiver. As illustrated by many of the examples in the following sections, both the production and perception of signals may be too direct to plausibly involve mental representations or advanced cognitive processing. According to some authors, it is therefore best to avoid the notion of information and to define communication as the process of altering other's behavior via evolved mechanisms [4,13]. For example, piercing shrieks draw attention and increase arousal simply because of their acoustic properties [14], leading the authors to propose a distinction between direct and indirect affect induction in the audience. This view of communication, with a focus on influencing instead of informing others, is a valuable contribution from biological research and a reminder that language is not the only possible form of communication. On the other hand, the existence of "direct" signals does not necessarily mean that they carry no information. If the effect—or meaning—of a signal depends on the receiver's set of sensory organs, cognitive architecture, and unique life history, the informational content of a signal is best treated not as an intrinsic property of the signal itself, but as a product of its interaction with a particular receiver in a particular context [15]. Once we acknowledge that the informational content of a given signal is not constant under all conditions, even "direct" signals such as startling shrieks can be accommodated by an information-based view of communication, which can then be defined as exchange of information via an evolved (for biological systems) or designed (for artificial systems) mechanism. It makes sense here to also contrast the biological perspective with Floridi's (see, e.g., [16,17]) notion of "true semantic content." Factual semantics is necessary when an agent needs to acquire knowledge about the world. This is particularly the case when that knowledge is needed as a means to an end, such as finding the solution to a problem, or finding the path to some goal. In situations when an agent cannot observe the world first hand, but is dependent on a third party for getting information, the veracity of that third party's account is critical. This state of affairs is common in human society, but does have analogues among animals as well. An example of this may be the case of bees reporting on the suitability of found hive migration sites (see, e.g., [18]). Yet, as will be further explicated below, in cases such as mate attraction, it may not make sense to speak of, e.g., a colorful plume, or towering antlers as conveying facts about an individual's fitness. Rather, it may be more plausible to understand the colors, or the antlers as inducing reward processes in the observer. These processes again facilitate and motivate approach behavior by reflex mechanisms.

The "conduit metaphor," however, is not compatible with a biological account of communication. As we argue below, the language-inspired notion of communication as the process of intentionally transferring a mental representation from the sender to the receiver via a symbolic code represents only the tip of the iceberg—a highly specialized and rather unusual example of communication in the biological world. Instead, a useful starting point in studying communication may be to specify the various cognitive mechanisms involved in the production and perception of different signals, from fairly "direct" to the most cognitively sophisticated. In other words, given that animal's innate capabilities are formed by evolutionary processes, that tend to vary widely, each particular species' capabilities and prerequisites are crucial to take into account. A failure to acknowledge, for example, non-linguistic or modal limitations risks missing essential communicative features. Therefore, an animal's cognitive faculties, as well as developmental factors, are potentially interesting.

3. Knower

To gain an overarching perspective of communication, we regard it conducive to link our discussion of communication to cognitive psychology (see, e.g., [11,12,19–22]). We will argue that both communication and knowledge feature parallel gradient properties that can be usefully compared. In doing so, we hope to be able to ground communication in a way that is elucidating.

All organisms have been formed by adaptations through evolutionary processes. They thus in a certain sense match their environment and so, through their structure, embody a form of "biological knowledge" [23]. What this means is that each organism has a specific set of ways to interact with the world and a specific set of faculties to perceive the world. In other words, an organism's structures and capabilities enable and delimit its impressions (actions) on the world and its fellow creatures, as well as its interpretations (perceptions) of incoming stimuli. Since different species live under very different circumstances and in different environments, they subsequently have been shaped to appear, act, and perceive very differently.

Focusing on cognitive capabilities, memory is central for knowledge. Memory is often divided into long-term memory and working memory.[1] Long-term memory can, in turn, be divided into (non-declarative or implicit) procedural memory governing actions, skills, and an animal's ability to tackle practical obstacles, as well as (declarative or explicit) semantic memory governing pattern recognition and categorizations, and episodic memory governing that is closely tied to remembrance and language. Working memory consists of a central executive that works as a decision making and conscious control station, a phonological loop governing internal linguistic sequences, a visuospatial sketchpad governing visual semantics and mental images, and an episodic buffer that binds information into episodes. Working memory, together with episodic long-term memory, governs reflection [20].

A way to facilitate our discussion is offered by the cognitive psychological Dual Process Theory (see, e.g., [21,22,24–27]). Sidestepping a number of details, Dual Process Theory divides mental processes into two kinds: non-conscious and automatic Type 1 processes and conscious and reflecting Type 2 processes. We will use these process forms, and the aforementioned memory forms, as a background framework to engage the following discussion of communication.

3.1. Reflexive Knowledge

From a cognitive psychological perspective, reflexive knowledge relies heavily on purely embodied structures. As mentioned, evolved structures provide limits and affordances for an organism's interactions with the world. This has led to vast differences in size, form, and capability. Moreover, different organisms rely differently on their various senses. For example, to some, tactile or olfactory stimuli are essential, whereas others primarily depend on visual or auditory input. It should be pointed out that it is important to understand both what form of stimuli an organism is capable to register, as well as to what degree it relies on a particular form of stimuli (or a particular weighted combination). These limitations will naturally also constrain that organism's abilities for communication with its environment, in particular its abilities to receive signals.

Concerning cognitive capabilities, reflexive knowledge also relies on non-conscious Type 1 processes [21]. Such reflexive knowledge is based in procedural and semantic long-term memory [11]. These processes are thought to primarily be implicit, intuitive, and automatic. They are linked to motor skills and abilities, such as bodily movement and the ability to vocalize. Moreover, the ability to categorize, and to associatively learn, comprise a conceptual form of reflexive knowledge. Procedural memory governs motoric, reflexive and perceptual pathways, whereas semantic memory governs, for example, associative pathways (see, e.g., [28,29]).

Reflexive knowledge may also include propensities for reward and aversion. These propensities can be seen as innate or learned reflexive associations between the sensory apparatus and the motivation and motor pathways [30]. These pathways enable quick and effortless reaction to rewarding or threatening stimuli, but often at the cost of accuracy [31]. Typically, rewarding stimuli motivates approach behavior, while aversive stimuli tend to motivate avoidance [32].

---

1 We will largely ignore the discussion of short-term memory and instead view it as subsumed in working memory.

Reflexive knowledge is spread throughout the animal kingdom, even though it can look very differently depending on the relevant context various animals find themselves in. Reflexive knowledge thus enables eliciting automatic signals, as well as automatic reactions to signals.

3.2. *Reflection*

Reflective abilities—although they are unusual in the totality of biological knowledge—rely heavily on Type 2 processes, involving capabilities such as language and mental representation. Type 2 processes are based in episodic long-term memory and working memory—albeit intertwined with procedural and semantic memory.

It is worth pointing out that "the brain mechanisms subserving episodic-like memory are highly conserved among mammals" [33] (p. 10373), and so many animals have some capacity for episodic long-term memory [34], such as rats [35], corvids [36], and primates [37]. Including all organisms, different species can be seen to range from having no such capabilities, or being capable of some rudimentary forms, to having open-ended general intelligence.

Although it is no trivial matter to dissociate short-term memory from working memory, it is also well established that many animals, such as mice, rats, dogs, and monkeys, have some form of working memory [38,39]. This involves being able to chunk information and to resist interfering stimuli. However, it remains unclear exactly to what extent animals have such working memory capabilities. This might in part depend both on that it is hard to measure such capabilities with certainty, and in part depend on that the research area is still relatively overlooked [33]. Working memory is involved in enabling domain-general information processing [20,40], where particularly the episodic buffer works as a link between the central executive in working memory and episodic memory in the long-term memory system.

Now, reflective capabilities involve, for example, rule-based explicit reasoning, mentalizing (mindreading), mental time travelling, hypothetical thinking, and language abilities (see, e.g., [28,41]). Working memory is divided into a number of discrete components (see, e.g., [20]). The phonological loop governs internal monologues and speech as well as interpretation. The visuospatial sketchpad governs visual and spatial information. The central executive governs cognitive control and executive functions. The episodic buffer governs mediation between memory systems, especially between the central executive and episodic long-term memory, integrating relevant information needed for planning and executive control. The episodic long-term memory governs integration of sensory streams, together with the episodic buffer, encoding and reconstructing episodes. In summary, reflective knowledge is also spread throughout the animal kingdom, even though it is much less clear how, and to what degree. It affords communicating about events and situations that are temporally and spatially distanced from the immediate environment.

3.3. *Grounding Communication*

In order to ground communication in memory, we link reflexive communication to embodied and innate aspects of an animal, as well as to reflexive behaviors. Indeed, this is plausible given that Type 1 processes rely on procedural and semantic long-term memory, thought to be instantiated in many animals. Importantly, as previously pointed out, such aspects can take very different forms depending on species. Moreover, reflexive signaling *can* involve conscious aspects, but are nonetheless seen to principally make up non-conscious functions. Reflective communication is linked to mentalizing abilities and Type 2 processes that rely on episodic memory and working memory. These competences are also widely spread although it is unclear to what extent they are found to a high degree.

Briefly put, and illustrated in Figure 1 below, procedural communicative features involve, for example, somatic, innate and inflexible responses and behaviors. Semantic features involve, for example, learned inflexible behaviors, whereas episodic features involve, for example, learned flexible behaviors as well as mental simulation, planning, and language.

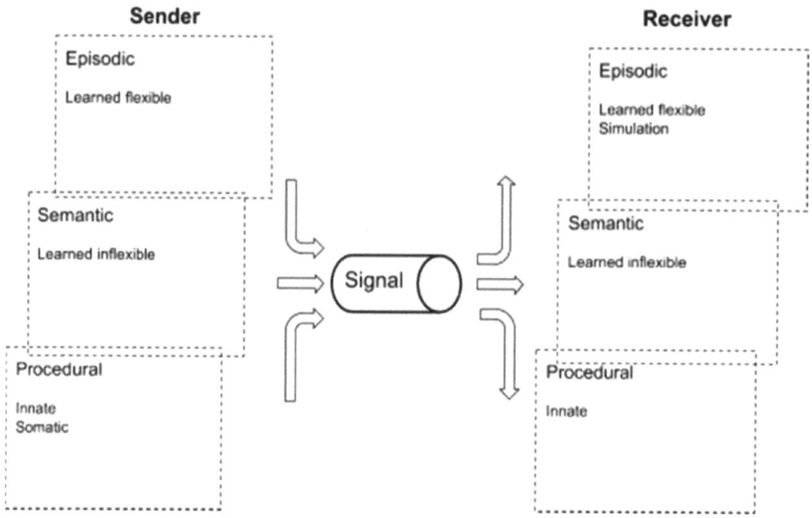

Figure 1. Schematic sketch of the communicative sender and receiver.

In the following two sections we will use this grounding of knowledge and communication in memory as a background framework to stepwise investigate communication, first from the perspective of the sender and thereafter from the perspective of the receiver.

4. Sender: Reflexive and Reflective Communication Production

In this section, we will focus on the communicative sender. This role can be filled by an animal, human, or AI system, potentially involving a wide range of communicative possibilities.

The least cognitively demanding production mechanisms rely on innate neural circuits and require neither learning nor conscious access to the communicated signal. In fact, the production of many signals does not even involve the brain. Signals of this type, which we refer to as 'somatic,' are long-term modifications of the signaler's body that evolved in order to inform other organisms about the fitness, age, sex, and social status of the signaler. For example, males of many animal species possess ornate and seemingly useless features: antlers in deer, large tail feathers in peacocks, brightly colored spots in fishes, and so on [42]. These decorations are thought to evolve due to sexual selection driven by female preferences. Sexual selection in humans is an object of continuing debate and speculation. For example, it is possible that a descended larynx and beard in males are examples of somatic features whose evolution was driven by female preferences and male competition in the context of attempting to exaggerate the apparent body size [43,44]. If that is true, these human peculiarities can be regarded as somatic communicative signals. It is also important to emphasize that the operation of sexual selection is not limited to somatic signals. Complex behavioral traits, such as songs of oscine birds or roaring contests of male deer [45], also evolve to regulate mating. There are even speculations that such uniquely human abilities as music and language [46] were affected by sexual selection.

Moving on from somatic features to signals whose production is rapid and controlled by the brain, there are many examples of communicative signals that are fully or largely innate in terms of both form and context of production. For example, worker ants returning from a food site lay down a pheromone trail, which helps to recruit and guide other workers, who in turn strengthen the trail with fresh pheromone markers until the food supply is exhausted. By using several types of attractant and repellent pheromones with varying half-life, ants can coordinate the behavior of the entire colony in an adaptive and highly flexible manner [47]. However, the form of signal (the choice of a particular pheromone) and the timing of its expression appear to be determined by simple 'if-then' rules, leaving limited room for learning, broader context, or conscious intentions. This may be obvious in the case of ants, but innate and relatively inflexible signals are by no means unique to invertebrates. On the contrary, a very large proportion of animal signals falls into this category. For example, the basic structure of nearly all primate vocalizations and many gestures is genetically determined [48], and each is associated with a range of typical eliciting contexts. In humans, congenitally deaf infants learn to laugh normally [49], which indicates that the appropriate motor programs (a coordinated activity of the diaphragm and muscles of the larynx) are species-typical behaviors that mature without auditory feedback and are triggered in a predetermined eliciting context (social play, tickling), again without the need for environmental input. Nor do we grow out of such innate signaling as adults: if suddenly frightened, most people will scream and display the classical primate "fear face" before being able to monitor or suppress this involuntary reaction. As demonstrated by this example, neural circuitry for the production of species-typical signals in relatively narrow, predetermined contexts remains operative in organisms endowed with a strong capacity for social learning and intentional control, including humans.

In contrast to ants laying pheromone tracks or deaf infants laughing when tickled, many animals deploy species-typical signals with a considerable degree of flexibility. In many cases, learning has only a limited role in determining the context of production. A well-known example is the alarm call that vervet monkeys use to alert members of the group to the presence of an aerial predator. While young monkeys initially produce the eagle alarm call to things such as falling leaves and harmless birds, they gradually learn which species of raptors are particularly dangerous and call only when they spot those [50]. The acoustic structure of the call itself is innate; further, there is a strong predisposition to apply this call type to threats from above rather than to terrestrial predators such as leopards or snakes, for which vervet monkeys use different alarm calls. Learning serves to fine-tune the eliciting context, but the production of alarm calls remains rather predictable.

At the opposite extreme of flexibility, calls of chimpanzees are much less context specific, even if their acoustic structure is innate, and some calls may even be produced with intention to inform. For example, chimpanzees appear to produce more alarm calls when other animals are not aware of the threat [51], and they may be able to inhibit the production of food grunts when it would be disadvantageous to disclose this information to others [52], although this inhibition appears to be effortful and is not always successful [53].

It is also important to point out that the same signal can be produced with varying degrees of flexibility or intentional control. The question of intentionality in animal communication is fraught with difficulty [54,55], but human emotional expressions are a clear case in point. Non-verbal vocalizations and facial expressions can be produced spontaneously, as when laughing at something amusing or showing a genuine, Duchenne smile [56], but they can also be used in a more controlled fashion, as when smiling or chuckling politely on social occasions. Interestingly, different neural circuits appear to be involved depending on whether an emotional expression such as a laugh is produced spontaneously or volitionally [57], which demonstrates that the same communicative signal can be generated by different cognitive mechanisms. In addition, there are detectable differences between spontaneous and volitional facial expressions [56] and vocalizations [58], indicating that markers of genuine affect are hard to fake and thus relatively "honest." The crucial point is that this honesty stems precisely from lack of intentional control. The less the context of production is open to manipulation, the more reliably the signal expresses the true mental state of the sender. As the amount of flexibility increases, the signal can potentially express a wider range of meanings [55], but it also places a greater burden on the receiver, who now has to take into account the broader context, and possibly also the reputation of the sender, since the "honesty" of communication is no longer guaranteed.

Finally, some aspects of language itself also appear to belong in the category of innate signals with relatively flexible usage. Emotional prosody in spoken language shows strong regularities around the world [59,60], making it straightforward to determine whether a speaker of an unfamiliar language is angry, happy, or sad. The changes in voice quality, rate of speaking, intonation and other acoustic features appear to stem from the even more universal nonverbal emotional vocalizations [61,62], which are in turn traceable back to the vocalizations of the great apes and other primates [63,64]. In addition to emotional prosody, spoken language utilizes a number of largely universal grammatical markers, such as rising intonation in questions [65] or simple interjections such as "Huh?" [66]. While their usage is flexible and subject to intentional control, the form of these signals is thus strongly constrained by the need to conform to the repertoire of vocal and gestural communicative signals that humans are genetically endowed with.

Signals with a completely arbitrary, purely learned form are not common in the natural world. The most obvious exception is human language, although even language is now regarded as less arbitrary than originally claimed by Saussure [67] due to the widespread presence of onomatopoeia and other forms of sound symbolism in basic vocabulary [68,69]. Among animals, the form of signals is normally either wholly or partially innate, but there are interesting exceptions to this rule. The gestural repertoire of great apes is generally considered to be more flexible than their vocalizations [70,71]. Furthermore, all species of great apes can be taught to understand and produce hundreds of signs from the American sign language (ASL). While the grammatical structure of their sentences remains relatively impoverished [72], rigorous testing has confirmed that they do understand the meaning of the signs and can produce them appropriately, not only to obtain reward but also to request information, inform others of their intended course of action, and so on [73].

The work with language-trained apes probably constitutes the most convincing example of intentional use of symbolic signals by any non-human animal, but signals with non-innate form do exist in the natural world. Vocal dialects are common among songbirds and have been reported in some mammals such as whales [74,75] and bats [76]. Learning plays an important role in the acquisition of such signals, which makes them more similar to human language than to human emotional expressions. Once learned, however, these signals may well be produced without intention to inform and with only limited sensitivity to context, placing them closer to the relatively inflexible signals discussed above.

A number of species have varying degrees of reflective communicative competencies, based in episodic memory and working memory, although testing these abilities are made difficult by the fact that it is sometimes possible to explain these same abilities in ways more in line with reflexive behavior [33,36]. At lower levels, capabilities can include any communicative behavior indicating recall of past events. At higher levels, reflective communication involves competencies such as rudimentary symbolic language and a sense of time. In its most advanced forms open-ended language abilities are tied to a developed general intelligence involving the ability to communicate through speech, writing, sign, or gesture, where arbitrary symbols are used as representations in socially agreed upon manners.

There are various theories concerning why such abilities might have developed. Examples include that it is in order for groups to plan for the future [77]. By playing out and discussing long-term future scenarios, rather than actually carrying them out, efficiency and survival can be increased by a large degree. For example, instead of going into a dark cave to explore, it is safer to first think through and discuss various scenarios and thereafter take relevant precautions beforehand. Such abilities offer enormous survival benefits.

By forming complex syntactic and semantic structures, communication can be both powerful and efficient, involving, for example, mental imagery, recollection, inner speech, reflective awareness, willed action, deliberation, and planning [78]. Such purpose-driven and intentional abilities of communication enable a highly flexible form of communication in large social groups, referred to by Hockett and Hockett [79] as "design features" involving 'displacement' (ability to tend to things not immediately present), 'productivity' (ability to understand new utterances), 'cultural transmission' (language learning in social groups), and 'duality' (meaningful language, made up by meaningless parts).

5. Receiver: Reflexive and Reflective Communication Perception

In this section, we will focus on the communicative receiver. As before, we will consider this role possible to be filled by an animal, human, or AI system.

The most direct effect of a signal on a receiver—in the sense of involving the smallest amount of neural processing—is largely determined by the properties of peripheral receptors. An example already mentioned in the Introduction is the generally aversive effect of harsh and loud shrieks on listeners [14]. It is also possible that the cries of infants in humans and other mammalian species are under selective pressure to (i) maximize their subjectively experienced loudness by carrying a significant amount of energy in the range of frequencies to which adults are particularly sensitive [64], potentially causing pain and even hearing loss in the listener [80] and (ii) prevent habituation by means of introducing frequency modulation, non-linear vocal phenomena, and other acoustic irregularities [64,81]. The aversive effect of such sounds is not mediated by learned associations, but basically stems from excessive stimulation of cochlear hair cells in their most sensitive frequency range. Some minimal degree of neural processing is still necessary, so there is arguably no absolute divide between receptor-driven and other innate responses discussed below. However, such "direct" signals are interesting theoretically since they highlight the danger of approaching all biological communication with a toolkit borrowed from linguistics. The informational content of these stimuli, if any, is clearly very different from that of a verbal utterance.

In many cases, the receiver's response is not predicated on the physical properties of the signal, but it is nevertheless innate—that is, largely predictable based on the characteristics of the signal and the genetic makeup of the receiver. A good example of such inflexible innate response is the startle reflex—a rapid, spontaneous defensive reaction to a threatening stimulus such as a sudden loud noise. The response does not have to be completely impervious to contextual effects. For instance, in humans the eyeblink to a sudden noise is attenuated by positive and enhanced by negative affective states [82]. Non-associative learning in the form of habituation can also play some role in modulating the response. However, the basic pattern of the eliciting stimulus and response are "hard wired" rather than learned.

In the animal world, innate responses are extremely common and crucial for survival. To refer back to the example of somatic signals that regulate mating, female preferences for features such as bright plumage or long tail feathers are not the product of associative learning, but rather innately specified responses to the appropriate triggering stimulus. In other words, a female peacock does not learn by observation that males with large tails produce healthy offspring; instead, their brain is predisposed to respond favorably to a particular combination of visual features on a large tail (see, e.g., [46]). Innately specified responses can persist not only without a chance to learn the meaning of the signal through previous exposure, but without even a theoretical possibility of such exposure. For instance, moths that migrated to Pacific islands relatively recently continue to drop to the ground upon hearing an ultrasound, although this defensive measure against bats is meaningless in their bat-free environment. In contrast, this motor response has been decoupled from the detection of bat cries in species endemic to the islands, although their ears are still sensitive to ultrasounds.

A well-documented example of an innate response in humans is rapid detection of threatening stimuli by subcortical circuits centered on amygdala, which orchestrates a reflexive fearful response to pictures of snakes and spiders [83]. Interestingly, amygdala also appears to respond similarly to facial expressions of fear in other humans, or rather to the increased visibility of the sclera as the sender's eyes open wide in fear [84]. In this case, both the production of the facial expression of fear and its detection appear to be innate and relatively inflexible—that is, hard to control or inhibit intentionally. Revealingly, the responsible neural mechanisms are largely subcortical, which makes both production and response very fast, but also hinders intentional control.

When there is no innate predisposition to respond to a signal in a particular way, the receiver has to learn the signal's meaning from experience. In behavioral terms, it means observing what events tend to follow the detection of this signal—in other words, what the signal predicts in terms of environmental changes or the ensuing behavior of the sender. In neurological terms, learning the signal's predictive power (or, more generally, its meaning) requires some form of associative learning. Depending on exactly what is learned and how this information is processed, we propose the following three subtypes of learned responses, from least to most cognitively sophisticated.

The simplest strategy is to associate a signal with a single, standardized response that does not depend on the broader context. Learned, but inflexible responses of this kind appear to be relatively uncommon in the natural world. Overtrained operant conditioning in laboratory animals or household pets is a possible example, but such "mindless" conditioning is seldom advantageous in nature. There is, however, an interesting special case, namely behavioral programs with an innately specified response to a learned signal. Imprinting is popularly associated with the image of Konrad Lorenz followed by his goslings, who had taken him to be their mother. In more natural circumstances, however, imprinting has an important role to play in creating a powerful bond between the mother and her offspring. In highly vocal and colonial animals such as seals and walruses, the ability of the mother to learn the voice of her pup is crucial for them to reunite after the mother's hunting expeditions. The pup's calls—more specifically, the unique signature of frequency modulation in the pup's bark that enables individual recognition [85]—are thus learned signals that trigger innate nurturing behavior in the mother.

In the majority of cases, when a signal is not coupled with an innately specified response, the animal learns to extract the relevant information from the signal and to respond appropriately, taking into account additional factors such as the sender's identity, the history of previous interactions with the sender, the presence of other group members, and other contextual factors. As a result, there is no longer a one-to-one correspondence between the signal and the response. For example, vervet monkeys respond to alarm calls depending on their current position—that is, the response is not stereotypical. An animal who hears an eagle alarm call while on the ground will rush up into the branches, whereas an animal who is already high up will descend from the exposed treetops [50]. Furthermore, if an alarm call is followed by the sound made by the actual predator, this otherwise frightening sound no longer provokes a strong response. For all practical purposes, it appears that an eagle alarm call evokes the mental representation of an eagle in the audience, a snake alarm call brings to mind the image of a snake, and so on [55].

The idea of signals evoking mental representations remains a controversial, but parsimonious explanation for flexible responses [54,55] to context-specific, or functionally referential, signals such as alarm calls. Whether or not mental representations are involved, highly flexible cognitive processing is required when the same signal can be produced in a broad range of contexts. For example, people can laugh with each other or at each other, and the meaning of a laugh can vary accordingly, from benign amusement to malicious taunting [86]. Likewise, chimpanzees who hear a sequence of screams from two familiar individuals seem to be able not only to determine who is the aggressor and who is the victim, but also to judge whether these roles conform to their expectations based on the existing social hierarchy [87]. In cases such as this, it becomes increasingly natural to describe animal communication in terms of the inferences that receivers make on the basis of the information that they extract from a signal.

This view aligns closely with the pragmatic approach to human communication. One implication is that the distinction between human language and animal communication has become increasingly blurred on the receiver's side, whereas the production of signals in the animal world is usually—but not always—restricted to species-typical displays [88]. Characteristically, comprehension far outstrips production both in human infants and in language-trained animals [73], again suggesting that the capacity for highly flexible, context-dependent interpretation of learned signals is more widespread and less cognitively costly than the corresponding production skills.

Phenomena such as gaze following, present in for example chimpanzees, indicate some form of ability of understanding mental states. Although, an alternative interpretation is that such competencies merely involve "goal-directed action and perception, common to all apes, (rather than a) sharing (of) psychological states with others in collaborative acts involving joint intention and attention, (which is) unique to the human." [89] (p. vii).

Now, communication perception can involve ascriptions of intentionality to others, where a subject can "see" others as intentional agents in their own right. This can be achieved by, for example, theorizing about other's mental states or by simulating them [41,78]. So, communication perception can involve the ability to take another's perspective as well as being able to discern their intent. In a social setting, such abilities enable complex social interactions, where long-term planning concerning goals in the distant future are made possible. By mental trial and error different predictions concerning perceptual input can then be made.

However, as far as we know, capabilities of such flexible mental representations, involving simulation and reflection on one's own and other's thoughts, are biological outliers.

6. Communication between Artificial Neural Networks

In this section, we describe how artificial neural networks can illustrate and lend support to the idea that different but similar experience enable communication between agents, irrespective of ontological commitments.

Artificial systems may presently communicate with humans or other artefacts through a variety of means: voice and natural language, by recognizing human facial expressions, prosody, and body language. So far, the communication of non-linguistic information has been mainly from humans to machines, as machines so far lack proper emotional systems and designers have to make do with simple theatrics. As for recognition in general, deep neural networks have afforded improvement both in inference of human emotional state, as well as generation of very natural-like language. The latter has in fact become good enough to be indistinguishable from the real thing, which in the context of phone calls raises ethical issues of subterfuge, and a call for artificial systems to be made to identify themselves when interacting with humans.

Deep learning neural networks have become one of the most essential computational engineering methods used today. These networks often consist of input, output and hidden neurons that learn to detect features of the signal in the data. In the early days of what became known as 'computer vision,' the first pattern recognition algorithms were developed that consisted of multiple layers of hand-coded feature detectors using fixed network weights [90,91]. Over time, new methods were developed in order to learn the weights of the hidden units that have state invisible to the observer—that is why a neural network with intermediate feature detecting hidden units is typically referred to as a 'black box.' One of the first methods to self-adapt their own weights were invented in analogy with evolution. By randomly sampling the parameter-space and preserving those weight parameters that returned the best results while dismissing those with poor results, the network started to learn its own weights [92]. A different approach came through the discovery of the backpropagation algorithm that allowed to learn neural networks more effectively [93,94]. Even though some theoretical neuroscientists are convinced that backpropagation is also a reasonable possibility of how the neural cortex learns to adapt synaptic strengths [95], many remain suspicious about the method. Alternatively, predictive coding schemes have been proposed suggesting that higher-cortical cells predict the neural activity of lower cortical cells. In a complex interaction between information flowing upstream towards and downstream away from higher cortical regions, the system begins to learn based on local rules of computation [96]. This is illustrated in Figure 2, where we see from the view of one neuron how information flows in both directions in order to gain an internal representation.

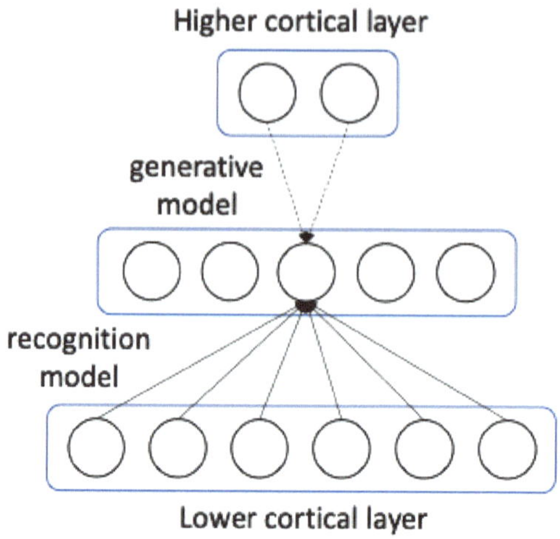

Figure 2. Every neuron—here only illustrated by the one central neuron—is bidirectionally connected to some or all the neurons in the higher and lower cortical layer. The generative and recognition models are indicated for a single neuron.

The downstream flow of information is given by a generative model that generates images from the internal representation in order to predict the activity of lower cortical regions. This allows a human-like communication process of two agents A and B to be modelled in a computer simulation as illustrated in Figure 3 [97]. Both agents are represented by a neural network that was trained on similar data, in this case a set of images of pears. Each of these networks includes a recognition as well as a generative model that allows the agents to build internal representations of pears and generate images of these. Given this setup, agent A can generate an image of, e.g., a pear that is fed as an input image into the network of agent B. This network then correctly recognizes the image and classifies it as a pear. Running this simulation back and forth closely resembles a dialog between two distinct human agents.

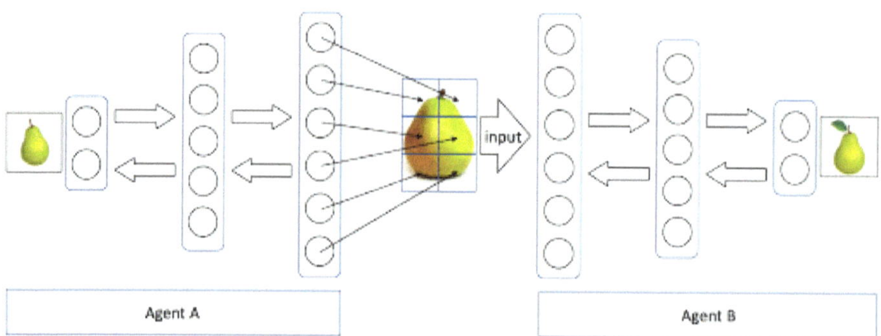

Figure 3. Two communicating (**Agents A,B**) are modelled using two independent neural networks, each of which was trained on different but similar data (here: images of pears). By simple forward computation from higher to lower cortical level, (**Agent A**) generates an image of a pear that is fed into the input layer of (**Agent B**). (**Agent B**) can then recognize the pear, resulting in an effective communication amongst the agents.

Traditional accounts of communication require the existence of an external material object x to express the perception of an agent ('A perceives x') and an expression for informing B about x, oftentimes by ostension ('A tells B about x by pointing at x'). The way human perception and action operate is closely related to the two-way process illustrated above. Thus, what we learn from such simulations is that no reference to any further objective reality must be made besides there being 'data' available. A communication process is perfectly expressible without relying on the individuals of an external reality. In other words, it is not required that the world appears to our senses in preparcelled form consisting of given objects. The world presents itself not in a veridical but instead in a way that is useful for preserving homeostasis. This is a view that found support by many scholars, including those in support of 'Evolutionary Epistemology' (see, e.g., [98,99]). This view was abandoned from mainstream thought for some decades before being resurrected with a new face in the light of modern neuroscientific findings [100]. Running a computer simulation of this model imitates a functioning communication process without committing to an ontology of a structured external world. Even if the communicators in the real world are humans rather than artificially intelligent agents, there is no particular commitment to an ontology about the real world. This is in line with contemporary constructivist and Kantian understanding of perception [101–104]. It is a view that challenges traditional semiotic accounts that assumes the existence of mind-independent objects with certain features that are signified by signs. It further challenges modern science oriented approaches that conceive information as veridical (see, e.g., [17]).

There are multiple theoretical advantages of this view. First, traditional problems concerning the inscrutability of reference (see, e.g., [105,106]) disappear since the commitment to a notion of reference is not required. A stronger claim, motivated by neuroscience research is made by Rosenberg [107,108] (Ch. 8). He says that even if introspection tells us that our thoughts are about something in the world, this is just an illusion. Advocating scientism, he says that science gives us no reason to believe that neural circuits are about something. In the end, neural circuits are nothing more than matter, and matter can never be intrinsically about anything at all. If our thought could be said to be about something in the world then we were to expect to find a sort of 'map' of the world in the brain. However, all we find is a neuronal network with altering connections. According to the sciences, mental states are not about the world but the neural structure is physically isomorphic to the world. Whether Rosenberg is right in his judgement is controversial and even though some others such as Kenny [109] (ch. 9) or Bennett and Hacker [110] share his type of sentiment, they draw different conclusions. According to them, when it comes to cognitive processes, we should not only consider the brain in isolation but the person as a whole. In any case, while we are not committed to Rosenberg's claim that the manifest concept of reference is *necessarily* defective, we do support the idea that reference is not *required*. Thus, we suggest that network modelling and simulations show how the notion of reference is not *required* in a science oriented perspective on communication. This is a weaker and less controversial position to be in.

Second, the representational account of perception supported by our scientific framework manages to avoid traditional problems of perception. These problems mostly concern the nature of how it is that we see an object and typically emerge once the mind-independent object is presupposed. However, not much is left of these problems within a framework that does not assume a world that is shaped prior to human perception. The problems simply disappear if objects are conceived of as constructions in the Kantian framework [97].

Third, communicative processes very often refer to fictive objects, such as unicorns or the characters in novels, rather than presumably existing ones. If there is no commitment to objects in the real world then there is also no need to make a distinction to fictitious objects. The process alone is what distinguishes the one from the other. While philosophers have pondered about the metaphysical distinction between real and fictional objects, an account of communication such as ours draws a line between the two types of objects by referring to the underlying cognitive process involved rather than the objects themselves. While the reference to real objects relies on a successful 'downstream' recognition process, the fantasizing of Sherlock Holmes depends on an 'upstream' information flow driven by the active generative network (see Figure 3).

Interestingly, this account of communication helps explain the phenomenon of humans communicating about fictitious, or made up, worlds and agents.

7. Conclusions

In this article, we have elucidated limitations of the classic conduit metaphor by investigating how various species use different forms of communication, both as senders and receivers, in order to successfully interact, survive, and mate in their respective environments. In particular, we have looked at communication as a biological natural phenomenon, being grounded in an organism's embodied structures and memory system, where specific abilities are tied to procedural, semantic, and episodic long-term memory as well as to working memory. In doing so, gradient reflexive and reflective properties, parallel to knowledge and reflection, have emerged. Finally, the account has explicated differences between non-verbal and verbal communication and shown how artificial neural networks can communicate by means of ontologically non-committal modelling. Hence we have attempted to sketch a picture of communication as a natural phenomenon where intentional communication between agents by means of structured language is but a special case of a larger continuity of communication which also includes biological signaling in the service of energy optimization. This means that the informational and the influencing approaches to biological communication are not necessarily in competition, but may best be seen as describing activity on different parts of a spectrum, and requiring different cognitive complexity to be supported.

Author Contributions: All authors have contributed to conceptualization, methodology and writing—original draft preparation, review and editing. All authors have read and agreed to the published version of the manuscript.

Funding: The authors gratefully acknowledge support from Makarna Ingeniör Lars Henrik Fornanders fond, grant numbers FO2019-0005; FO2020-0004 and Stiftelsen Elisabeth Rausings minnesfond: forskning, grant number RFh2019-0226.

Acknowledgments: We thank our anonymous reviewers for valuable comments. Sections 4 and 5 based on A. Anikin's doctoral thesis *Human nonverbal vocalizations* (2020).

Conflicts of Interest: The authors declare no conflict of interest.

References

1. Reddy, M.J. The conduit metaphor: A case of frame conflict in our language about language. In *Metaphor and Thought*; Ortony, A., Ed.; Cambridge University Press: Cambridge, MA, USA, 1993; pp. 284–310.
2. Lakoff, G. The contemporary theory of metaphor. In *Metaphor and Thought*; Ortony, A., Ed.; Cambridge University Press: Cambridge, MA, USA, 1993; pp. 202–251.
3. Andrews, K. Animal cognition. In *The Stanford Encyclopedia of Philosophy (Summer 2016 Edition)*; Zalta, E.N., Ed.; Stanford University: Stanford, CA, USA, 2016; Available online: https://plato.stanford.edu/archives/sum2016/entries/cognition-animal/ (accessed on 1 November 2020).
4. Stegmann, U.E. Introduction: A primer on information and influence in animal communication. In *Animal Communication Theory: Information and Influence*; Stegmann, U.E., Ed.; Cambridge University Press: Cambridge, MA, USA, 2013; pp. 1–39.
5. Shannon, C.E. A mathematical theory of communication. *Bell Syst. Tech. J.* **1948**, *27*, 379–423. [CrossRef]
6. Dawkins, R.; Krebs, J.R. Animal signals: Information or manipulation? In *Behavioural Ecology: An Evolutionary Approach*; Krebs, R., Davies, N.B., Eds.; Sinauer Associates: Sutherland, MA, USA, 1978; pp. 282–309.
7. Wilson, E.O. *Sociobiology: The New Synthesis*; Belknap: Cambridge, MA, USA, 1975.
8. Morton, E.S. *Animal Vocal Communication: Assessment and Management Roles*, 2nd ed.; Cambridge University Press: Cambridge, MA, USA, 2017.
9. Frick, R.; Bich, L.; Moreno, A. An Organisational Approach to Biological Communication. *Acta Biotheor.* **2019**, *67*, 103–128. [CrossRef] [PubMed]
10. Gärdenfors, P.; Stephens, A. Induction and Knowledge-What. *Eur. J. Philos. Sci.* **2018**, *8*, 471–491. [CrossRef]

11. Tulving, E. Memory and consciousness. *Can. Psychol.* **1985**, *26*, 1–12. [CrossRef]
12. Tulving, E. Episodic Memory: From Mind to Brain. *Annu. Rev. Psychol.* **2002**, *53*, 1–25. [CrossRef]
13. Rendall, D.; Owren, M.J.; Ryan, M.J. What do animal signals mean? *Anim. Behav.* **2009**, *78*, 233–240. [CrossRef]
14. Owren, M.J.; Rendall, D. An affect-conditioning model of nonhuman primate vocal signaling. In *Communication*; Springer: Boston, MA, USA, 1997; pp. 299–346.
15. Fischer, J. Where is the information in animal communication. In *Animal Thinking: Contemporary Issues in Comparative Cognition*; Menzel, R., Fischer, J., Eds.; MIT Press: Cambridge, MA, USA, 2011; pp. 151–161.
16. Floridi, L. Outline of a Theory of Strongly Semantic Information. *Minds Mach.* **2004**, *14*, 197–221. [CrossRef]
17. Floridi, L. *The Philosophy of Information*; Oxford University Press: Oxford, UK, 2011.
18. Passino, K.M.; Seeley, T.D.; Visscher, P.K. Swarm cognition in honey bees. *Behav. Ecol. Sociobiol.* **2008**, *62*, 401–414. [CrossRef]
19. Tulving, E. Episodic Memory and Autonoesis: Uniquely Human? In *The Missing Link in Cognition: Origins of Self-Reflective Consciousness*; Terrance, H., Metcalfe, J., Eds.; Oxford University Press: New York, NY, USA, 2005; pp. 3–56.
20. Baddeley, A. *Working Memory, Thought, and Action*; Oxford University Press: Oxford, UK, 2007.
21. Evans, J.S.B.T.; Stanovich, K.E. Dual-Process Theories of Higher Cognition. *Perspect. Psychol. Sci.* **2013**, *8*, 223–241. [CrossRef]
22. Lizardo, O.; Mowry, R.; Sepulvado, B.; Stoltz, D.S.; Taylor, M.A.; Van Ness, J.; Wood, M. What Are Dual Process Models? Implications for Cultural Analysis in Sociology. *Sociol. Theory* **2016**, *34*, 287–310. [CrossRef]
23. Plotkin, H.C. *Darwin Machines and the Nature of Knowledge*; Harvard University Press: Cambridge, MA, USA, 1993.
24. Evans, J.S.B.T. In two minds: Dual-process accounts of reasoning. *Trends Cogn. Sci.* **2003**, *7*, 454–459. [CrossRef] [PubMed]
25. Evans, J.S.B.T. Dual-Processing Accounts of Reasoning, Judgment, and Social Cognition. *Annu. Rev. Psychol.* **2008**, *59*, 255–278. [CrossRef] [PubMed]
26. Frankish, K. Systems and levels: Dual-system theories and the personal–subpersonal distinction. In *Two Minds: Dual Processes and Beyond*; Evans, J.S.B.T., Frankish, K., Eds.; Oxford University Press: Oxford, UK, 2009; pp. 89–107.
27. Kahneman, D. *Thinking Fast and Slow*; Farrar, Straus and Giroux: New York, NY, USA, 2011.
28. Kandel, E.R.; Schwartz, J.H.; Jessell, T.M.; Siegelbaum, S.A.; Hudspeth, A.J. *Principles of Neural Science*, 5th ed.; McGraw-Hill: New York, NY, USA, 2013.
29. Kim, H. Default network activation during episodic and semantic memory retrieval: A selective meta-analytic comparison. *Neuropsychologia* **2016**, *80*, 35–46. [CrossRef] [PubMed]
30. Roitman, M.F.; Wheeler, R.A.; Carelli, R.M. Nucleus Accumbens Neurons Are Innately Tuned for Rewarding and Aversive Taste Stimuli, Encode Their Predictors, and Are Linked to Motor Output. *Neuron* **2005**, *45*, 587–597. [CrossRef] [PubMed]
31. Vida, M.D.; Behrmann, M. Subcortical Facilitation of Behavioral Responses to Threat. *Sci. Rep.* **2017**, *7*, 13087. [CrossRef] [PubMed]
32. Rutherford, H.J.V.; Lindell, A.K. Thriving and Surviving: Approach and Avoidance Motivation and Lateralization. *Emot. Rev.* **2011**, *3*, 333–343. [CrossRef]
33. Carruthers, P. Evolution of working memory. *Proc. Natl. Acad. Sci. USA* **2013**, *110*, 10371–10378. [CrossRef]
34. Allen, T.A.; Fortin, N.J. The evolution of episodic memory. *Proc. Natl. Acad. Sci. USA* **2013**, *110*, 10379–10386. [CrossRef]
35. Panoz-Brown, D.; Corbin, H.E.; Dalecki, S.J.; Gentry, M.; Brotheridge, S.; Sluka, C.M.; Wu, J.-E.; Crystal, J.D. Rats Remember Items in Context Using Episodic Memory. *Curr. Biol.* **2016**, *26*, 2821–2826. [CrossRef]
36. Clayton, N.S.; Griffiths, D.P.; Emery, N.J.; Dickinson, A. Elements of episodic–like memory in animals. *Philos. Trans. R. Soc. B Biol. Sci.* **2001**, *356*, 1483–1491. [CrossRef]
37. Rilling, J.K.; Barks, S.K.; Parr, L.A.; Preuss, T.M.; Faber, T.L.; Pagnoni, G.; Bremner, J.D.; Votaw, J.R. A comparison of resting-state brain activity in humans and chimpanzees. *Proc. Natl. Acad. Sci. USA* **2007**, *104*, 17146–17151. [CrossRef] [PubMed]
38. Barth, J.; Call, J. Tracking the Displacement of Objects: A Series of Tasks with Great Apes and Young Children. *J. Exp. Psychol. Anim. Behav. Process.* **2006**, *32*, 239–252. [CrossRef] [PubMed]

39. Brauer, J.; Call, J. The magic cup: Great apes and domestic dogs (Canis familiaris) individuate objects according to their properties. *J. Comp. Psychol.* **2011**, *125*, 353–361. [CrossRef] [PubMed]
40. Awh, E.; Vogel, E.; Oh, S.-H. Interactions between attention and working memory. *Neuroscience* **2006**, *139*, 201–208. [CrossRef]
41. Goldman, A.I. *Simulating Minds: The Philosophy, Psychology, and Neuroscience of Mindreading*; Oxford University Press: Oxford, UK, 2006.
42. Zahavi, A. Mate selection—A selection for a handicap. *J. Theor. Biol.* **1975**, *53*, 205–214. [CrossRef]
43. Fitch, W.T. The Biology and Evolution of Speech: A Comparative Analysis. *Annu. Rev. Linguistics* **2018**, *4*, 255–279. [CrossRef]
44. Puts, D.A. Beauty and the beast: Mechanisms of sexual selection in humans. *Evol. Hum. Behav.* **2010**, *31*, 157–175. [CrossRef]
45. Reby, D.; McComb, K.; Cargnelutti, B.; Darwin, C.; Fitch, W.T.S.; Clutton-Brock, T. Red deer stags use formants as assessment cues during intrasexual agonistic interactions. *Proc. R. Soc. B Boil. Sci.* **2005**, *272*, 941–947. [CrossRef]
46. Miller, G. *The Mating Mind: How Sexual Choice Shaped the Evolution of Human Nature*; Anchor Books: New York, NY, USA, 2011.
47. Jackson, D.E.; Ratnieks, F.L. Communication in ants. *Curr. Biol.* **2006**, *16*, R570–R574. [CrossRef]
48. Snowdon, C.T. Plasticity of Communication in Nonhuman Primates. *Adv. Study Behav.* **2009**, *40*, 239–276. [CrossRef]
49. Scheiner, E.; Hammerschmidt, K.; Jürgens, U.; Zwirner, P. Vocal Expression of Emotions in Normally Hearing and Hearing-Impaired Infants. *J. Voice* **2006**, *20*, 585–604. [CrossRef] [PubMed]
50. Seyfarth, R.M.; Cheney, D.L.; Marler, P. Monkey responses to three different alarm calls: Evidence of predator classification and semantic communication. *Science* **1980**, *210*, 801–803. [CrossRef] [PubMed]
51. Crockford, C.; Wittig, R.M.; Mundry, R.; Zuberbühler, K. Wild chimpanzees inform ignorant group members of danger. *Curr. Biol.* **2012**, *22*, 142–146. [CrossRef] [PubMed]
52. Zuberbühler, K. Linguistic capacity of non-human animals. *Wiley Interdiscip. Rev. Cogn. Sci.* **2015**, *6*, 313–321. [CrossRef]
53. Goodall, J. *The Chimpanzees of Gombe: Patterns of Behavior*; Harvard University Press: Cambridge, MA, USA, 1986.
54. Manser, M.B. Semantic communication in vervet monkeys and other animals. *Anim. Behav.* **2013**, *86*, 491–496. [CrossRef]
55. Wheeler, B.C.; Fischer, J. Functionally referential signals: A promising paradigm whose time has passed. *Evol. Anthr. Issues News Rev.* **2012**, *21*, 195–205. [CrossRef]
56. Ekman, P.; Davidson, R.J.; Friesen, W.V. The Duchenne smile: Emotional expression and brain physiology: II. *J. Personal. Soc. Psychol.* **1990**, *58*, 342–353. [CrossRef]
57. Scott, S.K.; LaVan, N.; Chen, S.; Mcgettigan, C. The social life of laughter. *Trends Cogn. Sci.* **2014**, *18*, 618–620. [CrossRef]
58. Anikin, A.; Lima, C.F. Perceptual and Acoustic Differences between Authentic and Acted Nonverbal Emotional Vocalizations. *Q. J. Exp. Psychol.* **2018**, *71*, 622–641. [CrossRef]
59. Banse, R.; Scherer, K.R. Acoustic profiles in vocal emotion expression. *J. Personal. Soc. Psychol.* **1996**, *70*, 614–636. [CrossRef]
60. Pell, M.D.; Paulmann, S.; Dara, C.; Alasseri, A.; Kotz, S.A. Factors in the recognition of vocally expressed emotions: A comparison of four languages. *J. Phon.* **2009**, *37*, 417–435. [CrossRef]
61. Anikin, A.; Persson, T. Nonlinguistic Vocalizations from Online Amateur Videos for Emotion Research: A Validated Corpus. *Behav. Res. Methods* **2017**, *49*, 758–771. [CrossRef] [PubMed]
62. Cordaro, D.T.; Keltner, D.; Tshering, S.; Wangchuk, D.; Flynn, L.M. The voice conveys emotion in ten globalized cultures and one remote village in Bhutan. *Emotion* **2016**, *16*, 117–128. [CrossRef] [PubMed]
63. Davila-Ross, M.; Owren, M.J.; Zimmermann, E. Reconstructing the Evolution of Laughter in Great Apes and Humans. *Curr. Biol.* **2009**, *19*, 1106–1111. [CrossRef]
64. Lingle, S.; Wyman, M.T.; Kotrba, R.; Teichroeb, L.J.; Romanow, C.A. What makes a cry a cry? A review of infant distress vocalizations. *Curr. Zool.* **2012**, *58*, 698–726. [CrossRef]
65. Ohala, J.J. An Ethological Perspective on Common Cross-Language Utilization of F_0 of Voice. *Phonetica* **1984**, *41*, 1–16. [CrossRef]

66. Dingemanse, M.; Torreira, F.; Enfield, N.J. Is "Huh?" a universal word? Conversational infrastructure and the convergent evolution of linguistic items. *PLoS ONE* **2013**, *8*, e78273. [CrossRef]
67. De Saussure, F. *Course in General Linguistics*; Bally, C., Sechehaye, A., Eds.; Harris, R., Translator; Open Court: La Salle, IL, USA, 1983.
68. Blasi, D.E.; Wichmann, S.; Hammarström, H.; Stadler, P.F.; Christiansen, M.H. Sound–meaning association biases evidenced across thousands of languages. *Proc. Natl. Acad. Sci. USA* **2016**, *113*, 10818–10823. [CrossRef]
69. Dingemanse, M.; Blasi, D.E.; Lupyan, G.; Christiansen, M.H.; Monaghan, P. Arbitrariness, Iconicity, and Systematicity in Language. *Trends Cogn. Sci.* **2015**, *19*, 603–615. [CrossRef]
70. Arbib, M.A.; Liebal, K.; Pika, S. Primate Vocalization, Gesture, and the Evolution of Human Language. *Curr. Anthr.* **2008**, *49*, 1053–1076. [CrossRef]
71. Hobaiter, C.; Byrne, R.W. The gestural repertoire of the wild chimpanzee. *Anim. Cogn.* **2011**, *14*, 745–767. [CrossRef] [PubMed]
72. Terrace, H.S.; Petitto, L.A.; Sanders, R.J.; Bever, T.G. Can an ape create a sentence? *Science* **1979**, *206*, 891–902. [CrossRef] [PubMed]
73. Rumbaugh, D.M.; Savage-Rumbaugh, E.S. Language in comparative perspective. In *Animal Learning and Cognition*; Mackintosh, N.J., Ed.; Academic Press: San Diego, CA, USA, 1994.
74. Deecke, V.B.; Ford, J.K.B.; Spong, P. Quantifying complex patterns of bioacoustic variation: Use of a neural network to compare killer whale (Orcinus orca) dialects. *J. Acoust. Soc. Am.* **1999**, *105*, 2499–2507. [CrossRef] [PubMed]
75. Rendell, L.E.; Whitehead, H. Vocal clans in sperm whales (Physeter macrocephalus). *Proc. R. Soc. B Biol. Sci.* **2003**, *270*, 225–231. [CrossRef]
76. Prat, Y.; Azoulay, L.; Dor, R.; Yovel, Y. Crowd vocal learning induces vocal dialects in bats: Playback of conspecifics shapes fundamental frequency usage by pups. *PLoS Biol.* **2017**, *15*, e2002556. [CrossRef]
77. Gärdenfors, P. Cooperation and the evolution of symbolic communication. In *Evolution of Communication Systems*; Oller, D.K., Griebel, U., Eds.; MIT Press: Cambridge, MA, USA, 2004; pp. 237–256.
78. Passingham, R. *What Is Special about the Human Brain?* Oxford University Press: Oxford, UK, 2008.
79. Hockett, C.F.; Hockett, C.D. The origin of speech. *Sci. Am.* **1960**, *203*, 89–96. [CrossRef]
80. Calderon, L.E.; Carney, L.D.; Kavanagh, K.T. The Cry of the Child and its Relationship to Hearing Loss in Parental Guardians and Health Care Providers. *J. Evid. Informed Soc. Work.* **2016**, *13*, 198–205. [CrossRef]
81. Koutseff, A.; Reby, D.; Martin, O.; Levrero, F.; Patural, H.; Mathevon, N. The acoustic space of pain: Cries as indicators of distress recovering dynamics in pre-verbal infants. *Bioacoustics* **2018**, *27*, 313–325. [CrossRef]
82. Lang, P.J.; Bradley, M.M.; Cuthbert, B.N. Emotion, attention, and the startle reflex. *Psychol. Rev.* **1990**, *97*, 377–395. [CrossRef]
83. Öhman, A. Face the Beast and Fear the Face: Animal and Social Fears as Prototypes for Evolutionary Analyses of Emotion. *Psychophysiology* **1986**, *23*, 123–145. [CrossRef]
84. Whalen, P.J.; Kagan, J.; Cook, R.G.; Davis, F.C.; Kim, H.; Polis, S.; Johnstone, T. Human amygdala responsivity to masked fearful eye whites. *Science* **2004**, *306*, 2061. [CrossRef] [PubMed]
85. Charrier, I.; Aubin, T.; Mathevon, N. Mother–Calf vocal communication in Atlantic walrus: A first field experimental study. *Anim. Cogn.* **2010**, *13*, 471–482. [CrossRef] [PubMed]
86. Szameitat, D.P.; Alter, K.; Szameitat, A.J.; Darwin, C.J.; Wildgruber, D.; Dietrich, S.; Sterr, A. Differentiation of emotions in laughter at the behavioral level. *Emotion* **2009**, *9*, 397–405. [CrossRef] [PubMed]
87. Slocombe, K.E.; Kaller, T.; Call, J.; Zuberbühler, K. Chimpanzees Extract Social Information from Agonistic Screams. *PLoS ONE* **2010**, *5*, e11473. [CrossRef] [PubMed]
88. Seyfarth, R.M.; Cheney, D. Pragmatic flexibility in primate vocal production. *Curr. Opin. Behav. Sci.* **2018**, *21*, 56–61. [CrossRef]
89. Tomasello, M.; Carpenter, M.; Hobson, R.P. The emergence of social cognition in three young chimpanzees. *Monogr. Soc. Res. Child Dev.* **2005**, *70*, 1–152.
90. Selfridge, O.G. Pandemonium: A paradigm for learning. In Proceedings of the Symposium on Mechanisation of Thought Processes, London, UK, 24–27 November 1958; HMSO: London, UK, 1959; pp. 511–529.
91. Rosenblatt, F. *Principles of Neurodynamics*; Spartan Books: Washington, DC, USA, 1962.
92. Jabri, M.; Flower, B. Weight perturbation: An optimal architecture and learning technique for analog VLSI feedforward and recurrent multilayer networks. *IEEE Trans. Neural Netw.* **1992**, *3*, 154–157. [CrossRef]

93. Bryson, A.; Ho, Y. *Applied Optimal Control*; John Wiley & Sons: New York, NY, USA, 1975.
94. LeCun, Y. Une procédure d'apprentissage pour réseau a seuil asymétrique (A learning scheme for asymmetric threshold networks). *Proc. Cogn.* **1985**, *85*, 599–604.
95. Sacramento, J.; Costa, R.P.; Bengio, Y.; Senn, W. Dendritic cortical microcircuits approximate the backpropagation algorithm. *Adv. Neural Inf. Process. Syst.* **2018**, *31*, 8721–8732.
96. Rao, R.P.N.; Ballard, D.H. Predictive coding in the visual cortex: A functional interpretation of some extra-classical receptive-field effects. *Nat. Neurosci.* **1999**, *2*, 79–87. [CrossRef]
97. Schwaninger, A.C. What Simulations Teach Us about Ordinary Objects. *Open Philos.* **2019**, *2*, 614–628. [CrossRef]
98. Popper, K. *Objective Knowledge: An Evolutionary Approach*; Oxford University Press: Oxford, UK, 1972.
99. Maturana, H.R.; Varela, F.J. *The Tree of Knowledge: The Biological Roots of Human Understanding*; Shambhala: Boulder, CO, USA, 1987.
100. Metzinger, T. *The Ego Tunnel: The Science of the Mind and the Myth of the Self*; Basic Books: New York, NY, USA, 2009.
101. Clark, A. Whatever next? Predictive brains, situated agents, and the future of cognitive science. *Behav. Brain Sci.* **2013**, *36*, 181–204. [CrossRef] [PubMed]
102. Gładziejewski, P. Predictive coding and representationalism. *Synthese* **2015**, *193*, 559–582. [CrossRef]
103. Hohwy, J. *The Predictive Mind*; Oxford University Press: Oxford, UK, 2013.
104. Rescorla, M. Bayesian Perceptual Psychology. In *The Oxford Handbook of the Philosophy of Perception*; Matthen, M., Ed.; Oxford University Press: Oxford, UK, 2013.
105. Quine, W.V.O. *Word and Object*; Wiley: London, UK, 1960.
106. Putnam, H. *Reason, Truth and History*; Cambridge University Press: Cambridge, MA, USA, 1981.
107. Rosenberg, A. Naturalistic Epistemology for Eliminative Materialists. *Philos. Phenomenol. Res.* **1999**, *59*, 335. [CrossRef]
108. Rosenberg, A. *The Atheist's Guide to Reality: Enjoying Life without Illusions*; WW Norton & Company: New York, NY, USA, 2011.
109. Kenny, A. *The Legacy of Wittgenstein*; Basil Blackwell: Oxford, UK, 1984.
110. Bennett, M.; Hacker, P. *Philosophical Foundations of Neuroscience*; Blackwell: Oxford, UK, 2003.

philosophies

MDPI

Article

A Constructive Treatment to Elemental Life Forms through Mathematical Philosophy

Susmit Bagchi

Department of Aerospace and Software Engineering (Informatics), Gyeongsang National University, Jinju 660-701, Korea; profsbagchi@gmail.com

Abstract: The quest to understand the natural and the mathematical as well as philosophical principles of dynamics of life forms are ancient in the human history of science. In ancient times, Pythagoras and Plato, and later, Copernicus and Galileo, correctly observed that the grand book of nature is written in the language of mathematics. Platonism, Aristotelian logism, neo-realism, monadism of Leibniz, Hegelian idealism and others have made efforts to understand reasons of existence of life forms in nature and the underlying principles through the lenses of philosophy and mathematics. In this paper, an approach is made to treat the similar question about nature and existential life forms in view of mathematical philosophy. The approach follows constructivism to formulate an abstract model to understand existential life forms in nature and its dynamics by selectively combining the elements of various schools of thoughts. The formalisms of predicate logic, probabilistic inference and homotopy theory of algebraic topology are employed to construct a structure in local time-scale horizon and in cosmological time-scale horizon. It aims to resolve the relative and apparent conflicts present in various thoughts in the process, and it has made an effort to establish a logically coherent interpretation.

Keywords: platonism; predicate logic; monad; homotopy; probability

check for updates

Citation: Bagchi, S. A Constructive Treatment to Elemental Life Forms through Mathematical Philosophy. *Philosophies* **2021**, *6*, 84. https://doi.org/10.3390/philosophies6040084

Academic Editors: Gordana Dodig-Crnkovic, Marcin J. Schroeder and Rossella Lupacchini

Received: 16 August 2021
Accepted: 6 October 2021
Published: 12 October 2021

1. Introduction

The process of axiomatization is ancient, with wide array of applications in mathematics and in philosophy to coherently establish a theory to reach out to truth. The ideal goal is to attain truth in invariant forms independent of any contextual variables. In other words, the axiomatic development of a theory or knowledge aims to establish or systematize a structural form without any ambiguity and inherent contradictions within it [1]. Euclid first developed the axiomatized geometry in *Elements* where he made attempts to distinctly separate the *primitive* and the *derived* [1]. It is important to note that the philosophy of formal logic of Kant enables one to study abstract mathematical objects in nature [2]. The formal logic of axiomatization is often not purely sufficient to eliminate inherent ambiguity because of the process of *axiomatizations of initials* prior the coherent structures being established. However, in later times, Frege successfully formulated the methods of deriving logical inferences through the axioms in mathematics, which has found suitable applications in philosophy. It should be noted that Natorp and Cassirer disagreed with the concepts proposed by Frege and Russell on the ground, nature and logical demarcation [2]. According to Cassirer the New-logic is more suitable to study mathematical concepts in nature through relational structures. The mathematical philosophy of New-logic vindicates the Neo-Kantian theory of space and time [2].

It is often argued that philosophy must employ the axiomatic and logistic methods of constructive inquiry [3]. The reason is that the logic-based constructive systems such as mathematics enable us to gain insight to a philosophical system and natural observable systems free from internal inconsistencies. For example, Alonzo Church, as a philosopher and mathematical logician, applied the method of *hypothetico-deductive-rationalism* while theorizing epistemological aspects of mathematics, logic and philosophy [3]. It is important to note that the method of mathematical and philosophical investigations made by Church can be viewed as a Platonic realism, and sometimes his opinion about realism conflicts with Frege. Moreover, the Platonic and Aristotelian philosophies include the elements of Pythagorean doctrines [4].

1.1. *Self as Life Forms and Transitions*

In the natural or material world, the concept of the self is evident through the interactions to other elements of nature actively or passively. In other words, the recognizing existence of self is a set of spontaneous computable actions in the nature or in the material world [5]. As a natural consequence, the following question appears: What is the natural world in this context? The answer was proposed by Anaxagoras in ancient times based on the multiplicity of basic particles. According to Aristotle, the "homoiomeroi" constitutes the universe, and the natural material world can be described by relational ideas [5]. However, it is noted earlier that Platonic realism, mathematical axiomatization and Neo-realism aim to understand the true essence of natural world as a set of mathematical principles, which are foundational. According to Plato, Parmenides and Leibniz, the transitions from the mathematical world to the natural (physical or material) perceivable world happen due to the limitation of computations, and as a result, spatiotemporal forms of the perceivable physical world appear [6]. It is important to note that, in general, it is thought that pure logism is not very suitable to purely understand abstract principles of the natural world; however, philosophers often apply inductive reasoning for drawing plausible conclusions and such inductive reasoning includes elements of probability theory [7].

1.2. *Motivation*

The mathematically abstract notion of space is elegantly explained by Kant. In view of Kant, space is an outer form derived from the perceptions by life forms and space can be infinite [8]. Note that the Euclidean construction of space does not conflict with the view of Kant. The mathematical philosophy of Leibniz tried to investigate the nature of mind–body relation in a space. This results in the two directions of related thoughts. In one direction, the solipsistic monad is conceptualized to represent existential knowledge, and in another direction, the monad is not necessarily solipsistic in nature considering the interactions [9]. In the second line of thought, the causal separation from the rest of the universe is inevitable. Note that, at the foundation of mathematics, the shapes of natural world forms and space-time follow the structures constructed by axioms, logic and geometry, which are topological in nature [5]. On the other hand, the Cartesian proposal of mind–body dualism in mathematical philosophy is experimentally reconstructed and presented in [10]. The experimentation on a human life form illustrates that the multiplicity of self in the mind exists, and more importantly, it is observed that, mathematically, the existences are a set of almost continuous functions in space-time controlling body (i.e., biophysical or materialistic form) dynamics. The almost continuous functions in the varying spaces of the life forms are separated by cuts. This motivates us to investigate the abstract mathematical theory of existential life forms in space-time through the birth–death process of a materialistic element (i.e., body as a biophysical entity) in view of mathematical philosophy.

According to Carnap, the general theory and analysis should be based on *intensional isomorphism* [3]. The concept of isomorphism and homeomorphisms are widely used in various domains of mathematics such as algebra and topology, to name a few. The intensional isomorphism helps in determining isomorphism between two or more than two structures or statements where the local equivalency between substructures exists and, as a result such, equivalent substructures can be replaced, maintaining the overall isomorphism. The combinations of selective ingredients of Spinoza and Leibniz become a motivating factor to constructively approach to the existential nature of life forms and associated dynamics. The monadism of Leibniz emphasizes upon the concept of mathematical representation, and this paper tries to employ such an approach. Furthermore, in this paper we follow the analytical method of Carnap in order to prepare a constructive structure of elemental life forms in view of mathematical philosophy. It is possible that the mixed elements of Cassirer philosophy of symbolic forms, elements of Hegelian idealism and the Platonic realism of Alonzo Church can be observed in the constructive approach.

The rest of the paper is organized as follows. The descriptions about existentialism, dualism and abstractionism are presented in Section 2. The probabilistic treatment is formulated in Section 3. Section 4 presents the homotopic analysis, and finally, Section 5 concludes the paper.

2. Existentialism, Dualism and Abstractionism

The three broad and generalized schools of thoughts in philosophy and mathematics in explaining the perceivable nature and space are existentialism, dualism and abstractionism. The Platonism as well as the existentialism of Husserl proposed that the external sensual world and the associated perceptions are in reduced forms through local interactions with environments. The absolute realism is in a pure state which can be understood by the time-invariant mathematical abstract principles of nature. The existence of mind–body dualism further strengthens the school of thought of Plato to understand the true nature of existence in universe through the lenses of mathematics. Moreover, the philosophies of Pythagoras, Descartes, Kant and Leibniz propel the validation of the fact that the philosophical understanding of the true nature of life needs the invaluable doctrines of abstract mathematical understanding of the working of universe as a whole. Hence, the vivid understanding about existentialism, dualism and abstractionism are necessary. These approaches take distinct directions of analyses and inferences. In this section, brief presentations are made about these diverse schools of thoughts, and similarities as well as differences are pointed out whenever possible.

2.1. Existentialism

The theory of existentialism is clearly explained by Husserl in *Lebenswelt* as a basic form [11]. According to Husserl the experience, sensation and perceptions of a life form in a day-to-day environment is a horizon of life, which is concrete as well as local, and it is very different from the world horizon which follows a set of pure scientific or mathematical principles. Interestingly, the existentialism of Husserl has similar perspectives as compared to Platonism and the noumenal (i.e., unknown realm of freedom) elements of Kant [11]. The use of absolute rationalism to explain the world as a phenomenon (local events) fails to explain many day-to-day observations in the world horizon. For example, the statement derived from absolute realism saying that "water boils at 100 °C temperature" is *not* correct in the world horizon because it depends on the location of water in space. Thus, the existential forms, time and position, are important parameters to determine the locally observable or sensually measurable properties of the forms. It is important to note that Jean-Paul Sartre provided a convincing argument in favor of the existentialism of Husserl, Heidegger and Kierkegaard [11].

The philosophical approaches of Spinoza regarding existentialism have few similarities and some differences with respect to the philosophy of mathematician Leibniz. The existentialism of Spinoza has a twofold meaning. Spinoza specifically stressed on the *principle of ground-consequent,* and in this relation, the causation is identical [12]. As a result, all determined parameters become transitory and a series of cause–effect relational chains are formed where all consequences can be derived from a set of primitive causes. It is important to note here that the Aristotelian method of logical deductions sometimes fails to achieve clear uniformity if the system under investigation is complex [12]. On the other hand, Leibniz philosophized that all matter can be properly understood through abstract mathematical entities embodying the properties of corresponding matter. According to Leibniz, the characteristic feature of every substance is unity, and that unity is not conceivable just by the appearance of the substance in its forms and an activity is required [12]. It is important to note that the monadism of Leibniz may have some Platonic elements, and the concept of continuous evolution is embodied within the mathematical philosophy of Leibniz.

2.2. Dualism

The mind–body relation and dualism are intricately linked where a body is considered to be the material in a form. The notion of Hegelian idealism is contrary to the neo-materialism and the idealism attracted support from Bradley, Sir Eddington and Sir Jeans in the field of science [13]. Apart from the Thomistic approach, the unambiguous dualism enables the investigation of nature by removing the forms and species, while formulating a set of scientific principles exposing the laws of nature [14]. Fortunately, mathematics always, from antiquity to today, plays an unavoidable and extremely crucial role in the process. The dualism transformed the Aristotelian understanding of matter or materials and various other approaches are revived, such as atomism, skepticism and nominalism [14]. As a result, nature, matter and forms appear highly mathematical, exposing inherent natural laws, principles and their accurate interpretations [14]. This effectively gives way to the process of mathematical abstractions to understand nature as a set of purely mathematical principles.

2.3. Abstractionism

In the field of mathematical philosophy (and philosophy of mathematics), it is a well-accepted concept that humans make references to and deal with a large set of abstract objects or entities in everyday life [15]. The ontological and epistemological misunderstandings arise if one argues that abstract objects are not located in physical space. However, in this case the basic questions are: What is a space? Is it not the fact that the mathematical concept of space is also an abstract entity? According to Boolos and Gödel, abstract objects are understood through perceptual contacts in the everyday life of human beings [15]. For example, human or even some other life forms in nature count by recognizing numbers. Note that numbers are abstract (invisible) Platonic entities and humans are making perceptual contacts with them cognitively to make everyday life fruitful, indicating that such abstract mathematical objects exist in the universe. It is mentioned earlier that the perceptual understanding and measurement of contacts to living environment is highly limited in materialistic senses. Frege suggested that the meaning of abstract identity statements should be understood through finding out a suitable equivalence relation between them. Suppose S, T are two abstract objects (while retaining the possibility of isomorphism). The function form $f(a)$ can induce an equivalence relation Φab as given in the following functional expression [15]:

$$\Phi xy \equiv [f(x) = f(y)] \tag{1}$$

This invites the problem for finding an equivalence relation, which can be hard. Moreover, if such equivalence relation does not exist, then what would be the solution? The Platonic Neo-Frege theory illustrates that by employing abstraction principles, one can make perceptual contact with abstract objects, which does not necessarily need to maintain reductionism because abstract mathematical realm is in pure state.

3. Random Choices for Existence

The constructive mathematical as well as philosophical analysis of elemental life deals with the life forms of elements in a natural environment. In human society, an elemental life form is an individual in space-time having a mind–body relation in a birth–death process where the society is viewed as an ensemble of structure(s) composed of elements and their mutual interactions. This perspective of elemental life forms and society allows the evolution of societal and cultural structures over time, although no guarantee is made to achieve optimality or perfection. As a result, the corresponding evolutions of elemental life forms appear to be a continuous process, although the changes are unnoticed in smaller time-scales (in local-time horizon). Note that the cosmological time-scale of the universe is extremely large as compared to the local-time horizon of the elemental life forms.

The complex behaviors of elemental life forms as the social interactive elements are theorized in view of social choice theory and game theory affecting economic structures [16]. In this paper, we take a different discourse, and we will try to constructively formulate a mathematically consistent model of an elemental life form and its dynamics in the birth–death process in cosmological space-time (i.e., universe–time horizon). Our effort in this paper will be to offer a constructive treatment to elemental life forms in view of mathematics and philosophy, which is coherent to the materialistic biophysical understandings and experiences. The basis of a materialistic element is its biochemical structures and their mutual reaction pathways determining the dynamics of life of an element. The materialistic formation of an element begins when two genetic materials combine in a suitable growth environment, which we call as a birth of an element in environment or nature. As a result, the dynamics and properties of the elemental life forms are pre-determined by the genetic materials in a combination. However, the limitation of this viewpoint becomes considerable due to the fact that the life forms and natural environments mutually influence each other in bidirectional manner [17,18]. Hence, the dynamics of elemental life and its evolution are not purely pre-determined in a constantly changing environment and evolutionary discourse appears over time. As a result, the evolution of elemental life forms and the dynamics are not deterministic in full and appear to be probabilistic in nature based on some purely random choices satisfying the instantaneous requirements for existence in an environment. This can be considered as a process of *random choices for instantaneous existences*. This motivates to search a new approach and perspective to understand and analyze the elemental life forms by combining both mathematical and philosophical constructivism. First, we define a set of concepts in view of mathematical philosophy as follows:

Definition (Birth–death process): *LetU be the entire universe of existence. A birth–death process of an existential life form (as an element p) in the cosmological time-scale is the local dynamics of a continuous function $f_p(t)$ between the two fixed points in the continuous interval $[t_B(p), t_D(p)]$ for the corresponding element. The dynamics of $f_p([t_B(p), t_D(p)])$ are in continuum, and the discourse is determined by a sequence of probabilistic events attached to p in U.*

Every existential life form finitely interacts to the environment in local time-scale in the universe, and the number of elements in the environment are also finite for the life form for interaction within $[t_B(p), t_D(p)]$. As a result, an existential life form makes a set of finite choices in the environment during the process of interaction, which is defined as follows:

Definition (Selectable finite choices): *For all existential element p there is a finite $E \subset U$ forming an environment of p from which the existential element makes finite choices in $[t_B(p), t_D(p)]$. The choices made by the existential life form are the selection set $E_p \subset E$ of that element.*

The choices made by an existential life form out of selection set are probabilistic as determined by the element, and it affects the dynamics of the birth–death process and a set of such processes form a homotopy:

Definition (Homotopy forms): *A set $\{f_p([t_B(p), t_D(p)]) : p \in U\}$ forms homotopy if $\forall p, f_p(t_B(p)) < f_p(t_D(p))$ in local time-horizon and $\forall p, q \in U, [f_p(t_B(p)) \approx f_q(t_B(q))] \wedge [f_p(t_D(p)) \approx f_q(t_D(q))]$ in the cosmological time-horizon which is a half-open infinite space of Sorgenfrey line.*

The cosmological time-horizon is much larger than the local time-horizon of a set of birth–death processes resulting into the formation of the homotopy in the cosmological time horizon.

3.1. Existential Formation of Elements: A Probability Chain

In view of mathematical constructivism, an element comes into a materialistic existence by following two randomized processes producing a set of probabilistic outcomes of events. Let a set of elements in a dynamic natural environment at time t be denoted by $X(t) \subset U$. The first random process in $X(t)$ is denoted as $P_M(p,r)$, where $\exists p \exists r \in X(t)$ are the elements. The process $P_M(p,r)$ signifies the probability of the pairing of respective elements in the presence of selectable finite choices. The second random process is denoted as $Q(\{p,r\},q)$ signifying the probability of forming an element $\exists q \in X(t_2)$ if and only if $P_M(p,r)$ and $P_M(r,p)$ are successful birth–death processes at $t_1 < t_2$. We will logically establish later that $P_M(p,r)$ and $P_M(r,p)$ need not be always exactly equal. Suppose $t_\alpha \in (t_1,t_2)$ is an uncertain time instant (i.e., not predetermined) within the open interval. Let us algebraically denote the formation of element $q \in X(t_2)$ as $P_B(q|\{p,r\})|t_2$. Accordingly, for clarity, let us denote $P_M(p,r)|t_1$, $P_M(r,p)|t_1$ and $Q(\{p,r\},q)|t_\alpha$ signifying that these probabilistic events are measured at specific time instants in local as well as cosmological time-scale horizons as indicated. Note that the local time-scale horizon is a continuum, and it is an uncountable subset of cosmological time-scale horizon. The probability of combined birth–death processes generating an element are given in the following equation computed under an abstract algebraic operation (i.e., not necessarily multiplication at this point; detailed analysis is presented later in this section considering deterministic algebraic operations):

$$P_B(q|\{p,r\})|t_2 = [(P_M(p,r)|t_1) \cdot (P_M(r,p)|t_1)] \cdot Q(\{p,r\},q)|t_\alpha. \tag{2}$$

It is important to note the following properties of these two processes while in a combination exerting mutual influences satisfying the stability of structures and evolution:

$$\begin{aligned}
&[(P_M(p,r)|t_1) \neq (P_M(r,p)|t_1)] \Rightarrow [(P_M(p,r)|t_1) < 1] \wedge [(P_M(r,p)|t_1) < 1],\\
&[(P_M(p,r)|t_1) \cdot (P_M(r,p)|t_1)] = [(P_M(r,p)|t_1) \cdot (P_M(p,r)|t_1)],\\
&[(P_M(p,r)|t_1) \cdot (P_M(r,p)|t_1)] \cdot (Q(\{p,r\},q)|t_\alpha) \neq\\
&(P_M(p,r)|t_1) \cdot [(P_M(r,p)|t_1) \cdot (Q(\{p,r\},q)|t_\alpha)].
\end{aligned} \tag{3}$$

From the aforesaid properties, we can conclude that the random birth–death processes of pairing of elements are in a commutative relation; however, the combined processes for the generation of an element are not associative in nature (i.e., the abstract algebraic operation is commutative but not associative). It indicates that the operation generating the relation is at least not an algebraic division because, in that case, commutativity will not be valid (because division is not commutative). On the contrary, if the commutativity is valid with respect to the algebraic division operation, then it results in the conclusion that $(P_M(p,r)|t_1) = (P_M(r,p)|t_1) = 1$. However, in this case, the associativity will be achieved violating the principle in the local time-scale horizon of birth–death processes. Moreover, the non-associativity of the algebraic operation indicates that it is not an addition or a multiplication operation. Thus, the abstract algebraic operation generating such relation (commutative and non-associative) is sensitive to instantaneous time of measurement and cannot guarantee associativity under future projection in time, where the time is considered as a half-open infinite space of the *Sorgenfrey line*. In other words, the proposed construction includes the uncertainty in the time-scale horizon about the formation of events as presented in the following equation, *considering that the abstract algebraic operation is a multiplication*:

$$\begin{aligned}
&[(P_M(p,r)|t_1) = (P_M(r,p)|t_1) = 1] \Rightarrow\\
&[(P_M(p,r)|t_1) \cdot (P_M(r,p)|t_1)] \cdot (Q(\{p,r\},q)|t_\alpha) =\\
&(P_M(p,r)|t_1) \cdot [(P_M(r,p)|t_1) \cdot (Q(\{p,r\},q)|t_\alpha)].
\end{aligned} \tag{4}$$

It is important to note that if the abstract algebraic operation is multiplication and the probabilities $P_M(p,r)$ and $P_M(r,p)$ are each unity *signifying certainty*, then Equation (2) generates the absolute probability of the generation of a new element and Equation (3) becomes transformed into a commutative as well as associative under such certainty of pairing of elements. However, even in this case there is *no guarantee* that $(Q(\{p,r\},q)|t_\alpha) = 1$ will be achieved in Equation (4). Furthermore, Equation (2) exposes the fact that if $[(P_M(p,r)|t_1) < 1] \wedge [(P_M(r,p)|t_1) < 1]$, then $P_B(q|\{p,r\})|t_2$ cannot determine with absolute certainty whether a new element can be formed or not in the universe although commutativity is preserved (In this case associativity is not preserved in projection at different instants of time).

Interestingly, it is well known that the stable existence of $q \in X(t_2)$ is a highly probabilistic event due to the presence of several natural environmental factors affecting the existentialism of materials. We need to emphasize that although $Q(\{p,r\},q)|t_\alpha$ generates a probabilistic event, the observation $H(q)$ of event of formative existence of an element $q \in X(t_2)$ is a discrete function, as follows:

$$\begin{aligned} &(Q(\{p,r\},q)|t_\alpha) \in [0,1], \\ &H(q) \in \{0,1\}. \end{aligned} \tag{5}$$

Note that the theory of natural selection approximately determines the formation of the material form of an element and its evolving existence in a natural environment. It is relatively straightforward to observe that the theory of *natural selection* is maintained by the proposed constructive mathematical formalism.

3.2. *Relations to Empty, Entire and Observations*

One can analyze two cases of existential formation of the elemental life forms in a probability chain. The proposed analytical discourse combines the ingredients of mathematics, philosophy and social sciences. Let us first consider that the relation of an element with the *empty*, ϕ. Let us consider the related algebraic structure which can be denoted as $P_M(p,\phi)|t_{+\infty}$, where $t_{+\infty} \in [t_1, +\infty]$. If we analyze the extremely limiting but stable choice of the element $p \in X(t)$, then we can deterministically conclude that $[(P_M(p,\phi)|t_{+\infty}) = 1] \wedge [(P_M(\phi,p)|t_{+\infty}) = 1] = 1$. This directly results in the following logical conclusion in a birth–death process due to the degeneration of probabilistic measurement into the binary-valued determinism:

$$(P_B(q|\{p,\phi\})|t_2) \vee H(q) = 0. \tag{6}$$

On the other extreme end, let us consider the structure $P_M(p, X(t = t_n))|t_{+\infty}$ encompassing the *whole*. Clearly, this structure is not stably sustainable due to scale, and as a result, we can infer an equivalence relation between the *empty* and the *whole* as follows, following the degenerative probabilistic measurement into binary-valued determinism:

$$(P_B(q|\{p, X(t_n)\})|t_2) \vee H(q) = (P_B(q|\{p,\phi\})|t_2) \vee H(q). \tag{7}$$

Interestingly, the probabilistic outcome of both opposite extremes are the same, and the proposed construction also represents that once the probabilistic existential events are successfully chosen randomly (i.e., not 0) in time, then the corresponding observations are determined. If the probabilistic existential conditions of an element q are stable and satisfied with high probability, then $H(q \in X(t_2)) = 1$, reaching certainty. Otherwise, the model predicts that $H(q \in X(t_2)) = 0$ is a natural consequence in the natural environment. Furthermore, it exposes two different pathways of an elemental life form to evolve over time by following the homotopy theory of algebraic topology [19]. The topological homotopy theory appears to be a close fit to establish and explain the dynamics of elemental life forms in the cosmological time-scale of the universe in terms of constructive mathematical as well as philosophical principles.

4. Homotopic Existential Analysis of Elements

In the homotopic analysis of the existence of elements, we will consider two different time-scales: local and cosmological. The local time-scale is a subset of cosmological (universal) time-scale, where both are continuous in nature. We represent cosmological time scale as T_C, and we consider that $T_C \cong R_{+\infty}$, where $R_{+\infty} = [0, +\infty]$, is the subset of extended real numbers. Evidently, the local time-scales of an elemental form and its evolution is much shorter than the cosmological time-scale, where the cosmological time-scale is equivalent to a half-open (lower-limit) infinitary Sorgenfrey positive array [20]. Note that, in this case, the Minkowski space-time structure of topological space is decomposed and the time-like space is considered in our constructions and analysis. This observation leads to a set of interesting analytical insights in mathematical philosophy. In this section the emphasis is made on the homotopic analysis, and it omits the specified time instants within the local time-scale horizon for easy representation.

4.1. Homotopic Existence for $H(q) = 0$

The application of homotopy theory in analyzing the existential conditions of an element form for $H(q) = 0$ reveals a new philosophical inroad of understanding. In view of homotopy theory, we can derive two different existential conditions of an element $q \in X(t)$ by considering the birth–death process, where $t_B(q), t_D(q)$ represents the time of birth and death of the corresponding element in the local time-scale. These conditions are presented as follows:

$$\begin{aligned} &\forall p \forall r \exists q \in X(t), \\ &[Q(\{p,r\},q) = 0] \Rightarrow [H(q) = 0], \\ &[H(q) = 0] \wedge [Q(\{p,r\},q) \in [0,1)] \Rightarrow [t_D(q) \approx t_B(q)]. \end{aligned} \tag{8}$$

Hence, according to homotopy theory, this can be viewed as a formation of a trivial fundamental group denoted by $\pi_1(X(t), q)$ such that the homotopy class $[q]$ becomes a left as well as a right identity as a single element in $\pi_1(X(t), q)$ due to the fact that $([t_D(q) - t_B(q)]/T_C) \rightarrow 0$. In other words, $[q]$ is the nullhomotopy class representing $\pi_1(X(t), q)$ of a birth–death process in the cosmological time-scale horizon. This mathematical constructive approach validates the associated philosophical understanding of elemental life in a birth–death process in the observable local time-scale and its duality in the universe in the cosmological time horizon.

4.2. Homotopic Existence for $H(q) = 1$

This is another extreme where the probabilistic outcomes of events are deterministically observable in a natural environment. It is relatively straightforward to observe that, in this case, the following properties are maintained in the universe:

$$\begin{aligned} &\varepsilon(q) = t_D(q) - t_B(q) > 0, \\ &\exists F \exists K \in R_{+\infty}, F << +\infty, K << 1, \\ &\varepsilon(q) \in (0, F], \\ &\tfrac{F}{T_C} \in (0, K). \end{aligned} \tag{9}$$

Thus, one can infer from the materialistic as well as mathematical philosophy standpoints that the birth–death process of an observable element is deterministic with the corresponding two fixed points: $t_B(.), t_D(.)$. As a result, an n-dimensional path-homotopy structure is formed in local time-scale horizon within the cosmological time-scale horizon for all existential and observable elements of life forms, which can be represented as follows:

$$\begin{array}{l}|X(t = t_m)| = N(m) \in Z^+, N(m) \in (0, +\infty), \\ \forall p \in X(t \in [0, t_m)), f_p : [0,1] \to ([t_B(p), t_D(p)] \subset T_C) \times (S^{n-1} \subset R^{n-1}_{+\infty}), \\ \forall p \forall q \in X(t \in [0, t_m)), |[\varepsilon(p)/T_C] - [\varepsilon(q)/T_C]| \to 0, \\ ((|f_p(0) - f_q(0)|/T_C) \approx 0) \wedge ((|f_p(1) - f_q(1)|/T_C) \approx 0) \Rightarrow \\ [f_p(0) \approx f_q(0)] \wedge [f_p(1) \approx f_q(1)].\end{array} \tag{10}$$

Note that the path-homotopy expression is considering a generalized n-dimensional world horizon and an n-dimensional local horizon of perception. If we consider perceivable space-time as $n = 4$, then it is transformed into a four-dimensional perceivable space-time environment of the existential life forms. The observation $\varepsilon(p) \approx \varepsilon(q)$ derived from the above equations may appear surprising, but it is valid because it is always true that $\forall p \in X(t), 0 < (\varepsilon(p)/T_C) << 1$. It is important to note that each $f_p(.)$ is continuous representing dynamics of elemental life form of each discrete element in a local time-scale, and the path-homotopy functions of all elements are between the two fixed points in the cosmological time-scale of universe. However, this constructive approach does not violate another important material property, such as the actual local time-span of an elemental form in a homotopy of birth–death processes. If L_p represents the local time-span of an elemental life form in a path-homotopy, then one can compute L_p as follows by using line integral in n-dimensional space:

$$\begin{array}{l}L_p = \int_l \dot{f_p}(t)dt, \\ (L_p/T_C) \in (0, K).\end{array} \tag{11}$$

The line integral is computable because path-homotopic functions are continuous. However, one can conclude further that for any two elements $\exists p \exists q \in X(t)$, the values of L_p, L_q may be equal or not equal depending on the individual elemental dynamics in the local time-scale. As a result, one can infer that $\forall p \forall q \in X(t), \exists a \in R_{+\infty}, (L_p/L_q) \in (0, a)$, where $0 < a << +\infty$. This validates that the proposed constructive mathematical formalism maintains the corresponding philosophical interpretations of elemental life forms in local time-scale and also in the cosmological time-scale.

4.3. *Consistency of Transitions*

In the previous sections, we have presented two different abstractionisms, namely, probabilistic existentialism and homotopic existentialism. These two abstract forms are mutually consistent and there is no logical conflict between the two. In this section, we present a brief analytical outline illustrating that consistency of an equivalence class $S(t) = \{p, q, r\}$ is independent of nature of existentialism, where $S(t) \subset X(t)$. Moreover, the consistency of $S(t)$ is maintained under observable transitions in the cosmological time-scale horizon T_C. Suppose we consider a path-homotopy of elements in $S(t)$ in cosmological time-scale horizon T_C where the path-homotopy is maintained within local time-scale horizon $[t_a, t_b]$ such that $0 < ((t_b - t_a)/T_C) << 1$. It indicates that from the observational point of view one can conclude that $\forall t \in [t_a, t_b], \exists x \in S(t), H(x) = 1$. Note that this is a highly relaxed version considering individual elements at any instant of time within the path-homotopic time interval in local time-scale horizon. A stronger form of observation can be abstracted as: $\forall t \in [t_a, t_b], H(p) \vee H(q) \vee H(r) = 1$. An equivalence class $S(t)$ is called stable at a time $t \in [t_a, t_b]$ if, and only if, $\exists t \in [t_a, t_b], H(p) \wedge H(q) \wedge H(r) = 1$. It is relatively easy to infer that $\forall t \in [0, t_a) \subset T_C, \forall t \in (t_b, +\infty] \subset T_C, H(p) \vee H(q) \vee H(r) = 0$. Thus, there is a 0-0 existential transition in the cosmological time-scale horizon for every element in $X(t)$ except the path-homotopy interval, which is in a local time-scale horizon. In other words, the existential life forms are in an equivalent class of the 0-0 state everywhere outside of the corresponding local path-homotopy interval. Moreover, it indicates that there is an extended interval $t_{ae} < t_a, t_{be} > t_b, [t_{ae}, t_{be}]$ where almost every existential life form in universe U will preserve the 0-0 transition in the cosmological time-scale horizon, which is in line with the theory of evolution and disappearance of species in nature.

5. Conclusions

It is generally an accepted theory in evolutionary biology and social choice theory that life forms in nature follow the dynamic trajectories highly influenced by the probabilities of various events in nature. However, in the field of philosophy, it raises debate mainly between Platonic and realism schools of thoughts, which is further detailed towards resolution by neo-realism and New-logic. The proposed structure in light of mathematical philosophy applies the probability theory and predicate logic to illustrate that it is possible to establish a coherent structure of understanding. Moreover, the applications of topological homotopy theory nearly accurately encapsulate the conceptual and constructive understanding of existential life forms in nature without any observable contradictions. The cosmological time-scale representing the world or universal horizon plays an important role to formulate homotopic structures of dynamics of existential life forms in nature. The constructive mathematical treatment proposed in this paper enables us to coherently amalgam the structural notions of the local time-scale horizon and cosmological time-scale horizon of space-time through the lenses of mathematical philosophy.

Funding: The research is funded by Gyeongsang National University, Jinju, Korea.

Data Availability Statement: Not applicable.

Acknowledgments: The author would like to thank the anonymous reviewers and editors for their valuable comments and suggestions.

Conflicts of Interest: The author declares no conflict of interest.

References

1. Parsons, C. Informal axiomatization, formalization and the concept of truth. *Synthese* **1974**, *27*, 27–47. [CrossRef]
2. Heis, J. "Critical philosophy begins at the very point where logistic leaves off": Cassirer's response to Frege and Russell. *Perspect. Sci.* **2010**, *18*, 383–408. [CrossRef]
3. Anderson, A. Alonzo Church's contributions to philosophy and intensional logic. *Bull. Symb. Logic* **1998**, *4*, 129–171. [CrossRef]
4. Netz, R. The Pythagoreans. In *Mathematics and the Divine: A Historical Study*; Elsevier: Amsterdam, The Netherlands, 2005; Chapter 3; pp. 79–97.
5. Igamberdiev, A.U. Time and life in the relational universe: Prolegomena to an integral paradigm of natural philosophy. *Philosophies* **2018**, *3*, 30. [CrossRef]
6. Bekenstein, J. Universal bound on the entropy to energy ratio for bounded systems. *Phys. Rev. D.* **1981**, *23*, 287–298. [CrossRef]
7. Thagard, P. Naturalizing logic: How knowledge of mechanisms enhances inductive inference. *Philosophies* **2021**, *6*, 52. [CrossRef]
8. Parsons, C. Infinity and Kant's conception of the "Possibility of Experience". *Philos. Rev.* **1964**, *73*, 182–197. [CrossRef]
9. Russell, L.J. Some problems in the philosophy of Leibniz. In *Meeting of the Aristotelian Society*; Oxford Academy: Oxford, UK, 1923; Chapter XII.
10. Santos, T.J. On a multiplicity: Deconstructing Cartesian dualism using mathematical tools in performance. *Liminal. J. Perform. Stud.* **2014**, *10*, 1–28.
11. Wild, J. Existentialism as a philosophy. *J. Philos.* **1960**, *57*, 45–62. [CrossRef]
12. Hicks, G.D. The "Modes" of Spinoza and the "Monads" of Leibniz. In *Proceedings of the Aristotelian Society*; Oxford Academy: Oxford, UK, 1918; Volume 18, pp. 329–362.
13. Forster, E. Twenty-five years of philosophy. *Nature* **1935**, *136*, 332. [CrossRef]
14. Balz, A.G.A. Dualism and early modern philosophy. II. *J. Philos. Psychol. Sci. Methods* **1918**, *15*, 225–241. [CrossRef]
15. Heck, R.G. The existence (and non-existence) of abstract objects. In *Abstractionism: Essays in Philosophy of Mathematics*; Oxford University Press: Oxford, UK, 2016; Chapter 3; pp. 50–78.
16. Maschler, M.; Solan, E.; Zamir, S. Social choice. In *Game Theory*; Cambridge University Press: Cambridge, UK, 2013; pp. 853–883.
17. Rozzi, R. The reciprocal links between evolutionary-ecological sciences and environmental ethics. *BioScience* **1999**, *49*, 911–921. [CrossRef]
18. Corris, A. Defining the environment in organism-environment systems. *Front. Psychol.* **2020**, *11*, 1–13. [CrossRef] [PubMed]
19. Jardine, J. *Local Homotopy Theory*; Springer: New York, NY, USA, 2015; ISBN 978-1-4939-2300-7.
20. Douwen, E.K.V.; Pfeffer, W.F. Some Properties of the Sorgenfrey Line and Related Spaces. *Pac. J. Math.* **1979**, *81*, 371–377. [CrossRef]

MDPI AG
Grosspeteranlage 5
4052 Basel
Switzerland
Tel.: +41 61 683 77 34

Philosophies Editorial Office
E-mail: philosophies@mdpi.com
www.mdpi.com/journal/philosophies

www.ingramcontent.com/pod-product-compliance
Lightning Source LLC
Chambersburg PA
CBHW041409210926
51850CB00013B/3223
9783725832149